DEVELOPMENTS IN
ARID ZONE ECOLOGY
AND ENVIRONMENTAL QUALITY

DEVELOPMENTS IN ARID ZONE ECOLOGY AND ENVIRONMENTAL QUALITY

PROCEEDINGS of an INTERNATIONAL MEETING
12th Scientific Conference of the Israel Ecological Society
at the Hebrew University of Jerusalem, Israel, May 12–13, 1981

Editor
Hillel I. Shuval, The Hebrew University of Jerusalem

*1981 **BALABAN INTERNATIONAL SCIENCE SERVICES** Rehovot / Philadelphia*

Distributed by

International Science Services
POB 2039
76360 Rehovot, Israel

USA

S. Pomerantz
2242 Mt. Carmel Ave.
Glenside, Pa. 19038

Developments in
Arid Zone Ecology and Environmental Quality
Edited by H.I. Shuval,
Philadelphia and Rehovot

ISBN 0-86689-005-X

PREFACE

This monograph represents an effort in interdisciplinary scientific cooperation between ecologists and experts in environmental quality studies to present an overview of their most recent findings in relation to the ecology and environment in a unique ecosystem combining arid zones, Mediterranean climates and the problems of modern urban-industrialized society as found in Israel and the Eastern Mediterranean. While about three quarters of the papers represent the work of Israeli scientists, the volume is enriched by numerous excellent papers by scientists from other countries who have contributed balance and international perspective to this effort.

The Israel Ecological Society has strived since its founding to serve as a forum for scientific exchange and interaction between the community of ecologists and the environmental scientists who examine and study the same ecosystem from different angles. Participating in this effort are ecologists, botonists, zoologists, microbiologists, chemists, epidemiologists, engineers, atmospheric scientists and experts in environmental management. The meetings of the Society attempt to provide a common meeting ground aimed at encouraging the interdisciplinary approach to the problems of our ever more complex environment.

Thus the study of the fauna and flora and their environmental interactions becomes interdigited with the study of man's interactions with the environment and the detrimental effects on himself and on the plants and creatures that surround him.

This book containing 69 original papers selected by a panel of judges has been divided into ten sections covering most of the areas of ecological - environmental interest as follows:

I. Arid Zone Ecology
II. Environmental Quality Management
III. Aquatic Ecology
IV. Waste Recycling and Utilization
V. Environmental Microbiology
VI. Water Quality Management
VII. Wastewater Treatment
VIII. Environmental Health and Epidemiology
IX. Air Quality Management
X. Ecology of Streams in Arid and Semi Arid Areas

The 12th Scientific Conference of the Israel Ecological Society as presented in this volume represents the first effort to open the meeting to the international community of scientists in general and includes for the first time work representing important ecological and environmental studies carried out in Egypt concerning the Aswan High Dam. It is hoped that this first international conference of the Society will open the way for more active exchange with scientists in the Middle East who share a joint concern in preserving the ecological balance and environmental quality of this historic corner of the world.

Jerusalem

Hillel I. Shuval, Chairman
Israel Ecological Society

SPONSORS OF THE INTERNATIONAL MEETING

Bar Ilan University
Ben Gurion University of the Negev
Environmental Protection Service
Haifa Oil Refineries
Hebrew University of Jerusalem
Israel Electric Company
Israel National Committee on Problems of the Environment
Ministry of Health
National Council for Research and Development
Technion - Israel Institute of Technology
Tel Aviv University

CONTENTS

AQUATIC ECOLOGY

WASTE RECYCLING AND UTILIZATION

ENVIRONMENTAL HEALTH AND EPIDEMIOLOGY

AIR QUALITY MANAGEMENT

ECOLOGY OF STREAMS IN ARID AND SEMI-ARID AREAS

Developments in Arid Zone Ecology and Environmental Quality
Edited by H. Shuval

1

ECOLOGY OF THE NEGEV DESERT, A CRITICAL REVIEW OF OUR KNOWLEDGE

Michael Evenari

Department of Botany, Hebrew University, Jerusalem and The Blaustein Desert Research Institute, Sde Boqer, Israel.

ABSTRACT

The environmental parameters of the Negev desert with which desert organisms have to cope in order to survive, are described, as well as the main autecological adaptational features of the plants and animals living there, with emphasis on those characteristics which are common to all hot deserts. Convergent adaptations of plants and animals are compared. The main characteristics of the Negev ecosystem are discussed as well as the various possibilities of using the Negev for practical purposes without damaging or destroying its ecosystem.

A. DEFINITIONS

The great biologist Ernst Haeckel coined the term "Ecology" in 1866, deriving it from the Greek "Oikos" (house, home, domicile) and "Logos" (word, message). The literal meaning of "ecology" is therefore: The Science of the habitat. Haeckel already enlarged this simple sentence in defining ecology as the "gesamte Wissenschaft von den Beziehungen der Organismen zur umgebenden Aussenwelt, wohin wir im weiteren Sinne alle Existenzbedingungen rechnen koennen" (1866, p. 286). The more modern definitions do not differ much from Haeckel's. The simplest one could read: Ecology is the science of the relationship between the organisms and their abiotic and biotic environment. Most authors use this or a similar definition, and in all something important is missing. Instead of saying the "... the relationship between organisms and ..." the definition should read: "... the relationship between organisms in their living and dead state and ... . It may seem paradoxical to talk about a relationship between dead organisms and their environment but exactly this addition to our definition is of vital importance for understanding ecology in general and desert ecology in particular. As there is no life without death and no death without life, no ecological system would function without animals or plants dying in toto or in parte and abiological or biological decomposition of this dead material and recycling of its elements, a process deeply affected by environmental conditions. But in deserts especially as far as plants are concerned, the dead organisms also play a different, more direct ecological role which we could characterize as "the dead serving the living". We give only a few examples. The dead mother plant of the amphicarpic Gymnarrhena micrantha, through a complicated, purely physical mechanism, regulates over years the dispersal of its above ground telechoric dispersal units, synchronizing their dipersal with environmental conditions. The same mechanism prevents the simultaneous dispersal of all the units, spreading dispersal over years, which in deserts is a phenomenon of great survival value. At the same time the dead mother plant keeps its set of underground a-telechoric fruits (units of "non-dispersal") in place providing them with a habitat in which the survival probability of the seedlings is considerably higher than for the telechoric above ground dispersal units (Evenari 1963, Koller and Roth 1964, Evenari et al. 1971). The dead mother plant of Emex spinosus plays a similar role (Evenari et al 1977). In the rose of Jericho (Anastatica hierochuntica), a plant growing in the most extreme dry habitats of the Southern Negev, the dead mother plant persists for many decades and by hydrochastic "unfolding" movements of its dead seed carrying branches periodically disperses part of its seeds whenever rainfall conditions are favourable for germination. After a rain, when the atmosphere becomes dry, the branches "fold in" preserving the remainder of the seeds for a subsequent rainfall (Friedman et al. 1978). In both cases the "skeletons" (roots, stems and branches) of the dead mother plants must have special chemo-structural qualities preserving them from decomposition over a long period of time, a problem worth investigation. There are many other desert plants where "the dead are serving the living", such as Asteriscus pygmaeus, Plantago coronopus etc. (Evenari et al. 1971).

The "survival by partial death" (Evenari et al. 1971) so typical for many desert plants is another phenomenon where death as a function of environmental conditions has a survival value. We will deal with this later. All this is reason enough to stress the biological importance of dead organisms in the framework of any ecological system.

The science of ecology has been subdivided into Aut- and Synecology. As far as I know, both terms were first used by Schroeter and Kirchner (1902). In autecology we describe the relationship between the structural (morphological and anatomical), behavioural and physiological attributes and processes of the individual organisms and their environment. The ecologist considers some of these relationships as adaptations of the organism to its environmental conditions because they seem to him to be constitutional features which are advantageous for the continuity of the species in its struggle for existence. In its most literal sense the term "adaptation" describes therefore those specific organismic characteristics which according to our judgement enable it to survive in its given environment. It is obvious that, since a "value judgement" is involved (a fact which many ecologists are apt to forget) any statement concerning "adaptations" is and will always be of a subjective descriptive nature. There is always a danger, therefore, that "there will be constant lapses from description into explanation" as the modern philosopher Wittgenstein (1956,II, 85) stated of mathematics. The term "subjective" which we used above, does not relate only to the judgement of the individual ecologist. "Subjective" in this conjunction has a far deeper background. We must view the term "adaptation" also under its epistemological aspect. Our cognitive faculties of understanding nature are limited by the material composition and structure of our brain and of our sensor organs. It seems that we are able to perceive the organism-environment relationship only as an "adaptive" one so that the concept "adapatation" is also in that sense "subjective" that it relates to our innate cognitive abilities and not to the so called "real" or "objective" essence of nature which for us is not recognisable. There is a certain parallelism between this and the empiric hypothesis of language learning of the eminent linguistic philosopher N. Chomsky (1966, 1967). Chomsky claims that a language can be learned only because the human mind contains two innate ("a priori" ?) components enabling it to learn any language spoken by humans. One is a programmed function concerning certain structural qualities common to all human languages, the other is a value function evaluating the input during the learning process. Is not the same true for our conception of "adaptation" ? All this does not negate the necessity in ecology of using the terms "adaptation", "adapted to" etc.. The scientist should only be conscious of the basic "subjectivity" of, and the philosophical and semantic "pitfalls" contained in this and many other terms he is continuously using. Such another term used lately by nearly all ecologists is "strategies" or "survival strategies". This term is even more ambiguous than the term "adaptation" since its original meaning is the planning, directing and conducting of human operations. When applied to plants and animals, therefore, the term contains semantically a strong anthropomorphic element. I will therefore avoid its usage in this paper, although it is clear to me that it will be impossible to eradicate it altogether from ecology.

Synecology is the science of the relationship between biological communities and their environment. This aspect of ecology is many faceted. Population dynamics; ecosystemology (i.e. ecosystem research); ecology as applied to man-made (e.g. agricultural ecosystems, urban ecosystems etc.) and man-changed ecosystems, i.e. systems in which man is the dominant factor, are some of the aspects of synecology.

I stress these points because certain authors seem to equate synecology with ecosystemology, while for others ecology in general is synonymous with applied ecology.

B. MAIN ENVIRONMENTAL PARAMETERS AFFECTING LIFE IN DESERTS

The Negev is part of the belt of hot deserts composed of the deserts of North Africa and the Middle East. This belt is not only homo-climatic but also homo-ecological. Therefore, most of what we will say about the ecology of the Negev desert will be equally true for the North African and all Middle Eastern deserts. One can use various formulas to define these regions as deserts. The different maps based on the use of these different formulas, and showing the extreme arid, arid and semi-arid areas of the world, do not differ very much from each other. We will therefore use here only the formula of Thornthwaite, which is used by the new Unesco map (1979) of arid zones, and that of Oberlander (1979).

Thornthwaite's moisture index characterises the Negev as a desert because its values lie between -40 and -60. The Unesco map classifies the Northern Negev as semi-arid with $\frac{P}{ETP}$ values of 0.20-0.50 (P = mean value annual precipitation, ETP = mean annual potential evaporation), the main part of the Negev is shown as arid ($\frac{P}{ETP}$ = 0.03-0.20) and the southern part (including the Arava) as extremely arid ($\frac{P}{ETP} < 0.03$). According to Oberlander the Negev (and the whole Middle Eastern-North African desert belt) is an arid to extremely arid desert because its PE (potential evapotranspiration) lies between ca 1000 and 1500, its I_a (index of aridity) amounts to ca 80-98 and its I_b (index of humidity) values cover a range from slightly positive (ca +30) to very negative (Elath -82).

The Negev is a "hot" desert because the mean annual temperatures are between ca 18°C-26°C, the mean annual temperature of the hottest month is ca 25°C-39°C. This does <u>not</u> mean that in certain higher localities like Avdat the extreme minimum temperatures do not fall below 0°C, with sometimes even the appearance of snow. The values arrived at by using the different formulas and the figures of the various mean temperatures characterize the Negev as an area with a low amount of annual rainfall (the 19 year average for Avdat in the Central Highlands is 87 mm, see Evenari et al. 1980b), high evaporation (for Avdat an average yearly amount of 2611.3 mm as measured with a class "A" evaporation pan), high temperatures and a low relative humidity of the air (for Avdat an average annual value of 51.6%). But they do not inform us about a number of climatic parameters which, for the understanding of desert ecology, are of the utmost importance and are even more characteristic for hot deserts and of higher relevance than the "average" figures given above.

The most outstanding feature in this regard is what I would like to call the "rainfall instability syndrome" (RIS), i.e. the extreme quantitative, temporal and spatial irregularity, unreliability and variability of the rainfall which is therefore completely unpredictable. Of all environmental parameters affecting organisms living in the desert, this uncertainty factor is by far the dominant one and only those organisms which could develop ways to cope with the RIS can survive in the desert. The only exceptions are the few species whose water source is the much more reliable dew and/or fog which occur in certain deserts. The most simple ways to express the quantitative uncertainty of the rainfall are either the variability coefficient (maximum annual rainfall divided by its minimum, $\frac{M}{m}$) or the mean deviation as a percentage of the average annual rainfall. In non-extreme hot deserts, $\frac{M}{m}$ amounts to about 6-20, with values of 100 or more in extreme deserts. The corresponding values for the mean deviation are 35%-70% and more than 150%. In general the variability is inversely correlated to the amount of precipitation.

Because of the great variability of the annual precipitation, the isohyetes of the deserts shift from year to year over large distances. In the Negev, according to our estimates, the isohyete of 25 mm moved in 1962/63 at least 100 km to the North. Monod (1981) reports for the Sahel a 600 km move of the 100 mm isohyete in 1951/52.

Another typical rainfall feature is the irregular occurence of sudden showers - extreme rain pulses - which produce large quantities of water falling in a short time at high intensities. On October 5th, 1979, 14 mm of rain fell in the Mashabe Sadeh area of the Negev within 7 minutes, equalling an intensity of 120 $mm \cdot hr^{-1}$! The temporal aspect of the RIS can best be demonstrated by showing the variability of the length of the rainy season and by giving the dates of the occurence of the first and last rains during various rainy seasons. The Negev, as all the hot deserts of Africa and the Middle East, has a pronounced winter rainfall season with few rainy days (in Avdat an average of 16-20 days with a rainfall of more than 0.1 mm). In 1960/61 the rainy season in Avdat lasted for four months, while in 1963/64 it lasted for seven. During a 20 year observation period at Avdat, the earliest first rain fell on October 5, 1965 and the latest first rain on January 17, 1963. The dates for the earliest and latest last rains of the season were February 28, 1961 and May 16, 1967

(Evenari et al. 1980a).

Another biologically important aspect of the temporal variability of desert rainfall concerns the occurrence of "effective" rains which, according to Monod (1973), are rainfalls which wet the soil to such an extent that seed germination and seedling survival is possible. In deserts, the time of year of the efficient rainfalls and the time gaps between them are most variable. In extreme deserts the time gap between efficient rainfalls may be years, in less extreme deserts days or months. The date of efficient rainfalls are biologically very important because, although water is the main life limiting factor, in deserts, desert organisms are also subjected to the control of photo- and thermoperiodism which have to be "right" to permit germination, growth and flowering of plants and mating, reproduction and development of animals. Since both periodisms are date-dependant, an otherwise efficient rain may become inefficient because it occurs at the photo- and/or thermoperiodically "wrong" date. The time gaps between the rainfalls are important because a rain which in itself is only sufficient to trigger germination - to take an example from the plant world - but insufficient to guarantee seedling survival, may become "efficient" if the time gap between it and the following rainfall is not too large. In this way two "inefficient" rains may in combination become "efficient".

The spatial aspect of the RIS is best documented by two typical observations. The 14 mm rain falling in 7 minutes, mentioned above, was localised in an area of 30-40 km^2. We once saw the main road near Avdat divided lengthwise, half dry and half wet, as a result of a spatially restricted rain storm.

In deserts, runoff, i.e. that percentage of the rainfall which does not penetrate where it falls but runs off, plays an important biological role. In the Central Negev, runoff amounts to 10-30% of the average annual precipitation if we exclude sandy soils and sand dunes where infiltration is high. (Evenari et al. 1971, 1980a, Shanan 1975). In specific cases, in very small watersheds and during rainstorms of very high intensity, runoff may be much higher (50-80%). Runoff collects in certain localities creating "run-in" habitats, where even in bad drought years water conditions can be most favourable for plants and animals. This is one of the reasons why the stone cover of the soil surface ("desert pavement"), so typical for all deserts, has a biological function. Stones retain part of the runoff which accumulates below the stones in water pockets, which are the water source for plant roots.

For man living in the deserts, runoff and the floods caused by it are also of high relevance, as will be shown later.

For poikilohydrous plants and certain animals, dew is a more important water source than rain. The Negev Highland belongs to the dew-rich hot deserts. We observed in Avdat, over a 17 year period, an average of 195 dewy nights with an $\frac{M}{m}$ quotient of only 1.4, showing that in contrast to rain, dew is a much more reliable factor although the average annual amount of dew is low, 33 mm (measured with the Duvdevany dew gauge).

High air temperatures and intense radiation are other biologically important parameters in hot deserts. In the Negev Highlands, the extreme absolute maximum air temperatures move between 40°C-43°C, but in other parts of the Negev, Middle Eastern and North African deserts they reach 45°C-47°C. In order to survive, desert organisms must find ways to cope with these high air temperature maxima. But soil temperatures are biologically no less important. On a normal summer day with a maximum air temperature of 32°C we measured in the upper 2 cm of the loess soil, a maximum temperature of 44°C at noon and a minimum of 16°C around midnight, but there are days when the upper 2 cm heat up to more than 50°C. On the same day of measuring, the maximum temperature of the same soil layer under a stone was only 36°C, showing another reason why in hot deserts a hypolithic habitat is a comparatively favourable one. In sharp contrast to the upper soil layers, the soil temperature at 50 cm depth was about 24°C with a diurnal change of not more than 0.5°C. At 100 cm depth the soil temperature remained at a nearly constant 20°C the whole year round. This means that underground organs of plants and burrowing animals reaching a a depth of about 30-100 cm, live in a nearly "air conditioned" environment.

Because of the large number of cloudless days (3000-3600 sunshine hrs • $year^{-1}$), radiation is very intense throughout the year, reaching, at the hottest time of the year, values of ca 750 Kcal • cm^{-2} • day^{-1}. This subjects exposed organisms like above ground parts of perennial plants to a high radiation load, another factor with which they have to cope.

Desert soils of arid and extremely arid regions have a number of common characteristics. They are immature, have only weakly developed or no profiles, are - with the exception of salines - dry for the most part of the year, contain little organic material (ochric epipedon), are poor in nitrogen, are - with the exception of sands - saline, are rich in calcium carbonate and gypsum concreted in calciform and gypsiferous hardpans (duripans, caliche) forming lime and gypsum concretions and crusts. Most soils are alkaline, predominantly quartzite sands; stony, gravelly, silty and sandy loams (including loess) and clays; rocklands with skeletal lithic soils; alluvium in fans and colluvium in talus; gravel and stones in wadi beds.

In trying to classify the Negev soils we oncounter some difficulty because there is an "old" classification, used by Dan et al. (1976) in their soil map, and a "new" one (7th approximation, Flach and Smith 1969, Soil Survey Staff 1975). According to the new soil taxonomy, the main orders represented in hot deserts of the Middle East and North Africa are entisols and aridisols and to some extent alfisols. Entisols of these regions belong mainly to the sub-order of fluvents, orthents and psamments. In these suborders xerifluvents (xeric alluvial soils), terriorthents and xerorthents (lithosols and regosols) and torri- and xero-psamments (arid and xeric shifting or stabilised sand dunes predominate. Loess must apparently be classified under orthents (?). Aridisol with their two suborders of orthids and agrids are found exclusively in arid regions, as their names indicate. Many aridisols contain duripans (see above) classified as durorthids, durargids and calciorthids. Aridisols with salic and natric horizons, so typical for playas and salorthids (formerly "solonchaks and natrorgids (formerly "solonetz"). Camborthids ("sierozems", "loessial sierozems", "gray desert soils" of the "old" classification) have a cambic horizon, no duripans and no salic or natric horizon. Soils once called "red and gray desert soils" are now classified according to their structure as camborthids, durorgids, haplargids or paleorgids. Paleorgids, with their petro-calcic horizon, date back to the pleistocene (?).

Each of these soil types has its own specific biological importance. Psamments have a high infiltration rate and a very low water holding capacity. Rain water is therefore taken up fast and there is no runoff. If psamments are stabilised in sandfields overlying water impermeable layers on which the infiltrated water accumulates and very little of it is lost by evaporation, they become most favourable habitats for plants whose roots reach this sub-psamment water table. Naturally, moving psamments cannot sustain vegetation and are most destructive when moving over vegetated (natural or man made, i.e. agricultural) areas. Duripans in durorthids, durargids and calciorthids are mostly impenetrable for plant roots and therefore limit plant growth. Camb-orthids carry comparatively dense desert vegetation. Loess soils, when deep enough, as for example on many flood plains, and when water is available (see below) have been and can be turned into good agricultural soils. Terri- and xer-orthents are shallow and, as far as vegetation is concerned, poor soils. Salorthids and natrorthids are according to their salt and water content either without vegetation or are, at least partly, vegetated by halophytes.

Desert soils which were once thought to be "dead" soils are not static entiti but open systems in dynamic equilibrium with the climate and with the organisms they sustain. Together with the plant roots growing in them, the rhizosphere around the roots, the animals living in them and the microorganisms they harbour (bacteria, actinomycetes, streptomycetes, fungi and algae), desert soils could be considered to be organisms with their own history and evolution. We know from recent investigations how "alive" the desert soils of the Negev Highlands are. Buyanovsky (unpublished communication) found $1\text{-}2 \times 10^6 \cdot g^{-1}$ heterotrophic microorganisms in the upper 10 cm of soil during the dry season. During the rainy season their number increased to $3\text{-}4 \times 10^6 \cdot g^{-1}$. Under the dominant dwarf shrubs their number was 1.5-2 times higher than in the interspace. These desert soil microorganisms start to be active a few hours after a rainfall which for higher plants may be "ineffective" (Funke and Harris 1968). We will return later to the importance of these decomposers in desert ecosystems. Algae and certain lichens form crusts on desert soils. Fungi forming mycorrhizas and rhizomorphs are abundant in desert soils (Went and Stark 1968, Khan 1974, own observations). They possibly play a role in the roots water supply (?). Roots with mycorrhyzas are apparently more drought resistant than roots without them (Harley 1969). Desert rodents, ants, termites and isopods burrow into the soil, changing its physico-chemical qualities and affecting the erosional cycle. Hemilepistus reaumuri, an earth eating isopod, has in the Negev Highlands an annual soil turnover of $28.5\text{-}105.7\ g \cdot m^{-2}$ (Shachak et al. 1976) and snails have a soil-crust turnover of $14.2\ g \cdot m^{-2}$ (Shachak and Steinberger 1980). Plants are not only part of the desert "soil organism" because of their underground parts, but affect it also in another way. Soil accumulates under many desert shrubs creating microhabitats richer in carbon, nitrogen and water than the interspaces and affecting nutrient cycling. "Nebkas" and "Rebkas" are the most conspicuous of such soil accumulations around plants ("phytogenic hillocks").

The landscapes most typical for hot deserts are the immense regs with their desert pavement, composed of black-brown (desert varnish !) gravel and small stones; the stony hammadas; the regoid and hammadoid slopes; the dry river beds (wadis, oueds, arroyo secos); the ergs (mobile sand masses and dunes); sand plains; inselbergs; floodplains; single or compound alluvial fans (bajadas, piedmonts) and hydromorphous salines (playas, chottes, sebkas, solars, salines). As far as living organisms are concerned, each of these landforms is a macrohabitat and a macroecosystem, with its specific plant and animal communities (which in this short paper we cannot describe) and its specific climatic, topographic, hydrological and edaphic conditions.

In deserts, microlandforms creating microhabitats are more crucial for survival than macrolandforms because they create a great diversity and heterogeneity of microniches without which life in deserts would be even more sparse than it is. The more extreme the desert, the higher the survival value of microlandforms. The smallest runnels, the slightest depressions, animal burrows, rock fissures, rock shelters, phytogenic hillocks and the smallest shade-providing areas below dwarf shrubs are only some of the many important desert microniches.

C. THE MAIN AUTECOLOGICAL ADAPTATIONS OF DESERT ORGANISMS

For lack of space we can here only tabulate the main relevant features and indicate the literature which relates to organisms living in the Negev. Each list will be followed by some remarks enlarging upon certain details of the lists which need explanation, and stressing some facts of a general nature. We will deal first with plants.

AUTECOLOGY OF PLANTS

C.I. Poikilohydrous plants (Lichens and algae, Evenari et al. 1971, Berner 1974, Lange et al. 1975, Berner and Evenari 1978, Kappen et al. 1979, Evenari et al. 1980a).

1) Capacity of their thallus to take up liquid water instantaneously from rain or dew, and water vapour from air with a relative humidity above 70%.

2) Capacity to equilibrate with hydration level of their environment and to survive extreme and prolonged desiccation.

3) Capacity to enter cryptobiotic, i.e. metabolically inactive, state when desiccating, and to be resistant in this state to extreme cold and heat.

4) Rapid switching-on of metabolic activity as a function of water availability and rapid switching-off when thallus begins to dry out.

5) Capacity to go through repeated cycles of crypto- and phainobiosis* (metabolically active state) without damage.

6) Relative high rate of photosynthesis at low temperatures (lichens) and low light intensities (algae).

Remarks: The facts given in this list are obviously autecological adaptations to the environment. As such they have a high survival value in the harsh desert environment, but automatically carry with them certain limitations.

---

* From Greek 'phainesthai' - to appear, to reveal.

These limitations are a consequence of the fact that each adaptation is the end effect of an intricate, interwoven chain of internal reactions of the organism to external environmental factors. Because of its nature, such a reaction is always a compromise, the "advantageous" nature of an adaptation is limited, i.e. has "positive" and "negative" aspects. This is true for all the adaptations of plants and animals we will discuss. We may be permitted to express the same anthropomorphically, by saying that the organisms involved must pay a price for the sum of their adaptations. In the case of the poikilohydrous plants this price is an extremely slow growth rate (Lange and Evenari 1971). There is also another point which has to be stressed in connection with the general question of adaptations. Each organism possesses per-force certain "constitutional" characteristics which are innate and are independent of its specific environment. These features are neither adaptations nor non-adaptations. They could, in relation to the environment, be designated as "neutral". Only two examples will be given. All lichens including the tropical ones can photosynthesize at low temperatures of -8 to -14°C, a fact which for the Negev lichens has certainly no adaptive value. The photosynthesis of fully saturated lichens (160% water content) is depressed in comparison with the photosynthetic rate of thalli with a lower water content (Lange 1980). Since in the Negev the water content of the lichens does not surpass 90%, this specific feature of lichen physiology is, for the Negev environment, "neutral".

C.II. Homoiohydrous plants

a) Arido-passive (metabolically non active during the dry season) perennials (geophytes and hemicryptophytes, Evenari et al. 1971, 1980a).

1) Capacity of their persistant organs to become and remain dormant during the dry season of the year and to remain completely dormant during extreme drought years.

2) Ability to modify the length of their metabolically active growth- and flowering period which is subjected to the control by photoperiod, thermoperiod and endogenous rythms, according to the availability of water.

3) Ability to form fast growing adventitious roots ("rain roots") a few hours after water becomes available in the soil layer in which their persistent organs are located.

b) Arido-passive Pluviotherophytes (Zohary 1937, Evenari 1963, Koller and Roth 1964, Gutterman et al. 1967, Datta et al. 1970, Koller 1972, Evenari et al. 1971, 1977, 1980a).

1) Dispersal units and dispersal

1.1. Formation of telechoric and a-telechoric and of morphologically and/or physiologically heterogeneous and heteroblastic dispersal units on one mother plant (see above).

1.2. Restriction of dispersal to periods with environmental conditions favourable to germination and seedling survival.

1.3. Mechanisms preventing the simultaneous dispersal of all dispersal units of one plant.

2) Germination

2.1. Longevity of seeds.

2.2. Mechanisms (including "internal water clocks" and annual (endogenous ?) rythmicity) limiting germination in time and space to environmental conditions favourable to germination and seedling survival. In extreme drought years seeds, with few exceptions, do not germinate at all.

2.3. Extreme heteroblasty of seed populations preventing simultaneous germination even under most favourable environmental conditions.

2.4. Structures and mechanisms ensuring good seed-soil contact.

3) Development

3.1. Large scale modification of the temporal amplitude of the photo- and thermoperiodically controlled life cycle as a function of water availability, leading in extreme cases to ephemerality.

3.2. Regulation of size of whole body and single organs as a function of water availability, leading in extreme cases to tiny dwarf plants with very few and very small leaves.

Remarks: All our lists of adaptations contain some features which are typical of all or most of the types of plants or animals listed, whereas others concern only some specific species. The adaptations, for example, listed under 2.2, 2.3, 3.1,3.2 are common to practically all pluviotherophytes, the ones under 1.1, 1.2 and 1.3 are specific. An additional interesting feature of hot desert plants is that some are facultative pluviotherophytes and become, under certain conditions, arido-passive biennials or even perennials. An important consequence of the facts given in 2.2 is that in extreme drought years the contribution of the pluviotherophytes to the overall phytomass is zero in contrast to good rainy years when it may amount to 50% (Evenari et al. 1976).

c) Arido-active (metabolically active during the dry season) perennials (Evenari et al. 1971, 1975, 1980a, Lange et al. 1974, 1975b, 1975c, 1978, Schulze et al. 1980).

1) Structural, spatial and developmental adaptations

1.1. Permanent (constitutional) xeromorphy and/or tendency to develop temporary xeromorphy induced by external stresses.

1.2. Reduced metabolically active surface either by constitutionally small surface/dry weight quotient and/or by seasonal surface reduction (leaf and branch shedding) and/or "partial death" as a function of increasing water stress.

1.3. High value of the ration: undergrounf (root)/above ground phytomass, and great flexibility of root system as a function of the availability of soil water.

1.4. Roots exploit a relatively large soil volume, have a large horizontal extension in the upper soil layers and, according to habitat, reach great depths.

1.5. Avoidance of heat loads by appropriate leaf movements, by hair cover and high reflectivity.

1.6. Suppression of flowering and fruiting in extreme drought years.

2) Physiological adaptations

2.1. Ability to remain metabolically active at very low soil and plant water potentials, high internal water deficits and high temperatures.

2.2 Gas exchange regulating stomatal mechanisms with high sensitivity to the following external and internal factors and their combination: water stress, light intensity, temperature, relative humidity of the air.

2.3. Increase of residual resistance to gas exchange with decrease of water potential.

2.4. Ability to control transpiration and CO_2 uptake independently, at least to a certain degree ("flexible coupling").

2.5. Ability to change diurnal pattern of photosynthesis as a function of decreasing accumulated plant water potentials ("bar-days" of Schulze et al. 1980).

2.6. High temperature optima and temperature acclimation of net photosynthesis.

2.7. High cuticular resistance to gas exchange. Stomatal movements (opening and closing) are therefore most efficient regulators of gas exchange.

2.8. Decrease of dark respiration with decreasing water potentials.

2.9. Capability to tolerate high internal concentrations of inorganic ions.

2.10. Capability to reduce salt load by salt excreting glands.

2.11. Capability to store relatively large amounts of water (succulance).

Remarks: The main survival problem facing the arido-active desert plants is the "photosynthesis-waterbalance syndrome" . CO_2 uptake, and therefore photosynthesis are only possible when the stomata are open but open stomata result, through transpiration, in loss of water, the one overall life-limiting factor during the dry season. Therefore, during the dry season the arido-active plants have continuously to navigate precariously between Scylla and Charibdis, i.e. the danger of death by too much loss of water and the danger of death by insufficient dry matter production. The facts listed under 1.1, 1.2, 2.2-2.8 are all means with which to"navigate"succesfully and solve the photosynthesis-waterbalance syndrome. In this connection the term "partial death"

(1.2) needs some explanation. Some mature desert dwarf shrubs can split their stems and thus create out of one original mother plant a consortium of independent clones. In bad years some of these clones die, thus enabling the remaining clones to survive. With the progression of the dry season, the solution to the Scylla-Charybdis problem entails that water loss as well as net photosynthesis, and therefore also dry matter production, be more and more restricted, leading to a continuous decrease of the relative day-time gain and maximal rate of photosynthesis as a function of the accumulated decrease in water potential (bar-days). This is aggravated by the increase of respiration with increasing temperature. At the end of the dry season, when water potentials are very low (down to -163 bar), plants still show small gains of dry matter for 2-3 Hrs in the morning (see under 2.1), but for the other 10-11 daytime hours photosynthesis is below the compensation point, i.e. dry matter is lost. All this together explains why the overall growth of arido-active desert plants is so slow and why in bad drought years they cannot invest and "waste" dry matter in flower and fruit formation (1.6) (an adaptation which is only possible because they are perennials which can do without sexual reproduction for one or two years). To repeat what was already given as a general maxim we may say that this is the price paid for the various adaptations. The severity of the photosynthesis-water balance syndrome is alleviated by the capability of the root systems to take up water fast from a large volume of soil, to reach the smallest water pockets (i.e. below stones) and, in some cases, to store water (1.3, 1.4, 2.11).

For lack of space we will not deal with the interesting problem of the arido-active biseasonal annuals which in the Negev are represented only by a few species. We will also not touch upon the question of the competitive advantages and survival value of species with different photosynthetic pathways (C_3, C_4, CAM) since this question is still hotly discussed. We can only state that the vegetation around Avdat contains only a very few CAM species and that C_3 and C_4 plants coexist there successfully.

AUTECOLOGY OF ANIMALS

Our knowledge of the autecology of wild desert animals of the Negev is more restricted that that of plants. Only a limited number of species of the various taxonomic orders have been studied in detail. In order to prevent unwarranted generalizations we will mention the species studied which we did not do for the various plant groups.

C.III. Snails (Sphincterochila boissieri, S. prophetarum, Trochoidea seetzeni, Yom-Tov 1971a, b, Shachak et al. 1975, Shachak and Steinberger 1980, Steinberger et al. 1981).

a) Activity pattern

1) Activity (phaenobiotic phenophase) restricted to rainy days with wet soil and low temperature (ca 20 days per year). Inactivity (cryptobiotic phenophase) between rainfalls and during rainless period of year.

2) Capacity during short periods of activity to take up sufficiently large amounts of water and food to enable survival during long cryptobiotic phenophase.

b) Heat balance

1) High heat tolerance of body and high radiation reflectance by shell.

2) Isolating air layer between body located in upper part of shell and hot soil.

c) Water balance

1) Low evaporative water loss during cryptobiosis.

2) Capability of keeping continuous high hydration during cryptobiotic phenophase by formation of metabolic water.

d) Reproduction and life span

1) Synchronisation of reproduction with favourable environmental conditions. Suppression of reproduction in drought years.

2) Oviposition in localities protected against overheating and desiccation.

3) Relative longevity (8-15 years) as counterbalance against high mortality of young.

Remarks: These desert snails can be compared with hyper-arido-passive plants. Their activity rythm is most drastically steared by the environment. The main enemies of snails are carnivorous rodents and migrating birds.

C.IV. Isopods (Hemilepistus reaumuri, Shachak et al. 1977, Dubinsky and Steinberger 1979, Shachak 1980).

a) Water and heat balance

1) Water sources: dew, water content of eaten plant material and water in eaten soil.

2) Formation of metabolic water.

3) Avoidance of water loss and overheating by staying, during hot and dry time of day, in burrows with high relative humidity and moderate temperatures (temperate climate).

4) Deepening of burrows according to season of year and phenophase of animal, thus keeping soil moisture in burrows at a range of 6-13% and keeping temperatures relatively low.

b) Synchronisation of activity outside burrows (surface activity, S.A.), during various phenophases of animal, with favourable environmental conditions.

1) During pair formation phenophase (Feb.-March), S.A. is from 12;00-16:00 Hrs (unimodal).

2) During gestation and hatching phenophase (Apr.-May), and during growth phenophase (May-Nov.), S.A. is from 7:00-9:00 and 17:00-18:00 Hrs (bimodal). At end of growth phenophase S.A. is from 5:00-7:00 Hrs (unimodal).

3) During inactive phase (Nov.-Jan.) no S.A. During this time animals need neither food nor water (cryptobiosis ?).

c) Reproduction and social life

1) Reproduction synchronised with rainy and cool season.

2) Large clutch size.

3) Monogamy, staying together of family and parental care of the young increase probability of individual survival.

Remarks: It is quite remarkable that Hemilepistus is day- and not night-active, since night activity would solve the problem of high water loss by evaporation and of overheating because the animal has no heat tolerance (it dies already at a body temperature of 35°-36^oC) and cannot suffer dehydration. The reason for its day activity could be constitutional (?). Nevertheless, this activity is restricted to a limited number of hours which could perhaps be explained by the fact that its main enemy is a day active scorpion. Perhaps this is also the reason for the winter inactivity of the Hemilepistus (the scorpion is winter active), or perhaps we are dealing here again with a constitutional characteristic. Another certainly non-adaptational peculiarity is the high mortality rate especially during the pair formation phenophase. This is apparently at least partly counterbalanced by the features given under 'c'. Because of its food sources (alive and dead plant parts, organic material in soil), the animal never suffers from lack of food.

We do not know whether Hemilepistus possesses some additional adaptative features which are characteristic of other Arthropods, like spiracular control of water loss, conservation of body water by low rates of integumentary evaporation, low metabolic rates, low values of respiratory Q_{10}, or low oxygen consumption.

C.V. Reptiles (Eremias guttulata, Acanthodactylus boskianus, Agame stellio, Gecko spec., Uromastix aegyptius, Shkolnik in Evenari et al. 1971, 1980a, Orr et al. 1979).

a) Behavioural avoidance of heat and water stress

1) Diurnal reptiles exposed to heat- and water-stress only for short periods, through high mobility and use of stress protected habitats.

2) Nocturnal reptiles avoid stresses by remaining in "climatised" deep burrows during daytime.

3) Standing higher on their legs away from hot soil.

b) Temperature regulation (reptiles are poikilothermous !)

1) Exposing body to sun for short lengths of time.

2) Temperature control by counter current heat exchange and cutaneaous vasodilation (with heating) and vasocontraction (with cooling).

3) Colour change from dark to light by contraction of melanophores.

4) Evaporative cooling by panting and gular fluttering.

5) Cryptobiosis during cold winter.

c) Water balance

1) Water source of carnivorous reptiles is prey and of herbivorous reptiles is succulent plants.

2) Conservation of body water by excreting semi-solid urine ("urate pellets").

C.VI. Birds (Corvus corax ruficolis, Ammomanes deserti, Oenanthe lencopyga, Athene noctua, Columba livia, Pterocles alchata, Pt. coronata, Pt. senegallus, Alectoris chukar, Orr 1970, Marder 1973, Shkolnik in Evenari et al. 1971, 1980a).

a) Adaptations to high environmental temperatures

1) Constitutional high body temperatures (2-4°C higher than in mammals).

2) Facultative hyperthermia.

b) Prevention of overheating

1) Protection against overheating by air enclosed in plumage.

2) Evaporative cooling by panting and gular fluttering.

3) Cooling by soaring at high altitudes and by special posture of wings when resting.

4) Early morning and evening activity of diurnal birds during the hot season and resting in shade-providing niches during the hottest hours.

5) Exclusive nocturnal activity.

c) Water balance

1) Dietary intake as water source for carnivorous and omnivorous birds.

2) Ability of herbivorous birds to fly large distances to the few sources of drinking water and to transport water in their plumage to their young.

3) Water conservation by low glomerular filtration rate (G.F.R.), reduced urine flor (U.F.R.) and excretion of almost soild uric acid.

d) Reproduction

1) Synchronisation of breeding with rainy season.

2) Adaptation of clutch size to rainfall conditions.

3) Selection and construction of stress protected nesting sites.

C.VII. Rodents (Acomys russatus, A. cahirinus, Gerbillus gerbillus, G. dasyurus, Psammomys obesus, Meriones crassus, Jaculus jaculus, Shachak and McPhee 1970, Shkolnik in Evenari et al. 1971, 1980a, Boruth and Shkolnik 1974).

a) Behavioural avoidance of heat and water stress

1) Day and night active species living in "climatised" burrows or caves.

2) Synchronisation of outside activity of day-active rodents with favourable environmental conditions (e.g. morning and evening activity during hot dry season, allday activity during cool rainy season).

b) Water balance

1) Prey water source of carnivorous and plants water source of herbivorous rodents.

2) Ability to tolerate high salt content of their food (e.g. halophytes).

3) Water content of vegetable diet stored in burrows increased by absorbing water vapour from hunid air in burrows.

4) Formation of metabolic water.

5) Water conserved by excretion of extremely dry faeces and highly concentrated urine.

Remarks: Acomys russatus is a good example for b.1. Its main food consists of snails containing 60-90% water, making the animal fully independant of any other water source. The habit of Meriones crassus of storing seeds (e.g. of Zilla spinosa) in its burrows illustrates well the point made under b.3. The seeds stored in the burrows (relative humidity of 60%) contain 30% water while the water content of the same seeds outside amounts only to 4-7%.

C.VIII. Large mammals (Gazella gazella, G. dorcas, Capra beden, Hyaena striata, Vulpes rupeli, Felis ocerata, F. caracal, Hystrix indica, Lepus capensis, L. sinaica).

a) Avoidance of heat and water stress

1) Diurnal activity of grazing animals as described for rodents under a.2.

2) Protection against both stresses by fleece.

b) Prevention of overheating

1) Evaporative cooling by panting and sweating.

2) Facultative diurnal hyperthermy and nocturnal hypothermy.

3) Counter-current heat exchange.

c) Water balance

1) Ability to survive long periods without drinking water.

2) Large drinking capacity enabling animals to restore large water loss in few minutes.

3) Capacity to tolerate large decrease of body mass through water loss without change in volume and composition of blood.

4) Relatively dry faeces.

5) Reduction of GFR and UFR (see Birds under c.3).

d) Energy balance

1) Relatively low metabolic rates and modest food requirements.

2) Building up of food reserves (fat!) during favourable season for use during meagre summer months.

Remarks: This list of adaptations is based in large parts on our rich fund of information on the ecology of domesticated desert animals, i.e. camels, Bedouin sheep (Awassi breed) and goats, and donkeys (see Shkolnik in Evenari et al. 1971, 1980a and literature cited there), since our knowledge of the ecology of wild large desert mammals of the Negev is limited. In general, all herbivorous large desert mammals need drinking water whereas the carnivorous can apparently (?) do without it. With the exception of the grazing species (gazelles, ibex etc.), most of the species are exclusively night-active.

As a conclusion to this chapter on autecology it is worthwhile to compare the autecological adaptations of plants to that of animals, in order to demonstrate their convergence. But before doing this, one general statement has to be made: To the best of my knowledge there is not a single adaptation of desert organisms to their environment, the constitutional basis of which is not found in taxonomically comparable non-desert counterparts. What is then, in this regard, the difference between a desert and a non-desert organism ? It is only the fact that during evolution, through the process of natural selection of the fittest, various constitutionally given capacities characteristic of a specific taxonomic group were intensified in desert environments, developed and their amplitude enlarged, enabling the roganism involved to survive in the desert, "adapted" to it.

C.IX. Adaptational convergence of desert plants and animals

Plants	Animals
a. Stress avoidance	
Arido-passivity of whole plants (geophytes, hemicryptophytes)	Inactivity in time and space (geozonts* in burrows, other animals in stress protected above ground shelters)

---

* I propose the term "geozonts" as a parallel to geophytes for those animals which live in underground burrows. According to Bodenheimer (1957), about 72% of the mammals of arid zones belong to this group.

Postural adjustments (e.g. orientation of leaves)	Postural adjustments (e.g. position of wings)
Cryptobiosis of whole plants	Cryptobiosis of mature animals (aestivation, hibernation)
Dormancy of seeds	Cryptobiosis of eggs ("shelled embryos", larvae)
Growing in stress protected micro-habitats	Permanent habitation or temporary use of stress protected microniches

b. Stress control by structural features and eco-physiological processes

1) Reduction of water loss

By stomatal movements and change in residual resistance	By concentrated urine, dry faeces, reduction of GFR and UFR
By xeromorphy (thick lipid cuticula, hair cover, sunken stomata, small surface/volume ratio)	By water impermeable integuments, cover by fleece and feathers, small surface/volume ratio (large mammals)
By phyto-succulence	By zoo-succulence

2) Maximalization of water uptake

Direct uptake of dew and water vapour	Direct and indirect uptake of dew and water vapour
Fast formation of "water roots" after first rain	Fast drinking of large quantities of water, uptake of water from wet soil
Halophytes (uptake of highly saline water, high salt tolerance, salt excreting glands)	Halozonts (uptake of highly saline water, high salt tolerance, salt excreting glands)

3) Prevention of overheating

Transpirational cooling	Evaporative cooling
High heat tolerance and capacity to function at high temperatures	As in plants
Mechanisms decreasing and/or dissipating heat load (structures causing formation of boundary layers, high reflectivity, insulating structures)	As in plants

4) Control of reproduction

"Water clocks" of seed dispersal and germination	Sexual maturity, mating and births synchronised with favourable environmental conditions
Suppression of flowering and sprouting in extreme drought years	Sterility in drought years

D. SYNECOLOGY

In this section we cannot discuss the population dynamics of the Negev because as far as I know there are only two papers dealing with several botanical aspects of this topic (Evenari and Gutterman 1976, Gutterman, Evenari and Noy-Meir in manuscript) and none on the animals aspects. But we have some knowledge on the Negev ecosystem. The term "ecosystem" is comparatively new. Woltereck (1928) was the first to speak about an "oekologisches system" and Tansley (1935) shortened Woltereck's expression to "ecosystem". But the term "biocoenosis", meaning a "community of organisms", is much older since it was already coined by Moebius in 1877. It could be considered to be a precursor of the term "ecosystem" or a first step in that direction. The term "biogeocoenosis", used by some authors, has more or less the same meaning as "ecosystem" but is a more cumbersome expression. Since a "system" can be defined as "an assemblage of interacting things forming a complex or unitary whole" the shortest definition of an ecosystem would be: "A community of organisms and their environment in which the biotic and abiotic components are interacting and interdependent". The "community of organisms" is normally composed of primary producers (i.e. the autotrophic plants), consumers (i.e. the heterotrophic plants and animals) and decomposers. But there are some strange desert ecosystems without direct primary producers (e.g. in the Namib) and theoretically there could be ecosystems without consumers.

The most outstanding specific features of the Negev ecosystem are:

D.1. Of all the abiotic and biotic parameters which control and determine the reaction of the whole system, water (rain, runoff, dew, atmospheric humidity) is by far the most preponderant and effective

D.2. The biotic components of the system in which the water source is directly or indirectly rain, react to the irregular, unpredictable "rain pulses" (RIS) with equally unpredictable "bio-pulses". This is most evident in respect to primary production and germination of homoiohydrous plants. It is not true for those components for which dew is the main water source.

D.3. The primary producers have a low standing biomass, low production but high relative productivity. The same is true for the standing biomass and production of the consumers.

D.4. The system has, in addition to its food web, a complex "water web" which plays a major role in the functioning of the whole system.

D.5. Natural and bio-made microhabitats are of the utmost importance for the survival of the whole system.

D.6. Since nitrogen is a limiting parameter in the system the N-cycling plays a major role in comparison with the cycling of the other nutrient elements.

D.7. The main energy flow goes directly from the primary producers to the decomposers and comparatively little flows through the consumers to the decomposers. This is only true with the wild consumers and not with the man-introduced ones.

Some of these points have to be specified:

a) Primary production (Evenari et al. 1976)

The average standing phytomass and the net annual production of the primary producers of all desert ecosystems is the lowest of all ecosystems known, whereas their relative primary production, i.e. the quotient net annual primary production/standing phytomass, is relatively very high. This is obvious if we compare the relevant figures of desert ecosystems with those of tropical forests, which have the highest phytomass of all ecosystems.

	Standing phytomass of mature stands $(t \cdot ha^{-1})$	Net annual primary production $(t \cdot ha^{-1} \cdot y^{-1})$	Relative primary production
Tropical forests	60-800	10-50	0.04-0.17
Deserts	1-4.5	0.5-1.5	0.33-0.5

The extremely low figures for standing phytomass and annual production of desert ecosystems are a consequence of the extreme environemental stresses, and especially the water stress, to which the desert primary producers are subjected. The low figures for the relative production of the tropical forests and the high figures for deserts result from the very different life-form composition of these ecosystems. The main primary producers of the tropical forests are trees. A large percentage of the phytomass of trees consists of photosynthetically unproductive tissue to which there flows annually a large percentage of the energy produced by the photosynthetically active organs. In the Negev (as in most but not all hot deserts) trees are very rare and grow only in special habitats with much available water during most of the year and even then they grow only in limited numbers and at large distances from each other. Apparently trees in deserts cannot be a dominant life form because on the one hand the desert conditions do not permit the "luxury" of investing a large percentage of the photosynthetically produced energy (i.e. dry matter) in photosynthetically inactive organs, and on the other hand, light in deserts is not a limiting factor as in tropical forests where trees are a life form adjusted to the competition for light. In deserts the dominant life forms are annuals, hemicryptophytes, nanophanerophytes, chamaephytes and geophytes, all of which are life forms which contain, above ground only a small percentage or no photosynthetically inactive tissues.

The facts mentioned above under D.1 and D.2 are best illustrated by the following figure showing primary production, for two consecutive years, of a Negev Hammadetum:

Year	Rain (mm)	Plant	Peak above ground phytomass ($t \cdot ha^{-1}$)	Peak below ground phytomass ($t \cdot ha^{-1}$)	Total peak phytomass ($t \cdot ha^{-1}$)
1971/72	163	Hammada scoparia	0.73	0.41	1.14
		Peganum harmala	0.24	0.08	0.32
		Carex pachystilis	0.53	2.30	2.83
		Annuals	2.70	1.25	3.95
		Total	4.20	4.04	8.24
1972/73	54	Hammada scoparia	0.29	0.34	0.63
		Peganum harmala	0.10	0.02	0.12
		Carex pachystilis	0.00	1.11	1.11
		Annuals	0.00	0.00.	0.00
		Total	0.39	1.47	1.86

The total amount of the phytomass (dry weight) changed from one year to the next from 8.24 to 1.86, a fluctuation factor of 4.43. In the worst drought year (1962/63, 25.6 mm rain) this factor was much higher (7-8). In the years 1971/72 and 1972/73 the fluctuation of the above ground phytomass amounted to 10.77 and of the below ground phytomass to only 2.75. These two years are only examples of the annual fluctuations of the phytomass we observed for many years in the Hammadetum and other Negev plant communities. The conclusions to be drawn are as follows: The main reason for the fluctuations of the total phytomass are the annuals. In one year their phytomass amounts to more than 50% of the total, in another it is zero. The fact that the above ground phytomass fluctuates more than the total phytomass is due to the annuals and the geophytes (in the above case Carex) which in bad years do not sprout but preserve a large percentage of their below ground phytomass. This is one of the reasons why the fluctuation factor of the below ground phytomass is smaller than that of the total phytomass and much smaller than that of the above ground phytomass. The phytomass of the arido-active plants (in the above case Hammada and Peganum) fluctuates much less. Since we always observed that the fluctuation amplitude of the above ground phytomass is much larger than that below ground, we may say that the phytomass of the persistent underground organs buffers the total overall fluctuation amplitude. A comparison between the fluctuation factor of the total phytomass (4.53) and that of the rain (3.02) shows that they are not identical. But when we calculated the quantity of floodwater received by the Hammadetum, which is a typical flood plain community, we found that this amounted in 1971/72 to 48.9 mm and in 1972/73 to 11.34 mm. The proportion of these two figures is 4.31, i.e. it is nearly identical with the fluctuation factor of the total phytomass. This fact shows very clearly that in reality the "floodwater pulses" control the phytomass production of the Negev ecosystem to a larger degree than the "rain pulses", whereby it is clear that the floodwater pulses are a function of, but not identical to the rain pulses. This is only true for the biomass production of the homoiohydrous plants. The biomass production of the poikilohydrous plants (lichens and algae), which amounts to only a small fraction of the production of the homoiohydrous primary producers (phytomass of all types of lichens estimated as $0.14t \cdot ha^{-1}$, of all types of algae as $0.02t \cdot ha^{-1}$), depends mainly on the much more regular dewfalls and the production fluctuations from year to year are therefore very small in comparison with those of the homoiohydrous plants.

We can now state for the primary producers of the Negev ecosystem (and this applies to all hot deserts) that they are typified by a very large fluctuation amplitude of their phytomass and production, around a very low mean value. This fact can be considered from two different aspects. On the one hand these fluctuations represent a permanent danger to the whole ecosystem, especially when during 3-4 consecutive drought years there are too many fluctuations on the minus side, causing accumulative production debits. On the other hand, these fluctuations are all special adaptations of desert plants to their general environment and the special environmental conditions of each year are integrated into and culminate in one final result: A dry matter production geared and adjustable to the availability of water.

b) Secondary production of consumers

We have only a very few figures for the biomass of Negev animals. The zoomass of Sphincterochila is estimated at 0.0002-0.0014 $t \cdot ha^{-1}$; of Hemilepistus at 0.014-0.045 $t \cdot ha^{-1}$; of Eremias guttulata at 0.0004-0.0005 $t \cdot ha^{-1}$; of all ants about 0.0016 $t \cdot ha^{-1}$. We have no overall figures for the gastropods, arthropods, reptiles, birds and mammals. We can make only two suppositions: 1) The overall zoomass and possibly (?) the annual secondary production figures of the wild consumers of the Negev are very much lower than the corresponding figures for the primary producers. 2) Annual zoomass fluctuations are in general only to a minor degree a function of the phytomass fluctuations of homoiohydrous plants. This is a certainty for consumers that live on poikilohydrous plants whose phytomass fluctuations are small. This would mean that for the population dynamics of the Negev animals the quantity of food derived from plants plays no role or only a minor controlling role, while other biotic factors like the quality of plant food, predation, competition, fertility, reproduction rates and mortality, interrelated in a complex "biotic web" are more important. This statement is questionable. But if it is true there is at least one exception: seed eating ants which in the Negev consume about 60-70% of the annual seed production.

Unfortunately we have no exact figures of biomass and production of the decomposers in the Negev. This is, for reasons to be discussed in the next section, the worst gap in our knowledge of the Negev ecology. We can only assume that since their activity is moisture dependant, there are large fluctuations of biomass and production.

c) Food web, water web and energy flow

Fig. 1 tries to demonstrate in a schematic, non-quantitative way the water web of the Negev ecosystem, which at the same time also represents, at least partially, the food web. A "dew web" and a "rain web" are interrelated in a complex feedback system. The omnivorous rodents, for example, are part of the dew- and rain-web through the algae eating snails and the homoiohydrous plants. Another interesting point is that as far as the animals are concerned, only the large mammals and seed eating birds are affected by the RIS as are the homoiohydrous plants and most possibly the decomposers. Those animals which are directly or indirectly part of the dew web are unaffected by RIS as are the poikilohydrous plants.

The point made under D.7 concerning the energy flow of the system is illustrated in Fig. 2. This figure represents only a hypothesis which we cannot

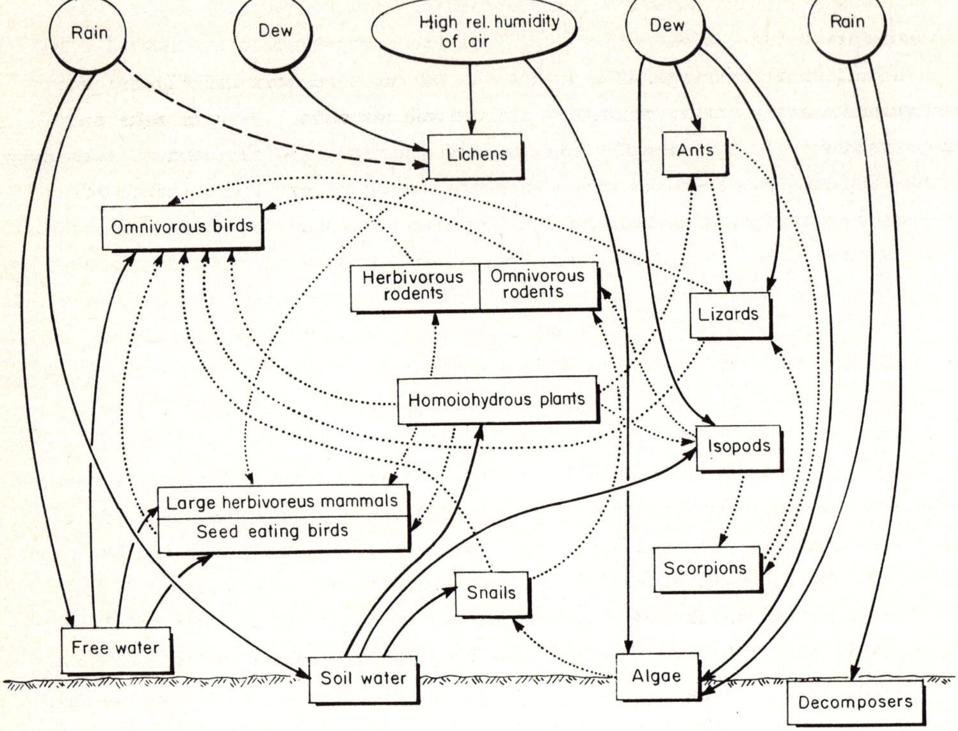

Fig. 1: The water web of the Negev ecosystem.

base on measurements. We can only state that in our ecosystem, on a rough estimate, not more than 1-5% of the organic matter produced and stored annually by the plants is eaten by animals (see also Remmert 1980, p.212), i.e. most of the plant produced energy does not reach the animals. Where does it go ? Every year, depending on the water conditions, the plants shed a large percentage of their biomass which reaches the soil (as measured by Evenari et al. 1976). Part of this is eaten by animals but the largest proportion of it remains in the soil, decays through weathering, is decomposed by soil microorganisms and reenters the mineral cycle. In other words, most of the energy flow bypasses the animals and reaches the decomposers.

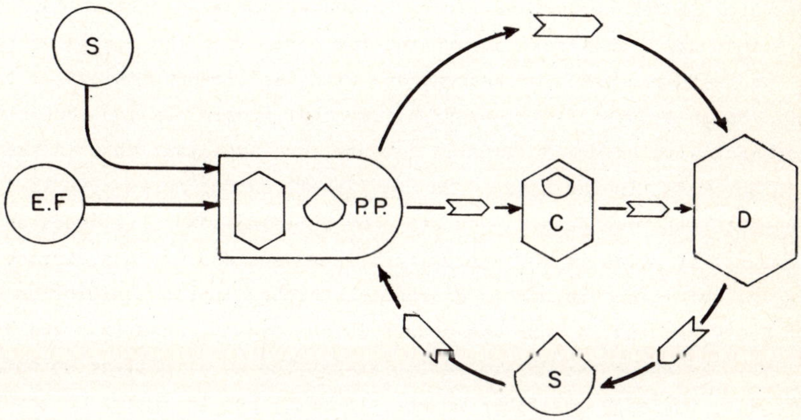

Fig. 2: A schematic drawing of the Energy flow through the Negev ecosystem (after Shachak).

S - sun, E.F. - external parameters affecting primary production, P.P. - primary producers, C - consumers, D - decomposers, S - soil. Hexagon symbol for respiration, Arrow symbol for energy transfer, Circle symbol for energy storage. The relative size of the symbols indicates the magnitude of the processes they represent.

E. CONCLUSION

Desert aut- and synecology is a highly fascinating research topic for the ecologist because the desert can be regarded as a monumental experiment of nature, in which, as in all well planned experiments, one paramount parameter is a limiting factor. But the results of desert research are not only of scientific interest. The fact that they help us to understand structure and functioning of the desert ecosystem is of vital practical importance in trying to fulfil the Messianic dream of making the desert productive and of ecological peace. And so it is written: "The desert and the parched land shall be glad and the steppe shall rejoice and blossom as the rose of Sharon, it shall blossom abundantly and rejoice" (Isaiah 35:1), "For in the desert shall waters break out and streams in the steppe, and the parched land shall become a pool and the thirsty ground springs of water" (Isaiah 35:6-7), "The beasts of the field shall honour me, the jackals and the ostriches, because I give water in the desert and rivers in the wilderness" (Isaiah 43:20), "And in that day I will make a covenant for them, with the beasts of the field, and with the fowls of heaven, and with the creeping things of the ground" (Hosea 2:20). Is it not high time that modern man too should make his covenant with nature ? This is nowhere more important that in the deserts and, in our case, in the Negev. Modern Israel, out of necessity, is trying very hard to make the desert productive by building towns, villages and settlements, by using it for agricultural purposes and by constructing factories, all of which are apt to disrupt the highly labile equilibrium of the desert ecosystem and to turn the desert into a dreary, cheerless and barren land. This is where the desert ecologist has the moral obligation not only to make his knowledge available to the planners, but to insist on being part of the planning and executive bodies in order to prevent ecological disasters. Doing this he must take into account not only his own knowledge of the Negev ecosystem, but also the fact that modern man is not the first to use the desert to his benefit.

In the Negev, the ancient Israelites (ca 900-600 B.C.) and the Nabateans, under their own rule and that of the Romans and Byzantines (ca 300 B.C. - 637 A.D.), erected towns and settlements in the Negev and carried out agriculture on a large scale (Evenari et al. 1971, 1980a). They did not only live <u>in</u> the desert, but lived <u>with</u> it and adapted all their pursuits admirably to desert conditions. This is not only true of their agriculture, in which they successfully imitated nature by collecting and using floodwater, creating an artificial desert ecosystem which fit perfectly into the natural ones. We could rightly call this "eco-agriculture". Their villages, settlements and houses were also built in harmony and conformity with the desert conditions, taking into account the Negev climate and collecting every drop of rainwater and runoff in rationally built cisterns for use as drinking water. The ancient city of Avdat can serve as a good example. Sometime after 637 A.D. the Bedouin became masters of the Negev and turned it into a grazing ground for their flocks. This certainly brought about changes in the ecosystem but after some time a new equilibrium was reached between the nomads, their flocks and the system. This new equilibrium because of overgrazing was most probably on a lowered production level of thc primary producers. We today, with all our modern technology, could learn much from the successes and mistakes of these ancient desert cultures. We could learn how to construct our houses, how to protect our roads and factories from floods, how to use runoff and how not to overgraze. Our modern knowledge of desert ecosystems can help us first of all to prevent self-destroying agricultural practices such as using marginal lands for dry farming, overirrigation with water brought in from the outside or diverted from "foreign" rivers flowing through deserts leading to irreversible salinization etc. Our research and practicel experience have taught us methods of controlled grazing preventing the destruction of the natural ecosystem by overgrazing, how to cope in general with the RIS etc. We know now what the main problems and their solutions are but our planners are far away from applying this knowledge. I feel that it is our duty to close this gap between research and its application.

This paper, covering as it does such a large topic, contains some contradictions, which the reader will easily discover, many question marks and much uncertainty. And so it is appropriate to end with a citation from Wittgenstein (1956, II:81): "For might we not possibly have wanted to produce a contradiction ? ... But why should people have tried to do this ? ... But why not e.g. in order to show that everything in this world is uncertain ?"

REFERENCES

Alcon P.U. (1974) Social behaviour of a population of chukar partridge (Alectoris chukar cypriotes) in Israel. Ph.D. Thesis, Cornell University.

Berner T. (1974) Studies on ecophysiology of the hypolithic algae from the flint stones of the Mashash formation in the Northern Negev, Israel. Ph.D. Thesis, Hebrew Univ. Jerusalem (Hebrew with English Summary).

Berner T. & Evenari M. (1978) The influence of temperature and light penetration on the abundance of the hypolithic algae in the Negev desert of Israel. Oecologia (Berl,) 33, 255-260.

Bodenheimer F.S. (1957) The ecology of mammals in arid zones. Unesco, Arid Zone Research VIII, Paris, 100-137.

Boruth A. & Shkolnick A. (1974) Physiological adaptations to desert environment. Environmental Physiol. (Edited by Robertshaw D.), Univ. Park Press, 185-229.

Chomsky N. (1966) Cartesian Linguistics. Harper and Row, New York.

Chomsky N. (1967) Receny contributions to the theory of innate ideas. Synthese 17, 2-11.

Dan J., Yaalon D.H., Koyumdjisky H. & Ras Z. (1976) The soils of Israel. Volcani Centre, Beth Dagan, Pamphlet 159.

Datta S.C., Evenari M. & Gutterman Y. (1970) The heteroblasty of Aegilops ovata L. Israel J. of Bot. 19, 463-483.

Dubinsky Z. & Steinberger Y. (1979) The survival of the desert isopod Hemilepistus reaumuri (Andouin) in relation to temperature (Isopoda, Oniscoidea). Crustaceana 36, 147-154.

Evenari M (1963) Zur Keimungsphysiologie zweier Wüstenpflanzen. Mitteil. Florist.-soziol. Arbeitsgemeinschaft N.F. 10, 941-944.

Evenari M., Shanan L. & Tadmor N. (1971) The Negev: the challenge of a desert. Harvard Univ. Press.

Evenari M., Schulze E.D., Kappen L., Buschbom U. & Lange O.L. (1975) Adaptive mechanisms in desert plants. Physiological adaptations to the environment. (Edited by Vernberg F.J.). Intext Educational Publishers, New York, 111-129.

Evenari M. & Gutterman Y. (1976) Observations on the secondary succession of three plant communities in the Negev desert Israel, I: Artemisietum herbae albae. Etudes de biology végétale, Hommage au Prof. P. Chouard. (Edited by Jacques R.). CNRS Gif sur Yvette, 57-68. (part II on Hammadetum scopariae in manuscript).

Evenari M., Schulze E.D., Lange O.L., Kappen L. & Buschbom U. (1976) Plant production in arid and semi-arid areas. Water and plant life (Edited by Lange O.L., Kappen L. & Schulze E.D.). Springer, 439-451.

Evenari M., Kadouri A. & Gutterman Y. (1977a) Ecophysiological investigation on the amphicarpy of Emex spinosus (L.) Campd. Flora 166, 223-238.

Evenari M., Lange O.L., Schulze E.D., Kappen L. & Buschbom U. (1977b) Net photosynthesis, dry matter production and phenological development of apricot trees (Prunus armeniaca L.) cultivated in the Negev Highlands (Israel). Flora 166, 383-414.

Evenari M., Shanan L. & Tadmor N. (1980a) The Negev: the struggle for survival in the desert. Mossad Bialik, Jerusalem. (In Hebrew, revised and enlarged edition of English book of 1971)

Evenari M., Masigh D. & Rogel A. (1980b) Runoff farming in the Negev desert of Israel. VIth Progress report. Dept. of Bot., Hebrew Univ. and Jacob Blaustein Inst. of Desert Res. Sde Boqer (Mimeographed).

Flach K.W. & Smith G.D. (1969) The new system of soil classification as applied to arid-land soils. Arid lands in perspective. (Edited by McGinnies V.G. & Goldman B.J.). AAAS and Univ. Press of Arizona, 59-73.

Friedman J., Gunderman N. & Ellis M. (1978) Water response of the hygrochastic skeleton of the true rose of Jericho (Anastatica hierochuntica L.). Oecologia (Berl.) 32, 289-301.

Funke B.E. & Harris J.O. (1968) Early respiratory response of soil treated by heat and drying. Arid lands ecosystems. (Edited by Perry R.A. & Goodall O.W.). Cambridge Univ. Press, I, 661-675.

Gutterman Y. (1980/81) Annual rythm and a position effect in the germinability of Mesembryanthemum nodiflorum. Israel J. of Bot. 29, 93-97.

Gutterman Y., Witztum A. & Evenari M. (1967) Seed dispersal and germination in Blepharis persica (Burm.) Kuntze. Israel J. of Bot. 16, 213-234.

Haeckel E. (1866) Generelle Morphologie der Organismen, 2 Bd. Georg Reimer (Berl.).

Harley J.L. (1969) The biology of mycorrhiza, 2nd ed. Leonard Hill, London.

Kappen L., Lange O.L., Schulze E.D., Evenari M. & Buschbom U. (1979) Ecophysiological investigations on lichens of the Negev desert, VI: Annual course of the photosynthetic production of Ramalina maciformis (Del,) Bory. Flora 168, 85-108.

Koller D. (1972) Environmental control of seed germination. Seed Biology. (Edited by Koslowsky T.T.), Acad. Press, London, Vol. II.

Koller D. & Roth N. (1964) Studies on the ecology and physiological significance of amphicarpy in Gymnarrhena micrantha (Compositae). Amer. J. Bot. 51, 26-35.

Lange O.L. (1980) Moisture content and CO_2 exchange of lichens, I: Influence of temperature on moisture dependant photosynthesis and dark respiration in Ramalina maciformis. Oecologia (Berl.) 45, 63 71.

Lange O.L. & Evenari M. (1971) Experimentell ökologische untersuchungen an Flechten der Negev Wüste, IV: Wachstumessungen an Caloplaca aurantia (Pers.) Hellb. Flora 160, 100-104.

Lange O.L., Schulze E.D., Kappen L., Buschbom U. & Evenari M. (1975a) Adaptations of desert lichens to drought and extreme temperatures. Environmental physiology of desert organisms. (Edited by Hadley N.P.). Dowden, Hutchinson and Ross, Stroudsburg, P.A., 20-37.

Lange O.L., Schulze E.D., Evenari M., Kappen L. & Buschbom U. (1974, 1975b, 1978) The temperature-related photosynthetic capacity of plants under desert conditions, I-III. Oecologia (Berl.) 17, 97-110; 18, 219-233; 34, 89-100.

Lange O.L., Schulze E.D., Kappen L., Buschbom U. & Evenari M. (1975c) Photosynthesis of desert plants as influenced by internal and external factors. Perspective of biophysical ecology. (Edited by Gates & Schmerl). Springer, 121-143.

Marder J. (1973) Body temperature regulation in the brown-necked raven (Corvus corax ruficolis), I, II. Comparative Biochem. and Physiol. 45, 421-440.

Monod Th. (1973) Les déserts. Horizons, Paris.

Monod Th. (1981) The Sahel. Ecosystems of the world, Vol. 12: Hot desert and arid shrubland. Elsevier (in press).

Noy-Meir I. (1973, 1974) Desert ecosystems. Ann. Rev. of Ecol. and Systematics 4, 25-52; 5, 195-214.

Oberlander Th.M. (1979) Characterization of arid climates according to combined water balance parameters. J. of Arid Environment 2, 219-241.

Orr Y. (1970) Temperature measurements at the nest of the desert lark (Ammomanes deserti deserti). The Condor 72, 476-478.

Orr Y. (1970) The underground burrow of Psammonys obesus. Teva Va'aretz 16, 280-284 (Hebrew).

Orr Y., Shachak M. & Steinberger Y. (1979) Ecology of the spotted lizard (Eremias guttulata guttulata) in the Negev desert (Israel). J. of Arid Environment 2, 151-161.

Orshan G. (1981) The deserts of the Middle East. Ecosystems of the world, Vol. 12: Hot deserts and arid shrubland. Elsevier (in press).

Remmert H. (1980) Ecology: A textbook. Springer.

Schroeter C. & Kirchner O. (1902) Die Vegetation des Bodensees. Ver. Gesch. Bodensees Bd. II

Schulze E.D., Hall A.E., Lange O.L., Evenari M., Kappen L. & Buschbom U. (1980) Long term effect of drought on wild and cultivated plants in the Negev desert, I and II. Oecologia (Berl.) 45, 11-18, 19-25.

Shachak M. (1980) Energy allocation and life history strategy of the desert isopod Hemilepistus reaumuri. Oecologia (Berl.) 45, 404-413.

Shachak M. & McPhee E. (1970) A comparative study of the ecology of two desert rodents (Acomys cahirinus and Gerbillus dasyurus) living in a rocky habitat. M.Sc. Thesis, Jerusalem.

Shachak M. & Steinberger Y. (1974) Field studies on the ecology of the small spotted lizard (Eremias guttulata guttulata) in the Negev desert, Israel. M.Sc. Thesis (Hebrew).

Shachak M., Orr Y. & Steinberger Y. (1975) Field observations on the natural history of Sphincterochila (S.) zonata (Bourgignat 1853) (=S. boissieri Charpentier 1847). Argamon. Israel J. of Malacology 5, 20-46.

Shachak M., Chapman E.A. & Steinberger Y. (1977) Feeding, energy flow and soil turnover in the desert isopod, Hemilepistus reaumuri. Oecologia (Berl.) 24, 57-69.

Shachak M. & Steinberger Y. (1980) An algae-desert snail food chain: energy flow and soil turnover. Oecologia (Berl.) 46, 402-411.

Shanan L. (1975) Rainfall and runoff relationships in small watersheds in the Avdat region of the Negev desert highlands. Ph.D. Thesis, Hebrew University (in English).

Soil Survey Staff (1975) Soil taxonomy. Agric. Handb. 436, Soil Survey U.S.D.A.

Steinberger Y., Grossman S. & Dubinski Z. (1981) Some aspects of the ecology of the desert snail Sphincterochila prophetarum in relation to energy and water flow. Oecologia (Berl.) (in press).

Tansley E.G. (1935) The use and abuse of vegetational concepts and terms. Ecology 16, 284-307.

Unesco (1979) Map of the world distribution of arid regions. MAB Technical notes 7.

Went F.W. & Stark N. (1968) The biological and mechanical role of soil fungi. Proc. Nat. Acad. Sci. 60, 497-504.

Wittgenstein L. (1956) Remarks on the foundations of mathematics. Basil Blackwell. Oxford.

Woltereck H. (1928) Ecologische Gestaltsysteme, Biol. Zentrbl., 48:521-551.

Yom-Tov Y. (1971a) The biology of two desert snails: Trochoidea (Xerocrassa) seetzeni and Sphincterochila boissieri. Israel J. of Zool. 20, 231-249.

Yom-Tov Y. (1971b) Body temperature and light reflectance in two desert snails. Proc. Malaco. Soc. London 39, 319-326.

Zohary M. (1937) Die Verbreitungsoekologischen Verhaeltnisse der Flora Palaestinas. Beih. Botan. Centralbl. 56, Abt. I, 1-155.

Zohary M. (1962) Plant life of Palestine. Ronald Press, New York.

Developments in Arid Zone Ecology and Environmental Quality
Edited by H. Shuval

2

INFLUENCE OF QUANTITY AND DATE OF RAIN ON THE DISPERSAL AND GERMINATION MECHANISMS, PHENOLOGY, DEVELOPMENT AND SEED GERMINABILITY IN DESERT ANNUAL PLANTS, AND ON THE LIFE CYCLE OF GEOPHYTES AND HEMICRYPTOPHYTES IN THE NEGEV DESERT

Yitzchak Gutterman

The Jacob Blaustein Institute for Desert Research, Sede Boqer Campus, and the Department of Biology, Ben-Gurion University of the Negev, Israel.

ABSTRACT

Large differences were found in the appearance of individual desert plants from year to year and within each year, depending on the distribution, quantity, and date of the rains. Survival of desert annuals, in particular, seems to be dependent on: (a) topochoric seed dispersal - near the mother plants, for determining suitable microhabitats; (b) germination at the right time, and of only some of the seeds at one time, even under optimal conditions (heteroblasty); heteroblasty is affected by environmental factors such as day length, temperature, seed position, and other factors. The influence of quantity and date of rain on the life cycle of geophytes and hemicryptophytes, in the hilly areas of the northern Negev desert of Israel is discussed.

INTRODUCTION

In the Negev deserts of Israel, the distribution, quantity and date of rains from year to year and within each year affect a great diversity of species of desert annual plants (Evenari et al., 1966). There also exists a relationship between the minimum amount of rain needed to cause germination and the mean daily evaporation (Evenari & Gutterman, 1976).

The survival of desert annuals seems to be dependent on topochoric behavior in seed dispersal (Zohary, 1962) for determining suitable microhabitats. Germination at the right time and germination of only some of the seeds at one time are other important survival mechanisms in desert annuals.(Gutterman 1980/81 a). Environmental factors such as day length, temperature, and seed position also determine seed germinability (Gutterman 1973, 1980/81 b).

In the deserts of Israel we find that, in some of the hottest and dryest areas, nearly all the winter annuals are ombrohydrochoric (dispersed by rain) (van der Pijl, 1972). There are species whose seed dispersal mechanisms are activated by only a few drops of rain (Gutterman, 1980/81a);and in these cases there is no connection between the rain that causes germination and the rain that disperses the seeds. On the other hand, there are other plants, such as *Blepharis sp.*, that need a long period of rain or wetness to trigger the dispersal mechanisms. In these cases there is a very close relationship between the time of seed dispersal and the time of germination. It is the same rain that causes both seed dispersal and germination (Gutterman et al., 1967). There are also other desert species whose behavior falls inbetween these two extreme categories. The geophytes and hemicryptophytes are perennials with a typical behavior also well suited to a desert habitat. They enter into the active stage according to the date of the beginning of the rainy season, which differs from year to year, and then enter into a dormant stage when the soil becomes dry (Evenari et al., 1971).

RESULTS AND DISCUSSION

1. Seed Dispersal and Germination of Hydrochoric Plants.

I. All the plants refered to in this part are desert winter annuals that dry out in the summer and remain straight with closed capitulum or capsules, and open rapidly after moistening.

A - Capsules that Open Rapidly and Immediately Disperse the Seeds.

Aizoon hispanicum L. need only 2-3 drops of rain for seed dispersal. The first drop opens the capsule and the second causes some seed dispersion and sticks them to the soil surface near the mother plant.

Some of those seeds germinate after other rains. In those plants there is not any connection between time of dispersal and time of germination.

B - Capsules Open Rapidly and Disperse Only Some of the Seeds in One Particular Rain.

Mesembryanthemum nodiflorum L. and *M. forsskalii Hochst.* need at least 2-3 drops of rain for seed dispersal. The first drop opens the capsule, the second drop of rain disperses some of the seeds.

In this case there is also very rapid seed dispersal and no connection between time of seed dispersal and time of germination. In *Mesembryanthemum nodiflorum* the seeds released first showed the highest germination percentages (Table 1). From observations on twenty capsules, average opening time of the capsule under moist conditions is 90 sec: 60 sec minimum, 123 sec maximum. The first seeds to separate do so after 132 sec, i.e. 42 sec after opening. The outer seeds of each capsule are first to separate. It is interesting to note that, of twenty capsules, which together contained 600 seeds, an average of thirty seeds per capsule, one third of the seeds are released during the first 15 min. Another third is released during the following 3 h and the last third is released during 2 more hours. Each lot shows different germination capacity. The effect of position on seed release and germination spreads the period of germination and prevents 100% germination even under optimal conditions (Table 1) (Gutterman, 1980/81 a).

TABLE 1. Time of wetting needed for *Mesembryanthemum nodiflorum* seed release, and percentage of seed germination after various times. (Whole dry plants, including capsules, were collected during summer of 1972. Germination experiment done on 14.1.1980.)

Duration of capsule wetting (min)	% germination after 12 days
15	61.0
200	5.5
320	1.0

Adapted from Gutterman (1980/81 a)

C - Plants that Open Their Capitula Very Rapidly and Disperse Some of the Achenes ("Seeds") that have been Disconnected from the Capitulum by Dew or by Previous Rains.

During the first summer the capitula of *Asteriscus pygmaeus* open with humidity above 85% or dew, and close during the day. The opening and closing releases some of the "seeds" from the periphery of the capitulum. Some of these "seeds" are dispersed by at least 2 drops.

The first drop opens the capitulum and the second drop may disperse the achenes ("seeds") that have previously been disconnected. There is no connection between the time of dispersal and the time of germination. All of these seeds have papus, and they either stay near the mother plant, are dispersed by the rain, or are dispersed by the wind after the rain.

II. Plants that disperse their seeds after about 1½ hours of continuous rain and are able to germinate after the dispersing rain.

Plants such as *Carrichtera annua* (Nordhagen, 1936) and *Anastatica hierochuntica* (Friedman et al., 1978).

After about 1½ hours of continuous rain, the dehiscence zone inbetween the halves of the pod is weakened, and then if a drop falls on the teaspoon-like extension of the Siliqua it opens and the mucilagenus seeds fall to the ground near the mother plant. They then stick to the soil surface by means of their mucilage.

In cases where the rain continues, these seeds can germinate a short time after dispersal. In this case the same rain both disperses and germinates the seeds.

III. Plants with a double safety mechanism of seed dispersal and rapid germination.

Desert plants of different species of *Blepharis* (*Acanthaceae*) have an east-Sudanian and east Saharo-Arabian distribution. The seeds remain undispersed on the dry mother plant, protected from weather and predators for many years, without lowering their full viability. Sepals enclosing ripened capsules remain closed until moistened either by rain (ombrohydrochory)(van der Pijl 1972) or by floodwater. After about 10-15 min of wetting, the sepals open and expose the capsule.

Tension which develops in the capsule during ripening and drying is released by penetration of water into the apex following a rain of about 1½ hours, and disperses some of the seeds from the dry mother plants.

This double safety mechanism ensures that seed dispersal takes place only after a heavy rain or flood (Gutterman et al., 1967; 1969). An exploding capsule throws out two dry seeds. Each dry seed bears numerous multicellular hairs which are part of the seed coat and are appressed to the seed (Gutterman et al., 1973). Upon falling on a wet substrate, the hairs hydrate rapidly, stand erect and reflexed, orient the seed at an angle to the soil surface with the root end in contact with the soil (Witztum et al., 1969; Gutterman et al., 1977), and the seed germinates immediately under a great variety of external conditions. Germination temperatures vary from 8°C to 40°C in the light and in the dark. The very fast root elongation (up to 48.7 mm after 24 h) ensures that the seedlings, which are not too drought-resistant, establish themselves before the upper layer of the soil has dried out (Gutterman, 1972). In *Blepharis* plants there is a close connection between the time of dispersal and the time of germination.

2. Quantity of Rain Required for Germination of Annual Winter Desert Plants During the Winter and Summer.

Dates and quantities of rainfall are unpredictable in the desert. In January, when evaporation rate is lowest in comparison to the rest of the year, rain of 10-14 mm brings about a good germination of annuals (Evenari & Gutterman, 1976).

In order to bring about the same germination at the same place during the summer, when temperatures and evaporation rate are highest, the addition of an equivalent of at least 200 mm is necessary. Plants such as *Schismus arabicus* and *Malva aegyptia* can germinate all year round depending on the quantity of water available.

3. Influence of Time of Germinating Rain on the Phenology of Desert Annual Plants.

Many of these plants belong to the group of long-day facultative plants for flowering. Those that germinate in short days, such as in December, develop a rosette with many large leaves and many lateral branches when soil moisture is sufficient. The flowering is delayed and there are many seeds.

Those that germinate when the germinating rain falls near the summer develop few, small leaves along a single long stem, and flower rapidly. They produce few seeds (Evenari & Gutterman, 1966).

4. Time of Seed Maturation and Germinability of Seeds.

The date of flowering of the plants will differ according to the time of the germinating rain and the soil moisture during the growth period. The flowering and the time of seed maturation depend on the amount and allocation of the rains. Different pods ripen under different day lengths.

As seen in *Ononis sicula* (Evenari et al., 1966; Gutterman & Evenari, 1972; Gutterman & Heydecker, 1973) and *Trigonella arabica*

(Gutterman, 1978), the last eight days of seed maturation have a great influence on seed coat color, structure, permeability to water, and germinability.

Seeds matured during short days germinate relatively earlier than those matured during long days.

Along the same stem, different pods develop seeds of different germinability according to the day length in the last 8 days of ripening. This mechanism spreads the germination over time. This phenomenon is very important for the survival of those species under desert conditions (Gutterman, 1973; 1980/81 b).

5. Geophytes and Hemicryptophytes - Rain and Life-Cycle

At the end of the dry season in 1980, when the first big rain was on December 11, 1980, which is relatively late in Har Hanegev area, many geophytes and hemicryptophytes such as *Erodium hirtum* (Willd), *Peganum harmala* L., *Eremostachys laciniata* (L.), *Urginea maritima* (L.) Bak., *Urginea undulata* (Desf.) Steinh., *Asphodelus microcarpus* Salzm et Viv., were found to start producing leaves during November 1980, i.e. before the first rain, and when the soil was still completely dry.

In most of them the leaves remain at a minimal size until the rainfall. These plants exploit their water and storage material that they store in organs below the soil surface. Shortly after the rain they produce a canopy of leaves, and cover their area before the annuals germinate and attain a certain size.

The annuals germinate about 10 days after the first big rain, and remain small for a number of weeks. The advantage of the geophytes in occupying an area is obvious.

It is interesting to note that, on the same date, geophytes and hemicryptophytes such as *Sternbergia clusiana* (Ker-Gawler) Spreng., *Narcissus tazetta* L., *Pancratium sickenbergeri* Aschers. et Schweinf., *Tulipa montana* Linde, and *Scorzonera paposa* DC. that were kept during the summer under laboratory conditions in paper bags, produced leaves at the same time as those same species produced leaves in the field.

If an area in the field is irrigated, as done by Prof. M. Evenari in Avdat during July (personal communication), and also as we observed on irrigated areas at Sede Boqer, a hemicryptophyte such as *Erodium hirtum* does not become dormant, and has leaves throughout the summer. It also flowers at times.

In an experiment we irrigated in November an area in which *Anemone coronaria* L. grows, i.e. one month before the first heavy rain. These plants grew leaves one month before the plants in areas that were not irrigated.

When irrigation was carried out in the middle of September in fields where *Iris sisyrinchium* and *Iris hieruchamensis* are found, these species produced early leaves in the middle of October, long before the rain. Apparently there are conditions whereby rain (or irrigation) brings to an early end the dormancy stage of hemicryptophytes and geophytes. Without rain, they enter the active stage at a more or less fixed time. They produce leaves that remain small. When the first heavy rain falls, the opening of the leaves is very rapid and a canopy of leaves is produced.

REFERENCES

Evenari M. & Gutterman Y. (1966) The photoperiodic response of some desert plants. *Z. Pflanzenphysiol.* 54, 7-27.

Evenari M. & Gutterman Y. (1976) Observation on the secondary succession of three plant communities in the Negev desert, Israel. I. *Artemisietum herbae albae*. Etudes de Biologie Végétale. Hommage au Prof. P. Chouard. R. Jacques ed. Paris, 57-86.

Evenari M., Koller D. & Gutterman Y. (1966) Effects of the environment of mother plants on germination by control of seed-coat permeability to water in *Ononis sicula*. *Aust. J. Biol. Sci.* 19, 1007-1016.

Evenari M., Shanan L. & Tadmor N. (1971) The Negev: The Challenge of a Desert. Harvard University Press, Cambridge, Massachusetts.

Friedman J., Gunderman N. & Ellis M. (1978) Water response of the hygrochastic skeletons of the true rose of Jericho (*Anastatica hierochuntica* L.). *Oecologia* (Berl.) 32:289-301.

Gutterman Y. (1972) Delayed seed dispersal and rapid germination as survival mechanisms of the desert plant *Blepharis persica* (Burm.) Kuntze. *Oecologia* (Berl.) 10, 145-149.

Gutterman Y. (1973) Differences in the progeny due to daylength and hormonal treatment of the mother plant. Seed Ecology. W. Heydecker ed. London: Butterworth 59-80.

Gutterman Y. (1978) Seed coat permeability as a function of photoperiodical treatments of the mother plants during seed maturation in the desert annual plant: *Trigonella arabica*, del. *J. of Arid Environ.* 1, 141-144.

Gutterman Y. (1980/81 a) Annual rhythm and position effect in the germinability of *Mesembryenthemum nodiflorum*. *Isr. J. Bot.* 29, 93-97.

Gutterman Y. (1980/81 b) Influences on seed germinability: phenotypic maternal effects during seed maturation. *Isr. J. Bot.* 29, 105-117.

Gutterman Y. & Evenari M. (1972) The influence of daylength on seed coat colour, an index of water permeability of the desert annual *Ononis sicula* Guss. *J. of Ecol.* 60, 713, 719.

Gutterman Y. & Heydecker W. (1973) Studies of the surface of desert plant seeds. I Effect of day length during maturation on the seed coat of *Ononis sicula*. *Ann. Bot.* 37, 1049-1050.

Gutterman Y. & Witztum A. (1977) The movement of integumentary hairs in *Blepharis ciliaris* (L.) Burtt. *Bot. Gaz.* 138, 29-34.

Gutterman Y., Witztum A. & Evenari M. (1967) Seed dispersal and germination in *Blepharis persica* (Burn) Kuntze. *Isr. J. Bot.* 16, 213-234.

Gutterman Y., Witztum A. & Evenari M. (1969) Physiological and morphological differences between populations of *Blepharis persica* (Burm.) Kuntze. *Isr. J. Bot.* 18, 89-95.

Gutterman Y., Witztum A. & Heydecker W. (1973) Studies of the surface of desert plant seeds. II Ecological adaptations of the seeds of *Blepharis persica*. *Ann. Bot.* 37, 1051-1055.

Nordhagen R. (1936) Über dorsiventrale und transversale Tangentballisten *Sven. Bot. Tidskr.* 30:443-473.

Pijl, van der, L. (1972) Principles of Dispersal in Higher Plants. Springer-Verlag, Berlin, Heidelberg, New York.

Witztum A., Gutterman Y. & Evenari (1969) Integumentary mucilage as an oxygen barrier during germination of *Blepharis persica* (Burm.) Kuntze. *Bot. Gaz.* 130, 238-241.

Zohary M. (1962) Plant Life in Palestine, Israel and Jordan. Ronald Press, New York.

Developments in Arid Zone Ecology and Environmental Quality
Edited by H. Shuval

3
CO_2 EXCHANGE AND GROWTH RATE OF THE LOESS SOIL CRUSTS ALGAE IN THE NEGEV DESERT OF ISRAEL

Yehuda Levi, Tamar Berner and Yigal Cohen

Department of Life Sciences, Bar-Ilan University, Ramat-Gan, Israel

ABSTRACT

The CO_2 exchange of the microbial population of loess soil crusts of the Negev desert was examined under various combinations of relative humidity, temperature and light intensity, using an Infra-Red Gas Analyzer (IRGA) and incorporation of labelled carbon. Maximal CO_2 assimilation measured by IRGA was recorded at 15°C and relative humidity of 100%. No CO_2 assimilation was observed but high values of CO_2 evoluaton were recorded at 35°C, probably due to high respiration rates. For any combination of temperature and light intensity, CO_2 exchange increased with increasing relative humidity. CO_2 assimilation by the crust algae was measured by ^{14}C uptake. At relative humidity of 100%, the optimal temperature for CO_2 fixation by the algae was 25°C. Optimal light intensity ranged between 280-400 $\mu E.m^{-2}.sec^{-1}$.

INTRODUCTION

Soil algae are widespread and are an important component of terrestrial ecosystems (Both, 1941). These algae exist even in desert habitats, mainly in soil crusts (Cameron, 1964; Shield, 1964). The algae improve water retention by the crusts (Fritch, 1922), increase soil aggregate (Bailey, 1973) and total nitrogen and carbon content of the soil (Lynn, 1973).

Loess crusts which cover vast areas of the northern Negev desert in Israel contain algae (Friedman & Galun, 1974), fungi, and bacteria (Borut, 1960). Much is known about the taxonomy of algae in crusts (Cameron, 1964), particularly in the Negev (Friedman et al., 1967), but little information is available concerning the relationship between their activity and environmental conditions. In this paper we report on growth and CO_2 fixation of soil crust algae under various combinations of relative humidity, light intensity and temperature.

STUDY AREA, MATERIALS AND METHODS

The study area is located about 30 km south of Beer-Sheva at 500 m above sea level. The average annual rainfall is 76 mm with a minimum of 23 mm and a maximum of 137 mm (Evenari et al., 1971). The average temperature in January (the coldest month) is 9°C, while in August (the hottest month) - 26°C (Israel Atlas, 1963). The average annual dewfall, as measured at the Avdat station 15 km south of the study area, is 33 mm condensing throughout 195 nights (Evenari et al., 1971).

Algal material - Soil crusts were collected from the research site and subsequently stored in the dark at room temperature. The amount of algae in the crusts was assumed to be proportional to the chlorophyll *a* content (Lynn, 1973) and expressed as μg chlorophyll $a \cdot cm^2$.

CO_2 exchange - CO_2 exchange of the crusts was measured using an Infra Red Gas Analyzer (IRGA) (Oras, Hartman and Brown, West Germany) in a closed system as described by Lange (Lange, 1972). Air was circulated in a closed system including the cuvette and the fluctuations in the CO_2 concentration detected by the IRGA were recorded.

Environmental control - Soil crust samples were illuminated by a 150w quartz halogen lamp . Light intensity at the sample surface was measured with the aid of LI-185 quantum meter (Lambda Instruments Corp., Lincoln, Nebraska). Temperature at the surface of the samples was measured with the aid of small thermopouples attached to the crust.

Air at various relative humidities was supplied to the crusts by pumping it through appropriate saturated salt solutions (Winston, 1960). CO_2 fixation was determined from the amount of ^{14}C incorporated by the crusts. $^{14}CO_2$ was generated by injecting 10% H_2SO_4 through a serum cap onto a $NaH^{14}CO_3$ solution placed in a 10 ml beaker inside a sealed system of a known volume. Crusts were incubated in a controlled temperature chamber ($\pm$ 0.2°C). After an incorporation period of 3h crusts were removed and placed in Soluene solution (Packard, USA) for about 24h at 57°C, and then centrifuged for 10 min at 10,000 rpm. Ten ml of Insta Gel (Packard, USA) scintillation cocktail were added to 0.2 ml aliquot of the supernatant, and counted in a Tricarb scintillation counter (Packard, USA).

Growth rate - Crusts were placed in petri dishes on moistened filter paper under controlled temperature and light for various periods of time up to 96h. The amount of chlorophyll *a* per unit area served as an index for algal proliferation.

RESULTS AND DISCUSSION

The combined effect of relative humidity, light intensity, and temperature on the CO_2 exchange of the microbial population of the crusts are presented in Fig. 1 (a,b,c). At relative humidity of 75% no CO_2 exchange was detected regardless of the temperature and light intensity (Fig. 1a).

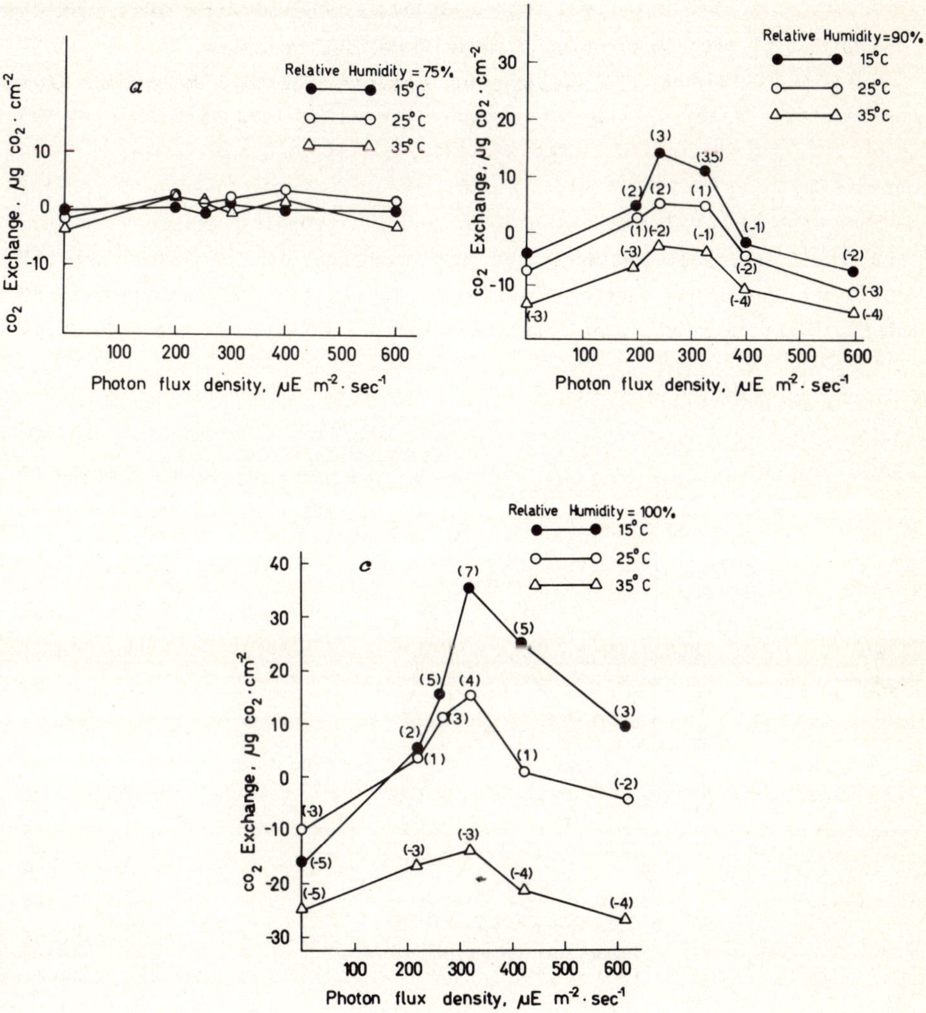

FIG. 1 (a,b,c). The combined effect of relative humidity, light intensity, and temperature on CO_2 exchange of the microbial population of soil crusts of the Negev.

At a relative humidity of 90% CO_2 fixation was recorded only at 15°C. At temperatures of 25°C and 35°C CO_2 evolution was measured indicating that respiration exceeded photosynthesis (Fig. 1b).

The highest level of CO_2 exchange was found at relative humidity of 100%. At temperatures of 15°C and 25°C fixation was 31, and 11μg $CO_2.cm^{-2}.h^{-1}$, respectively (Fig. 1c). At 35°C, under any light intensity, CO_2 evolution was noted probably due to the high rate of respiration of the whole microbial population.

^{14}C incorporation studies showed that the optimal temperature for net fixation is 25°C. At this temperature with light intensity of 400 $\mu E.m^{-2}.s^{-1}$ 35μg $CO_2.cm^{-2}.h^{-1}$ was fixed (Fig. 2). At 35°C and 15°C, maximal CO_2 fixation was 25 and 12μg $CO_2.cm^{-2}.h^{-1}$, respectively.

The discrepancy between the IRGA and ^{14}C results arises from the fact that the IRGA measures the CO_2 exchange of the whole crustal population including bacteria and fungi, while the ^{14}C method measures the CO_2 fixation of the algal population only. In both measurements

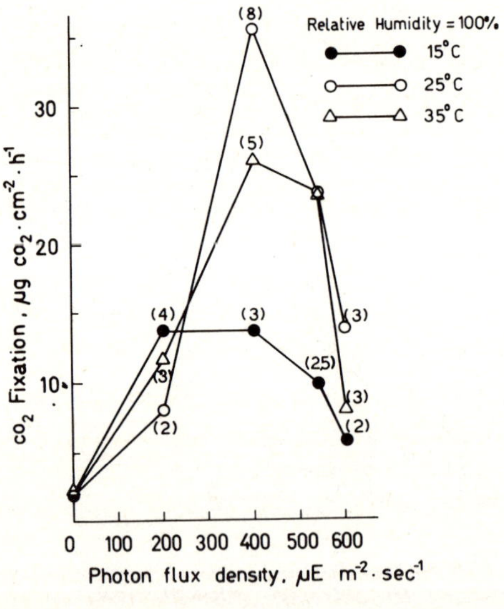

FIG. 2. Carbon fixation by loess crusts algae under various temperatures and light intensities.

at light intensities of > 500 $\mu E.m^{-2}.S^{-1}$, CO_2 fixation was partially inhibited, probably due to photooxidation of chlorophyll.

Moistening the crusts prior to their exposure to ^{14}C increased chlorophyll *a* content per unit area, while the amount of carbon fixed per unit weight of chlorophyll *a* remained constant (Table 1). It seems, therefore, that dewfall during the night, before sunrise, has no influence on the fixation potential of the algae per unit weight of chlorophyll, although it will significantly increase areal production.

TABLE 1. The effect of moistening the loess crusts prior to their exposure to $^{14}CO_2$.

Premoistening time, hours	Chlorophyll *a* content μg.Chl *a* cm^{-2}	CO_2 fixation per μg Chl *a*.h^{-1}
0	1.3 ± 0.08	32 ± 4
30	2.8 ± 0.12	30 ± 5
60	6.5 ± 0.88	33 ± 5
96	8.1 ± 1.1	31 ± 6

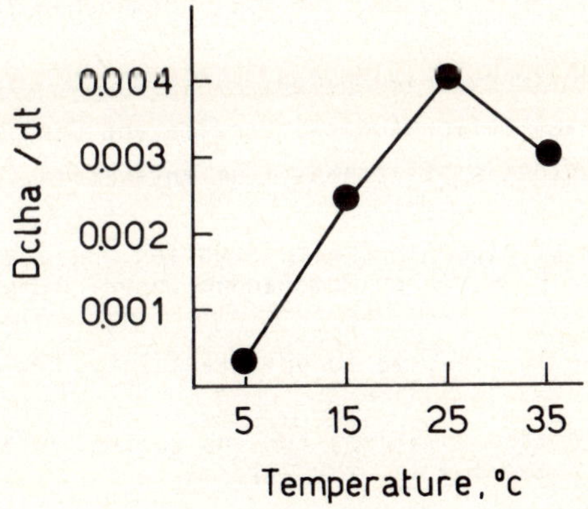

FIG. 3. The effect of temperature on chlorophyll *a* biosynthesis by algae of the crusts.

Fig. 3 shows that the rate of chlorophyll biosynthesis was highest at 25°C where it reached $4x10^3$ μg chlorophyll *a*.h^{-1}, whereas the lowest rate was measured at 5°C and reached only $3x10^{-4}$ μg chlorophyll *a* .h^{-1}.

Our results show that optimal conditions for photosynthesis by the desert crust algae are relative humidity of 100%, temperature of 25°C and light intensity of about 400μE.m^{-2}.S^{-1}. However, the optimal conditions in which a net increase in organic carbon loess crust occurs are a temperature of 15°C prevailing in nature during the early hours of the morning, when the soil surface is still wet as a result of dew condensation, and light level does not exceed 500 μE.m^{-2}.S^{-1}.

REFERENCES

Bailey D., Andrew P.M. & Rosowski J.R. (1973) Aggregation of soil particles by algae. *J. Phycol.* 9, 99-101.

Borut S. (1960) An ecological and physiological study on soil fungi of the Northern Negev. *Bull. Res. Counc. Israel* 80, 65-80.

Both W.E. (1961) Algae as pioneers in plant succession and their importance in erosion control. *Ecology* 18, 38-46.

Cameron R. (1964) Terrestrial algae of southern Arizona. *Trans. Amer. Microscop. Soc.* 83, 1212-1219.

Evenari M., Shanan L. & Tadmor N.H. (1971) *The Negev.* Harvard University Press.

Friedman I., Lipkin Y. & Ocampo-Paus R. (1967) Desert algae of the Negev. *Phycologia* 65, 185-200.

Friedman I. & Galun M. (1974) Desert algae, lichens and fungi. In: *Desert Biology* (Ed. Brown), Vol. II, pp. 166-212, London-New York, Academic Press.

Fritch F.E. (1922). The terrestrial algae. *J. Ecol.* 10, 220-236.

Israel Atlas (1963) Israel Meteorological Service and Bialik Foundation.

Lange O. (1972) Photosynthesis of lichens from Antarctica. *Antarctica Res. Ser.* 20, 83-95.

Lynn C. (1973) The role of algae in crust formation and nitrogen cycling in desert soils. I.B.P. Research Memorandum Desert Biome U.S. Internat. Biol. Program.

Shield L.M. (1964) Algae in relation to soil fertility. *The Botanical Review* 92-128.

Winston T.W. (1960) Saturated solutions for the control of humidity in biological research. *Ecology* 41, 232-237.

Developments in Arid Zone Ecology and Environmental Quality
Edited by H. Shuval

4

MOISTURE AS A FACTOR INFLUENCING SELECTION OF HIDING PLACES BY JUVENILE *SALAMANDRA SALAMANDRA* (L.) FROM SEMI-ARID HABITATS

G. Degani and H. Mendelssohn

Department of Zoology, Tel Aviv University, Tel Aviv, Israel

Summary

The response of juvenile *Salamandra salamandra* from semi-arid habitats to different depth of holes and moisture in them was studied shortly after metamorphosis.

The most important factor influencing the selection of hiding places of juvenile salamanders was soil moisture, this factor being more important than the size of hole opening and the depth of the hiding place.

According to this study it seems that the selection of moist soil in their hiding places enables the juvenile salamanders to survive long dry periods.

Introduction

Salamandra salamandra is found in different habitats in central and southern Europe, reaching the southern limit of its distribution in northern Africa and in Israel (Eiselt 1958; Gasser 1975; Degani 1979). The activity of *S. salamandra* takes place mostly in winter during periods of high relative humidity and of low temperatures (Joly 1968; Degani & Mendelssohn 1979b). The selection of moist substrates was found in Plethodontidae (Bateson 1965; Jaeger 1971), and *Ambystoma* (Marangio & Anderson 1977). Several studies established a relationship between moisture preference and water balance in Urodela (Taub 1961). In some winter ponds in semi-arid habitats the *S. salamandra* tadpoles completed metamorphosis just before the water source dried up in spring or summer (Degani & Mendelssohn 1979a; Warburg, Degani & Warburg 1979). Under semi-natural conditions juvenile salamanders hide in holes in the ground. There is no information available about the relationship between soil moisture and the depth and size of the hiding places of juvenile *S. salamandra*.

Materials and Methods

The juvenile salamanders were collected as tadpoles from the Gush Halav pond at the end of March and completed metamorphosis in the laboratory (see detailed description of Gush Halav pond, Degani & Mendelssohn 1979a).

Two experimental enclosures (A, A^1) were built on flat wooden boards (100 x 100 cm). A clay fence was built round the boards. The size of the enclosure was 80 x 80 cm and 6.5 cm high (Fig. 1). On each side of the fence two small cells were built (8 cm diam.). Enclosure A contained openings of different widths: two of 0.5 cm width, two of 1 cm., two of 1.5 cm., and two of 2 cm. Each opening was placed at random. The control enclosure (A^1) contained several openings of the same size of 1 cm.

A third enclosure (B) was made from panes of glass covered with black cloth and forming a box 90 x 35 x 40 cm. This box was filled with soil with two sides containing holes of different depths: 5 cm., 10, 15, 20, 25, and 30 cm at angles of 45° (Fig. 2). All the experiments were conducted at room temperature 20-25°C. The percentage of water in the soil was determined as previously described by Degani & Warburg (1979).

Experiments

Experiment 1: In order to test the influence of the width of opening on the response of juvenile salamanders to hiding places, enclosure A was used. Each cell in the enclosure contained the same amount of humid soil, containing 31.5% water. Enclosure A^1 was used as a control under the same conditions. The salamanders were tested in groups of 15 animals in the enclosures in the afternoon and examined after 8 hours.

Experiment 2: Enclosure A^1 was used for a comparison between saturated soil and dry soil. Moist soil containing 31.5% water was placed in 4 cells, and dry soil (dried at 105°C for 24 hours) in four other cells. The experiment was conducted in the same way as in experiment 1. The cells were selected at random.

Experiment 3: Using enclosure A^1 four cells were selected at random and moisture was maintained at approximately 4.65%, 13.60%, 22.55% and 31.5%. The experiment was conducted in the same way as experiment 1.

Experiment 4: In order to test the relationship between soil moisutre and depth of hole enclosure B was used. The 5, 10 and 15 cm depth holes were filled with 5 ml water and the 20, 25 and 30 cm depth holes were left dry. 8 salamanders were tested each night and examined 8 hours after placing them in the enclosure.

Experiment 5: Enclosure B was used. Six holes of different depths in one side were filled with saturated soil and the other six holes were left dry. The experiment was conducted as described for experiment 4.

Results

At the beginning of experiment 1 more juvenile salamanders entered the cells with the wide openings than those with the narrow openings. Moreover, after 8 hours 60% of the juvenile salamanders were found in the cell with a 2 cm wide opening and the cell with the 0.5 cm opening was empty. (Fig. 3). The difference between salamanders selecting holes with wide or narrow openings respectively, in enclosures A and A^1, was significant only when the holes contained moist soil. No significant difference was found, however, between the salamanders in enclosure A^1 and random distribution.

Most of the juvenile salamanders selected the moist soil rather than the dry soil (Fig. 4). The difference between the two possibilities was significant. Moreover, the selection of hiding place by juvenile salamanders in the cells with openings of the same width was according to the amount of moisture of the soil (Fig. 5). The difference between the Test (enclosure A^1 with different soil moistures in the cells) and the control (enclosure A^1 with 31.5% soil moisture) was not significant. No significance was found between the control and the random distribution (Fig. 5). The juvenile salamanders selected the holes according to the moisture of the soil and not according to the depth of hole. Most salamanders selected the shallower holes (5, 10, 15 cm) when they were moist, compared to deeper holes (20, 25, 30 cm) when these were drier (Fig. 6). The difference was significant ($P < 0.001$). A similar situation was found when a comparison was made of one side of enclosure B, where all the holes were moist, to the other side, where all the holes were dry (Fig. 7). There was a significant difference between the selection of salamanders of moist holes compared with the dry holes. No significant difference was found regarding the different depths ($P > 0.05$).

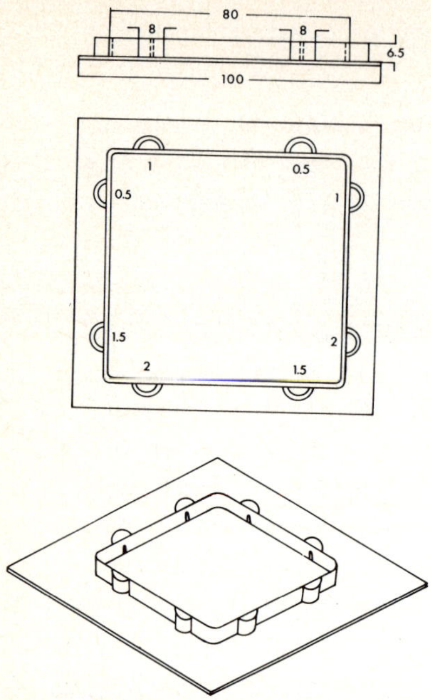

Fig. 1. Measurements of the enclosures (A, A^1)

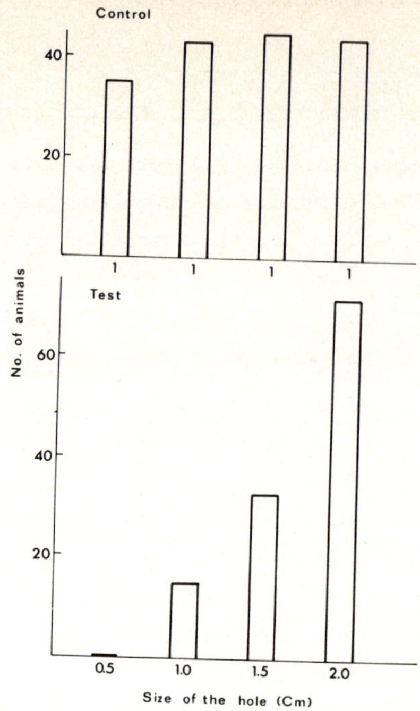

Fig. 3. Juvenile S. salamandra selection of hiding places by width of opening.

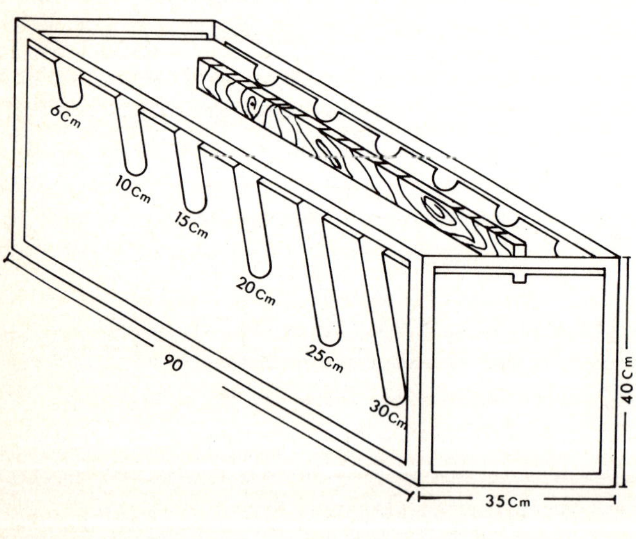

Fig. 2. Measurements of enclosure (B)

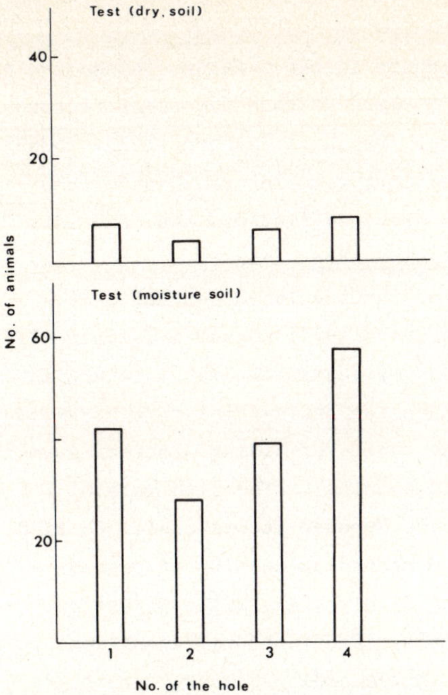

Fig. 4. Number of juvenile S. salamandra choosing saturated soil to those choosing dry soil (experiment 2).

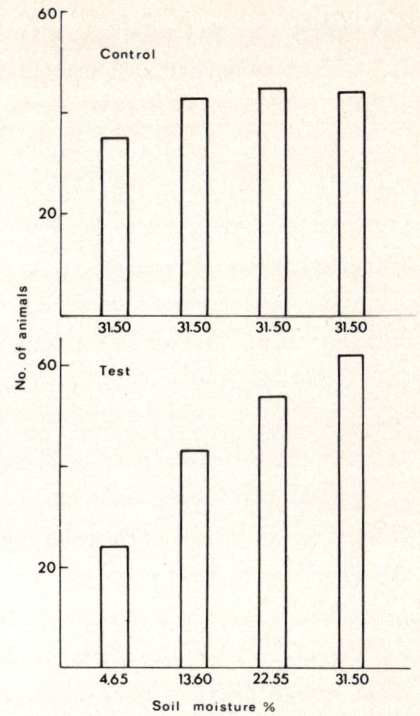

Fig. 5. The response of juvenile S. salamandra to hiding places with different degree of substrate moisture (experiement 3).

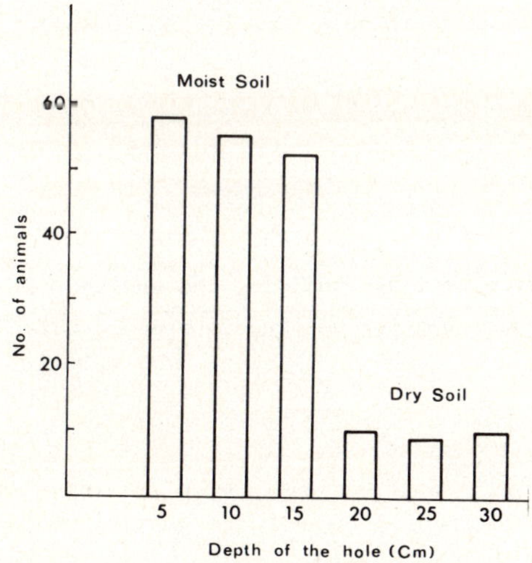

Fig. 6. The response of juvenile S. salamandra to hiding places with different depths of holes and varying degrees of substrate moisture (experiment 4).

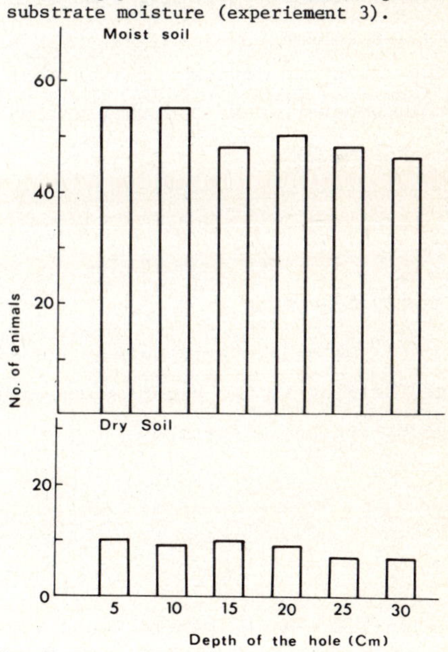

Fig. 7. The response of juvenile S. salamandra to hiding places with different depths of holes and varying degrees of moisture (experiment 5).

Discussion

The response of juvenile *S. salamandra* to the moisture inside the hiding places (cells and holes) was very strong: more so than any other factor such as depth of hole or width of opening. Degani & Warburg (1979) found that the juvenile *S. salamandra* responded more to moisture changes than did adults. This fact can be related to previous findings that the rate of dehydration is higher in juvenile *S. salamandra* than in adults. The response to substrate moisture was studied in Urodela (Rosenthal 1975; Bateson 1965; Jaeger 1971; Marangio & Anderson 1977; Degani & Warburg 1979). All those studied established a definite preference for high moisture in the substrate. This study shows the connection between moisture preference and two other contributing factors in the choice of hiding place of juvenile *S. salamandra* (depth of hole and width of opening). The high rate of selection of rooms with wide opening may be because it was easier to find the moist soil inside them. No selection according to size of opening was found when the rooms contained dry soil only. A similar situation was found in the selection by depth of hole. Under natural conditions these three factors were found together in the hiding places. In Israel *S. salamandra* are found in semi-arid habitats. In summer many spawning ponds became dry (Degani & Mendelssohn 1979a)and the salamanders find hiding places underground, where the soil remains moist. The depth of the holes and narrow openings that reduced evaporation from the soil did not, however, have a direct influence on their selection by the juvenile *S. salamandra.*

Acknowledgements

We thank Ravit Noiman and Sasson Lavi for building the enclosures and helping in the experiments, and Nomi Paz for helping to prepare this paper for publication.

References

Bateson, T.D., 1965. Some comparative behaviour studies on three genera of Salamandra. Kentucky Acad. Sci., 25:120-128.

Degani, G., 1979. Morphological and biochemical differences between the Salamandra salamandra (L) populations in Israel. Israel J. Zool. 28:54-55.

Degani, G. & H. Mendelssohn, 1979a. The food of Salamandra salamandra (L) tadpoles in Israel in different habitats. Israel Ecol. Soc., 19c-45c.

Degani, G. & H. Mendelssohn, 1979b. Seasonal activity of adult and juvenile Salamandra salamnadra at the limit of their distribution. Brit. J. Herpetol.

Degani, G. & M.R. Warburg, 1978. Population structure and seasonal activity of adult Salamandra salamandra (L) (Amphibia, Urodela, Salamandridae) in Israel. J. Herpetol. 12:437-444.

Degani, G. & M.R. Warburg, 1979. The response to substrate moisture of juvenile and adult Salamandra salamandra (L) (Amphibia, Urodela). Biol. Behav. (in press).

Eiselt, J., 1958. Der Feuersalamandra Salamandra salamandra (L). Beitrag zu einer taxonomischen synthese. Abh. Ber. Mus. Naturk. u. Vorgech. (Magdeburg) 10:77-154.

Gasser, F., 1975. Recherche sur le status microevolutif de deux amphibiens urodeles, l'espece pyreneene Euproactes asper (Duges) et l'espece palearctique Salamandra salamandra (L). proteins et groupes seriques, cycles sexuels femelles et morphologie. Presente l'universite Paul Sabatier de Toulouse pour obtenir le grade de docteur d'etat Mention. Sciences, 1-247.

Jaeger, R.G., 1971. Moisture as a factor influencing the distribution of two species of terrestrial salamanders. Oecologie 6:191-207.

Joly, 1968. Donnees ecologiques sur les salamandra tachetee Salamandra salamandra (L). Ann. Sci. Mat. Zool. 12^{e} ser. 10:201-366.

Marangio, M.S. & J.F. Anderson, 1977. Soil moisture preference and water relations of the marbled salamander Ambystoma opacum (Amphibia, Urodela, Ambystomatidae). J. Herpetol. 11:169-176.

Rosenthal, G.M., 1975. The role of moisture and temperature in the total distribution of plethodontid salamander <u>Aneides lugrubis</u>. Univ. Calif. Dupe. Zool. 54:374-420.

Warburg, M.R., G. Degani & I. Warburg, 1979. Growth and population structure of <u>Salamandra salamandra</u> (L) larva in different limnological conditions. Hydrobiologia.64:147-155.

Developments in Arid Zone Ecology and Environmental Quality
Edited by H. Shuval

5
A MICROBIAL COMMUNITY OF THE SAND SEASHORE OF ISRAEL

Tamar Berner*

**Department of Biology School of Education - Oranim, Haifa University, Kiryat Tivon, Israel*

ABSTRACT

A microbial community was found in the sand in several locations along the Mediterranean coast of Israel. This sand community formed a profile of stratified sand-grain layers, each several mm in thickness. The upper surface layer, grey in colour, was free from visible microorganisms; below this layer, in that order, were an algal, a purple bacterial and a dark silt layer. Two community types were found within the algal layer: a) epipelic motile algae, dominated by *Oscillatoria tenuis Ag.*, free living among the sand grains, and b) epipsammic, non-motile algae, dominated by *Microcystis reinboldii (Richt.) Forti*, attached to the grains. This community was found within a salinity range of 780-5680 mg $Cl \cdot l^{-1}$. The salinity of the sand increased from the bottom to the surface due to evaporation while the water content showed an opposed trend. Light intensity was quickly attenuated with depth. The presence of the algal community depended on a regular influx of organic matter, either by seepage from nearby fish ponds, or by infiltration of raw sewage.

INTRODUCTION

An algal layer, buried under an upper grey sand stratum, which covered a bacterial and a silt layer, has been reported in the North Sea (Warming, 1904; Skuja, 1924; Schulz & Meyer, 1939; Hoffmann, 1942), where it was named "Farbstreifen-Sandwatt", meaning "the coloured layer sand", and some other locations (Flint, 1955; Ganapati, 1959; Amos, 1965). A similar algal layer has been found by us in the sandy eastern Mediterranean coast of Israel.

The aim of this work was to follow the seasonal variation in distribution of the algal species and to study the environmental requirements of the community.

DESCRIPTION OF RESEARCH AREA

The study was carried out mainly at Ma'ayan Michael, Israel, on the eastern shore of the Mediterranean. The community occurred in the sand between the fish ponds on the east and the sea on the west. The water elevation in the fish ponds went from 6.2 m above sea level in the eastern ponds to about 3 m in the western ones. The width of the beach at this site was about 50 m. The water source for the ponds was the Taninim River, whose salinity ranged from 500 mg $Cl \cdot l^{-1}$ during the winter to 800 mg $Cl \cdot l^{-1}$ in the summer. Due to the ponds elevation and their porous bottom, the water, containing organic matter, seeps through the sand to the sea shore.

During the winter the upper layer of sand was blown about, causing instability of the algal layer and even its disappearance.

Another research area was at Kiryat Yam along the same coast, where organic matter source was the infiltration of raw sewage.

MATERIAL AND METHODS

Cores of sand were taken with glass cylinder, and the layers' thickness was measured through the glass. After extrusion the layers were separated. Chemical data are averages of all measurements taken.

Water content was determined as loss in weight of samples of each layer after 24 hr at 105°C and calculated as percentage of dry weight.

Cl determination - by $AgNO_3$ titration (Standard Methods, 1976) of the interstitial water, as mg $Cl \cdot l^{-1}$.

Nitrogen determination - by micro Kjeldahl (Steyermark, 1961) as mg $N \cdot l^{-1}$.

Areal chlorophyll a - by extraction in methanol 90% (Vollenweider, 1974), expressed as mg Chl. a · cm^2.

Light measurements - taken with Li-185 light-meter (Lambda Instruments) with quantum sensor, in $\mu E \cdot m^{-2} \cdot sec^{-1}$.

RESULTS

The algae, Chlorophytes, Euglenophytes and Cyanophytes, were identified at least to the genus level. They occurred either attached to the sand grains or among them. The algal layer was covered by an upper grey layer while underneath the algae, a purple bacterial layer, dominated by *Rhodospirillum rubrum*. At the bottom of the core, a dark silt layer was presented, dominated by *Beggiatoa alba*. (Fig. 1)

Table 1 shows the seasonal distribution of the algae. Three species only were consistently present whenever the algal community existed.

In order to determine the conditions upon which the algal community existence depended, three major environmental factors were examined:

TABLE 1. Distribution of the Algae in Ma'agan Michael

Algae	Total number of months the organism occurred
Attached	
Microcystis Reinboldii (Richt) Forti	10
Phormidium fragile Gom.	9
Anabaena "affixa"	7
Merismopedia affixa P. Richt (Lagerh)	5
"N. gen. Chlorophycedrum"	3
Non-attached, non-motile	
Anabaena sp. (1)	5
Gloeotheca sp.	6
Gloeocapsa sp.	5
Anabaena sp. (2)	3
Anabaena sp. (3)	2
Nodularia Harveyana Thur.	2
Non-attached, motile	
Oscillatoria tenuis Ag.	10
Chroococcus turgidus (Kütz) Näg var. submarinus	10
Spirulina maior Kütz	6
Euglena van-goori Defl.	6
Merismopedia sp.	3
Eutrptia viridis Perty	2
Pseudoanabaena Lauterb.	1
Oscillatoria sp.	1

TABLE 2. Water content (%) and salinity ($mg \cdot l^{-1}$) at different distances from the sea

Distance	5m		20m		30m	
Layer	Water content	Salinity	Water Content	Salinity	Water Content	Salinity
Upper grey	21.8±2.4	5790-5970	18.6±8.1	1570-1640	13.4±8.4	690-740
Algal	24.5±2.1	5580-5650	22.1±2.8	1470-1530	19.9±6.2	630-720
Bacterial and Silt	26.3±2.2	4930-5110	25.3±2.5	1420-1510	21.4±8.9	640-710

TABLE 3. Chlorophyll a content ($mg \cdot cm^{-2}$) and N concentration ($mg \cdot l^{-1}$) in the algal layer at different distances from the sea

Distance	Chl. a	N
5m	0.032	12.5±1.3
20m	0.295	50.5±5.0
30m	0.198	30.8±1.4
40m	0.038	9.2±1.1

r = .986

excreta, dead fish and plankton. The N source in the North Sea are the excreta of sea birds and organic matter washed ashore by the sea (Wohlenberg - personal communication).

Preliminary measurements showed photosynthetic activity. However, the existence of algae well below 1% light level (compensation point) may suggest that they benefit from the organic compounds by mixotrophy (Fogg, 1973; Stanier, 1973; Droop, 1974). The dominance of the blue-green algae suggests the possibility of anaerobic activity and photo-assimilate CO_2, with sulfid from the bacteria and silt layers as an electron donor (Stewart, 1973; Cohen et al., 1977).

It appears that a combination of environmental factors determines the stratificiation of the microbial community. Due to the wide fluctuations in hydration and salinity in the upper grey layer, as well as its exposure to high light intensity, most of the living organisms, certainly algae, are excluded from it . In the subsequent layer the light intensity is reduced so that chlorophyll is less prone to photo oxidation and the algae are able to photosynthesize or use other kind of nutrition. In the lower layer, as shown by the appearance of *R. rubrum* there is a decrease in oxygen concentration while the layer is still illuminated allowing bacterial photosynthesis. The dark silt underneath the purple bacteria is anaerobic and rich in H_2S as proved by the *B. alba* population. The fluctuations in layers' thickness is probably due to environmental factors affecting the activity of the living organisms in the profile. The occurrence of the microbial community depends on a regular influx of organic matter and the stratificiation -- on the relationship between the environmental gradients in this habitat.

REFERENCES

Amos W.H. (1965) The living sand. *Nat. Geogr. Mag.* 127, 820-833.

Anagnostidis K. (1961) *Untersuchungen über die Cyanophyceen einiger Thermen in Griechenland.* Inst. f. Syst. Bot. u. Pflanzen geographic. Univ. Thessaloniki. 7, 1-322.

Cohen Y., Krumbein W.E. & Shilo M. (1977) Solar Lake (Sinai). 2. Distribution of photosynthetic microorganisms and primary production. *Limnol. Oceanogr.* 22, 609-620.

Droop M.R.(1974) Heterotrophy of carbon, 530-559. *Algal Physiology and Biochemistry* (Edited by Stewart W.D.) Blackwell, Oxford.

Flint L.H. (1955) A green sand from the Gulf of Mexico. *Proc. Louisiana Acad. Sci.* 18, 22-24.

Fogg G.E., Stewart W.D.P., Fay P. & Walsby A.E. (1973) *The blue-green Algae*. Academic Press, London and New York.

Ganapati P.N. (1959) Tidal rhythms of some diatoms and dinoflagellates inhabiting the intertidal sands of the Wisakhapatnam beach. *Curr. Sci.* 28, 450-451.

Geitler L. (1932) *Cyanophyceae*. Akademische Verlagengezellschaft, Liepzig.

Hefer B. (1958) Chemical changes in fertilized and non-fertilized fish pond in the sub-tropical conditions of Israel. *Bamidgeh* 11, 3-17 (In Hebrew).

Hoffman C. (1942) Beiträge zur Vegetation des Frabstreifensandwattes. *Kieler Meeresforsch.* 4, 85-108.

Ramane A. & Schlieper C. (1971) *Biology of Brackish Water*. John Wiley & Sons Inc., New York.

Schulz E. & Meyer H. (1939) Watere Untersuchungen über das Farbstreifensandwatt. *Kieler Meeresforsch.* 3, 321-336.

Skuja H.(1924) Beitrag zür Algenflora des Rigaschen Meerbusens. *Acta Univ. Latviensis* 10, 24-36.

Standard Methods for Examination of Water and Wastewater (1976) 15th. Edition. American Public Health Association.

Stanier R.Y. (1973) Autotrophy and heterophy and unicellular blue-green algae 501-518. *The Biology of blue-green Algae*. (Edited by Carr N.G. & Whitton B.A.) Blackwell, Oxford.

Stewart W.D. (1973) Nitrogen fixation 260-278. *The Biology of blue-green Algae*. (Edited by Carr N.C. & Whitton B.A.) Blackwell, Oxford.

Steyermark A. (1961) *Quantitative Organic Microanalysis*. Academic Press, New York.

Umezaki I. (1961) *Marine Blue-green Algae of Japan*. Kyoto University, Japan.

Vollenweider R.A. (Ed.) (1974) *A Manual on Methods for Measuring Primary Production in Aquatic Environments*. IBP. Handbook No. 12. Blackwell, Oxford.

Warming E. (1904) Bedrag til vadernes, Sardenes og Marskens Naturhistorie. Klg. Dansk. Vidensk. Selsk. Skrift. 7. Raekke. *Naturv-math. Afd.* I 1-56.

Wohlenberg E. Marshenbanamt Husum, Germany (Personal communication).

Developments in Arid Zone Ecology and Environmental Quality
Edited by H. Shuval

6
BEHAVIORAL ECOLOGY OF THE MARINE SNAIL *TRUNCULARIOPSIS (MUREX) TRUNCULUS*

Ehud Spanier

Center for Maritime studies and the Department for Maritime civilizations, University of Haifa, Mount Carmel, Haifa, Israel.

ABSTRACT

Historical sources suggest that the marine snails, *Trunculariopsis (Murex) trunculus,* believed to be an important raw material for the ancient purple dye industry, were collected during limited seasons. Field and laboratory studies indicate that these snails are more abundant on and above the sediments in the colder months than in the summer, when most of the snails are burrowed in the sediment. This, together with sea-state considerations, explains the short collecting seasons. The fact that the southeastern Mediterranean has the warmest water within the distribution range of this species may be the reason for this burrowing behavior. It is suggested that the snails try to avoid the high temperature of the summer by burrowing in the sand, possibly in the form of aestivation.

INTRODUCTION

The marine snails of the species *Trunculariopsis (Murex) trunculus* (Prosobranchia, Muricidae) are believed to be one of the main sources from which the ancients obtained the Royal or Tyrian purple (e.g. Jensen, 1963). The purple dye's chemical constitution has been accepted as 6,6' - dibromoindigotin (Baker, 1973) and is obtained from a secretion of the hypobranchial gland of these snails (Hyman, 1967). The assumption that this species served as an important raw material for the purple dye industry is supported by historical records as well as by the large quantities of shells of this species found in various archaeological excavations around the Mediterranean Sea (Reese, 1980).

The historical sources suggest that these relatively shallow-water marine snails were not collected throughout the year but rather during certain limited seasons. Pliny, in his *Historia Naturalis* (Rackham, 1962) mentioned that the snails should be caught at the end of the summer or in early spring. The present study was initiated in order to decide whether the collecting seasons were limited due to oceanographic reasons (i.e. sea-state) or resulted from the ecology and/or behavior of these molluscs.

MATERIALS AND METHODS

A population of *Trunculariopsis trunculus* (the generic name has been substituted for the old nomenclature *Murex*; other recent combinations are *Phyllonotus trunculus* and *Muricanthus trunculus*) was studied in their natural habitat by SCUBA divers from May 1979 to September 1980. The site of the study was 30 m southeast of the "tower of flies" in the Akko Bay on the northern Mediterranean coast of Israel. Water depth at the site was 6 m. Bottom sediment was coarse sand covered with stones and small rocks. An area of 25x25 m at this site was fenced off with heavy ropes. This square was divided into five equal strips using lighter ropes. In May, 1979 all *T. trunculus*, collected in the fenced-off area, were marked with red polyurethane paint (resistant to sea water). The height (i.e., length of shell) of each snail was measured with a caliper. The marked, live snails were then returned to the center of the fenced area. The population was monitored almost every month in the subsequent period of 16 months. During each visit, several pairs of divers carefully inspected each strip within the fenced square and collected every *T. trunculus* detected on the bottom. These snails were counted, inspected for red paint, measured and returned to the center of the fenced area. Environmental conditions, such as water temperature near the bottom, visibility, flow and sea state were recorded during each dive. During the last dive, in September 1980, all *T. trunculus*, found burrowed in the sediments in a small area of 3x3 m within the fenced square, were dug up and recorded.

Snails of the same species collected in the same general area (but not close to the fenced square) were kept in glass tanks in the laboratory. The size of the snails was similar to that of those found in the field (average height - 47.9 mm). The water in the tanks was aerated and filtered through under-gravel and outside filters. The bottom of the tanks was covered with a 5 cm layer of coarse sand. *T. trunculus* were fed with live *Patella coerulea* (see Spanier and Karmon, 1980). During the preliminary observation, 40 muricid snails were kept in the winter in a 100x30x40 cm glass tank. The water of this tank was heated by a submerged heater. Additional winter experiments were carried out with 12 *T. trunculus* kept in a 60x30x40 cm glass tank. This time the temperature of the water was adjusted by a submerged heater as well as by an outside heater which warmed the whole room. Snails were scattered on the sand before each experiment. Water temperature was kept constant for at least 48 hours and the position of each snail in the tank was recorded. In this way the vertical distribution of specimens of *T. trunculus* in the experimental tank was recorded at various water temperatures ranging from 13°C to 29°C.

RESULTS

After several months of successive field dives it became apparent that the red paint, though it did not peel off, slowly disappeared because of marine fouling (such as barnacles, vermetid worms etc.) covering the shells of the municids snails. However, data obtained from those snails whose marking paint was still detectable revealed that the population seemed to remain in the fenced area. Therefore, the recording of the population throughout the year, under different environmental conditions was continued.

Water temperature was the factor which affected the abundance of *T. trunculus*. Fig. 1 clearly indicates that the number of snails was inversely proportional to the water temperature. The only exception is the spawning period associated with the crowding of the snails into reproductive mounds.

More than 300 snails were collected in the 25x25 m area during the cold months while this number dropped to less than 100 during the summer. However, during the last dive, when snails were collected by digging a small area within the fenced square (from a depth of less than 2 cm under the surface of the sand) the number was almost doubled.

The average and the range of the height of the shells collected were not significantly different at different water temperatures.

In the preliminary laboratory experiments snails climbed and adhered to the glass walls of the tank (which was colder than the water in the tank) when water temperature was raised to above 25°C. In the ancillary laboratory experiments, in which water in the tank and the surrounding air was kept at the same temperature, another behavior was observed. *T. trunculus* snails moved closer to the sediment and some burrowed shallowly in the sand (with siphon tips in contact with the water). This behavior was noticed in water temperatures of and above 25°C and also in the lowest water temperature (Table 1).

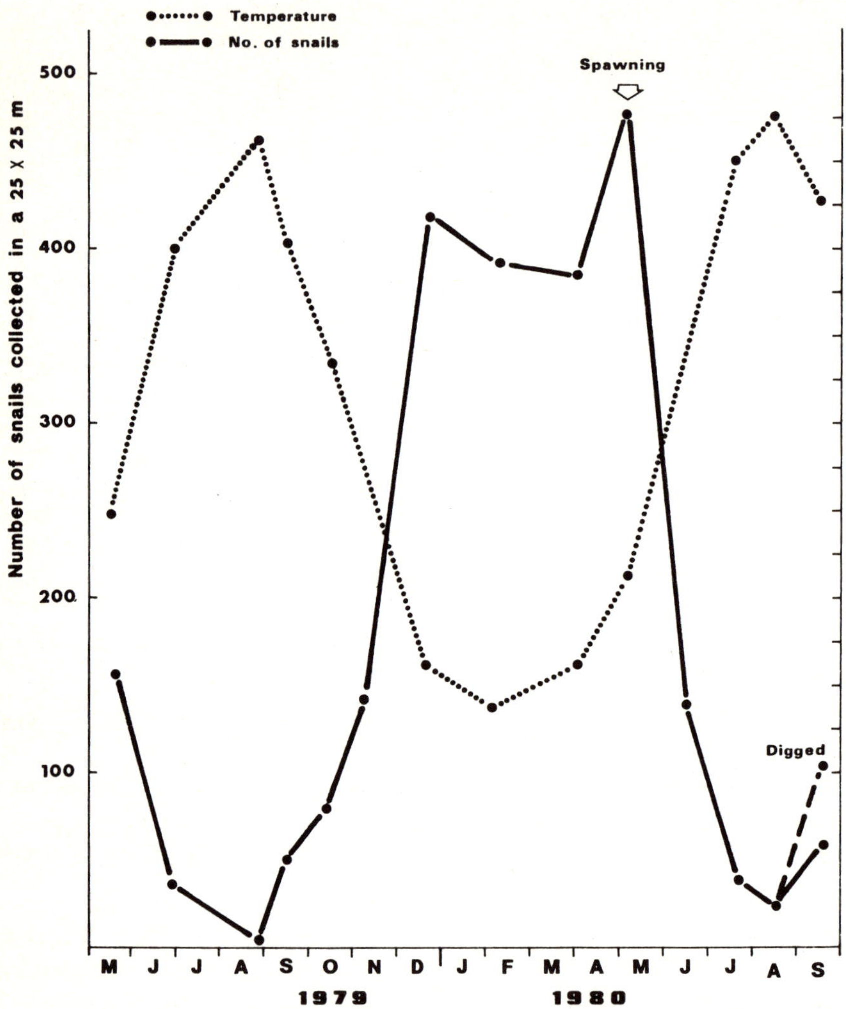

FIG. 1. Number of *T. trunculus* collected in the fenced square as a function of water temperature.

TABLE 1. Average vertical distribution of *T. trunculus* snails kept in a laboratory tank at different water temperatures.

water temp.	% snails burrowed in the sédiment	% snails on and in the sediment	% snails located up to 5 cm above the sediment	% snails located more than 5 cm above sediment
13°C	33.3	66.6	8.4	25
17°C	0	33.3	8.4	58.3
19°C	0	33.3	0	66.7
20°C	0	33.3	12.5	54.2
22°C	0	25	0	75
24°C	0	54.2	12.5	33.3
25°C	16.6	75	16.6	8.4
26°C	0	70.8	0	29.2
27°C	12.5	75	16.6	8.4
28°C	16.6	83.4	0	16.6
29°C	33.3	91.6	0	8.4

DISCUSSION

Both field observations and laboratory experiments indicate that *T. trunculus* is more abundant above the bottom sediment in the colder months. Consequently, they were more available for collectors in the eastern Mediterranean Sea from October to May (Fig. 1). The fact that these snails are common in relatively shallow water where breaking waves prevent collection, must also be taken into consideration. The sea state along the Mediterranean coast of Israel, as represented by mean wave height and standard deviation of wave height, (Goldsmith, Golik & Sofer, personal communication, 1980) is calmest in May and October with transient conditions in April and November. This and the above ecological considerations exclude the rest of the year as possible collecting seasons of *T. trunculus*. Thus, the short seasons suitable for the collection of considerable amounts of snails are, as Pliny recommended, at the end of the summer (October and part of November) and the early spring (April - May).

The present study also indicates where the snails disappear in the warm months -- they burrow shallowly in the sediments. This phenomenon may have zoogeographical and physiological explanations. This species is distributed along the Mediterranean and the Atlantic coasts of Europe,

North Africa and Asia. The water of the southeastern Mediterranean is the warmest in this range of distribution. They do not migrate to much deeper and colder water during the summer. This water, at least in the study area, was too far to reach by these slow-moving snails. It is possible therefore, that the snails try to avoid the unfavorable high water temperature in the summer by burrowing in the sediment. This may be described as a form of aestivation. Carriker & Van Zandt (1972) and Manzi (1970) described similar burrowing behavior at temperatures lower than 7°C in the Northeastern Atlantic muricid snail *Urosalpix cinerea* as hibernation. Thus, muricid snails hibernate in lower temperatures (such as 13°C for *T. trunculus* see Table 1) and aestivate in warm temperatures (25°-30°C). Only further physiological and physical studies may clarify this hypothesis and the mechanisms controlling this phenomenon.

ACKNOWLEDGEMENTS

I would like to express my gratitude to O. Weissmann, N. Karmon and R. Rosental for their assisstance in the laboratory experiments. Appreciation is directed also to many colleagues and students who provided useful help in the diving operation and especially to Y. Tur-Caspa and S. Breitstein.

This study was supported, in part, by a grant from the Israeli Academy of Sciences and a research grant from Dr. S.M. Edelstein, N.J.

REFERENCES

Baker J.T. (1973) Tyrian purple: an ancient dye, a modern problem. *Endevour* 33, 11-17.

Carriker M.R. & Van Zandt D. (1972) Predatory behavior of a shell-boring muricid gastropos. In *Behavior of Marine Animals* Vol.1 (Edited by Winn H.E. & Olla B.L.) pp. 157-244. Plenum Press, New York.

Hyman L.H. (1967) *The Invertebrates. Vol. 6, Mollusca 1.* McGraw-Hill, New York.

Jensen L.B. (1963) Royal purple of Tyre. *J. Near East Stud.* 22,104-118.

Manzi J.J. (1970) Combined effects of salinity and temperature on the feeding, reproductive and survival rates of *Eupleura caudata* (Say) and *Urosalpinx cinerea* (Say) (Prosobranchia: Muricidae). *Biol. Bull.* 138, 35-46.

Rackham H. (1962) *Pliny -- Natural History* vol. 3, p. 253. Harvard University Press, Cambridge.

Radwin G.E. & D'Attilio A. (1976) *Murex Shells of the World.* Stanford University Press, Stanford.

Reese, D.S. (1980) Industrial exploitation of Murex shells: purple dye and lime production at Sidi Khrebish, Benghazi (Berenice). *11th Ann. RP. Soc. Libyan Stud.*, 79-93.

Spanier E. & Karmon N. (1980) Notes concerning the predatory behavior of the purple snail *Murex trunculus. Levantina* 28, 321-323.

Developments in Arid Zone Ecology and Environmental Quality
Edited by H. Shuval

7

THE ECOLOGY OF THE CARPENTER BEE, Xylocopa sulcatipes Maa IN ISRAEL

Dan Gerling and Abraham Hefetz

Tel Aviv University, The George S. Wise Center for Life Sciences, Dept. of Zoology

ABSTRACT

The carpenter bee, Xylocopa sulcatipes occurs in Israel in the desert areas and in most of the Rift Valley. The bees overwinter, mostly as unmated males and females in old nests. After mating they establish new nests in hollow, canes, pithy stems or thin branches and raise up to 4 generations per season. Flight and flower visits begin after ambient temperature exceeds 22°C and continues during most of the day. X. sulcatipes differs in this detail from its sympatric bee, X. pubescens that starts to fly at 18°C but stops collecting during the hot hours of the day. Mating follows territorial flights that may be carried out at flowers or nest sites. Nests are guarded against invading parasites and conspecifics. In gregarious nests, bees feed each other through trophallaxis.

Xylocopa sulcatipes is one of the 5 species of carpenter bees living in Israel. All are characterised by digging tunnels in woody substances in which they nest. Their progeny develop in sealed cells within the tunnels, while feeding on a mixture of pollen and honey provided by their mother. They are characterised as solitary bees by some authors (Wilson 1971) but recent findings showed a high degree of mother-daughter cooperation (Michener, 1974, Gerling, Hurd and Hefetz, in preparation).

X. sulcatipes is distributed from Aden to Israel and Transcarpia (Maa 1970). In Israel it occurs through the Rift Valley from Eilat to the North of the Kinneret, and in most of the Negev. It overwinters as young males and unfertilized females in old nests and starts its activity with massive territorial flights about flowering plants in March. From then until September or October, X. sulcatipes nests actively, raising, on the average, 4 generations.

The nesting cycle includes nest digging in thin branches, canes, or other soft, slim woody substrates; provisioning with a mixture of pollen and nectar (bee bread), laying an egg and sealing the cell. There may be up to 8 such cells in a row, all are prepared within a short time. The bees within them develop to adulthood after about a month and emerge almost contemporaneously.

X. sulcatipes females need ambient temperatures of about 21°C in order to become active. Once active, they may continue to collect pollen and nectar from flowers incessantly throughout the day, even at the hot noon temperatures of the Arava summers. The flowers visited include local plants such as *Calotropis procera*, *Acacia* spp. and *Moringa peregrina*, as well as many introduced ornamentals. Some, such as *C. procera* serve only as sources for nectar and may depend upon bees of that genus for pollination, whereas *Acacia* serves as a source for both, nectar and pollen.

The fact that *X. sulcatipes* needs relatively high ambient temperatures for initiation of foraging, causes it to come second to many of the flowers. Other bee species, especially *X. pubescens* which starts foraging at an ambient temperature of 18°C have the advantage of visiting the flowers earlier in the day. As a result, *X. sulcatipes* has to expend more energy for collecting similar amounts of food for her progeny and herself and therefore is often seen laboring at flowers for longer hours than its sympatric bee, *X. pubescens* (Fig. 1a and b).

Mating, that occurs in the spring and, later, when each generation emerges during the summer, is associated with territorial behaviour of the males. The territories defended by the males and the strategies that they use vary. Massive territorial flights occur near flowering *Moringa* trees at Ein Gedi during the spring, when these constitute the main food source for the females and therefore are a convenient place for the males to find them. More localized, individual territorial flights of males take place where flower sources are more diffuse, like the Hatzeva Field School (Velthuis & Gerling, 1980).

Very active and aggressive territorial flights occur near nesting sites of the bees. These are less frequent than those near food sources and were sighted under 2 sets of circumstances: A dense concentration of nests in cane roofs on buildings whence there were scores of males flying about pursuing the females, and chasing each other; and a very sparse concentration of nests and food sources where apparently it was more profitable for a male to fly from one nest to another and take the slim chance of meeting a virgin female, than to look for her on one of the many, widely dispersed, possible food sources.

Once the male meets with a receptive female, they mate following the exchange of visual and pheromonal signals (Velthuis and Gerling 1980).

The number of available and easily obtainable nest sites is limited, and finding a suitable nest as well as provisioning it with food require much effort. Consequently, there are always bees especially young ones,

Captions for Figure 1

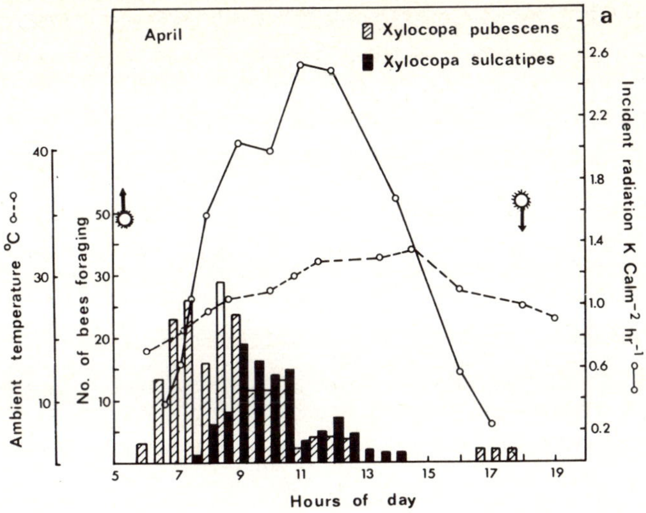

a) Daily foraging activity of X. pubescens and X. sulcatipes during the month of April, as related to averaged temperatures and incident radiation. Counts of bees were conducted either at flowering plants or nest entrances. Numbers presented are cumulatives of half hour counts.

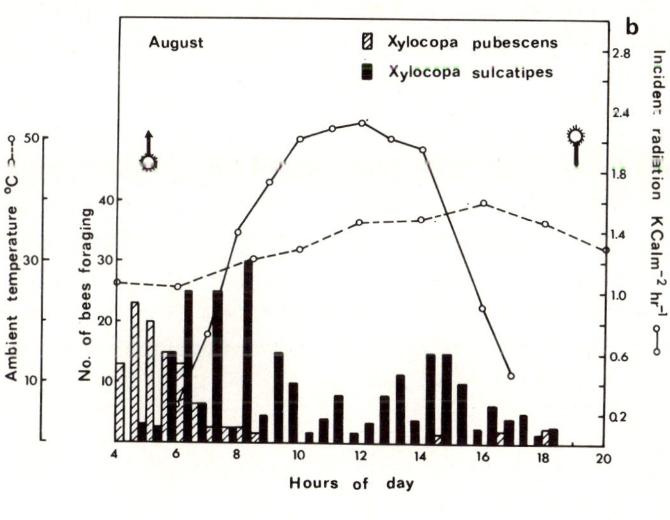

b) Daily foraging activity of X. pubescens and X. sulcatipes during the month of August, as related to averaged temperatures and incident radiation. Foraging was monitored as described in Fig. 1a.

that try to invade already established nests. As a counter measure, every established nest is guarded whenever possible, solitary nests are guarded whenever the bee is not out looking for food, and gregarious nests are guarded continuously. In the latter case, the guard female is fed by the forager through trophallaxis upon her return to the nest.

References

Maa, T.C. 1970. A revision of the subgenus *Ctenoylocopa* (Hymenoptera Anthophoridae). Pacific Insects 12(4):723-752.

Michener, C.D. 1974. The Social behaviour of the bees. Harvard University Press. 404 pp.

Velthuis, H.H.W. and D. Gerling 1980. Observations on territoriality and mating behaviour of the carpenter bee *Xylocopa sulcatipes*. Ent. exp. Appl. 28:82-91.

Wilson, E.O. 1971. The insect societies. Harvard University Press. 548 pp.

Developments in Arid Zone Ecology and Environmental Quality
Edited by H. Shuval

8

A SIMULATION MODEL OF THE NORTHERN LEAF BLIGHT OF CORN

Yehuda Levy*, Emanuel Noymeir* and Yigal Cohen+

*Department of Botany, The Hebrew University, Jerusalem and +Department of Life Sciences, Bar-Ilan University, Ramat-Gan, Israel

ABSTRACT

A simulation model for the development of the northern leaf blight of corn incited by Helminthosporium turcicum Pass. is presented. A high degree of similarity was found between the model-predicted and the actual development of the disease in growth chambers. Validation of the model under field conditions is yet needed.

INTRODUCTION

Northern leaf blight caused by Helminthosporium turcicum Pass. is a major disease of corn in Israel where it reduces yields substantially in the late growing season (Levy & Cohen, 1979). In a series of growth chamber experiments and field observations, the relationship of environmental and biotic factors to the development of the pathogen and of the disease were studied (Cohen, Levy & Eyal, 1978; Levy & Cohen, 1979, 1980a, b). The data collected were used to design a simulation model in order to provide a means for quantitative projections of disease epidemics.

In simulation models, a disease epidemic is broken down into a hierarchy of subsystems and the effect of different environmental and biotic factors on each subsystem is examined, as well as their effect on increase of the disease over a period of time. A simulation model attempts to represent biological information with symbols to achieve a better mimic of the natural system.

In this study a simulaton model of the northern leaf blight of corn was constructed according to Waggoner et al. (1972). The model was validated in growth chamber miniepidemics (Cohen & Rotem, 1971).

THE MODEL

The various stages in the life cycle of the pathogen were quantitatively characterized and calculated during the disease development.

The detailed functional relationships which characterize the host-pathogen-environment interactions were incorporated into the model (Fig. 1). Environmental data have to be incorporated before an epidemic can be simulated. The model is divided into three distinct subsystems:

1) Germination, appressorial formation and infection as a function of environmental and host conditions.

2) Reproduction of the pathogen, spore dispersal and spore deposit on vacant host as a function of environmental, host and pathogen conditions.

3) Actual environmental conditions to which the host- pathogen interaction is exposed.

The simulation of the model was done using the CSMP computer language. (For detailed program see Appendix No. 1).

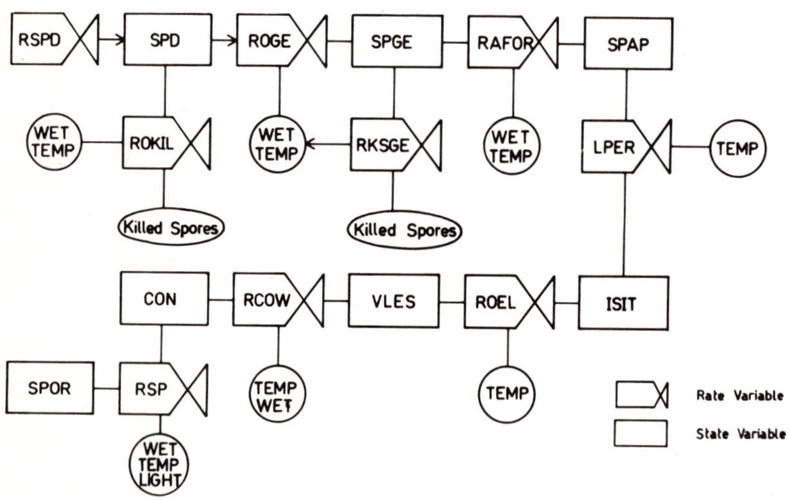

FIG. 1. The functional relationships of environmental and biotic factors in the development of Helminthosporium turcicum on sweet corn.

The environmental parameters in the model are: temperature in °C (TEMP), free leaf moisture (WET) and light (LIGHT). These were defined as a daily course repeated every day as a function of time. For light and free leaf moisture two conditions were distinguished: light (LIGHT=1) and darkness (LIGHT=-1), wet (WET=1) and dry leaves (WET=0).

We assumed that the crop growth is independent of the growth of the pathogen. Host leaf area was introduced as a function of time (LEA). The epidemic initiates by spores deposited on leaves (SPD) under appropriate conditions. Spores germinate (SPGE) at a rate (ROGE) dependent on temperature and wetness conditions. Some of the germinating spores produce appressoria (SPAP). Wetted, nongerminating and germinating spores which failed to produce appressoria are killed by a dry period of about 12 h (ROKIL, RKSGE) and, therefore, eliminated. The rate of appressorial formation (RAFOR) is dependent on temperature and wetness conditions.

Spores with appressoria penetrated into the leaf and produced infection sites (ISIT1). Infection sites became visible lesions (VLES) after an incubation period of 1-10 days, depending on plant and leaf age. The rate of growth of visible lesions (ROEL) is temperature-dependent. Under suitable conditions of wetness and temperature conidiophores grow on the infected leaves (CON). The number of spores produced by conidiophores depends on temperature, wetness and light conditions (SP). Conidiophores can sporulate up to three times (depending on temperature) with a maximal rate of sporulation at the first sporulation cycle (RSP1). Spores produced are dispersed by wind and, depending on host leaf area, are deposited on leaves (SPD).

MODEL VALIDATION

The model was validated by a series of growth chamber experiments in which environmental conditions prevailing in nature were mimicked.

Three environmental programs were selected: a) "Spring" (minimal night temperature 10°C, maximal day temperature 20°C, 14 h of wetting during each night; b) "Summer" (minimal night temperature 20°C, maximal day temperature 30°C, 10 h of wetting during each night),and c) "Autumn" (minimal day temperature 15°C, maximal day temperature 25°C, 12 h of wetting during each night). To initiate the disease, equal amount of spores was dispersed into each chamber. Fig. 2 shows the actual and model-predicted development of the disease in each chamber. The results show a close relationship between the actual and the model-predicted development of the disease. The results also show that the environmental conditions prevailing in nature in the autumn are most favourable for the development of the disease.

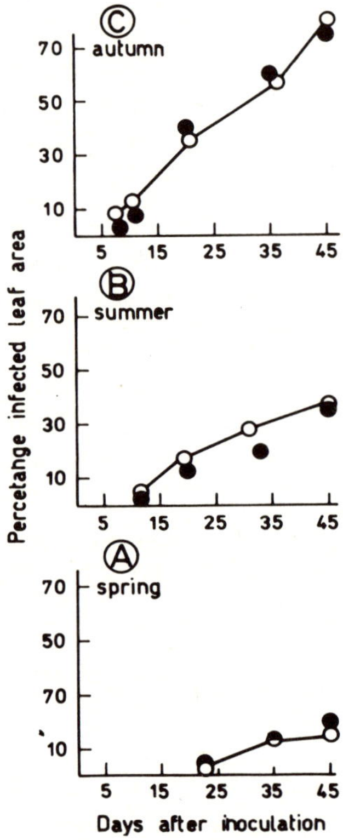

FIG. 2. The epiphytotics of Helminthosporium turcicum on sweet corn under three environmental regimes in growth chambers. ● - observed; o - model-predicted.

DISCUSSION

The simplified model established by Berger (1970), which has proven successful for three years in Florida, was found to be inefficient to forecast epidemics in Israel. Berger defined blight favourable hours (BFH) as hours in which humidity is near 100% and temperature above 15°C. A daily average of 6.5 BFH caused too little blight to justify fungicide application, while 11 BFH caused an epidemic regardless of fungicide application. Field observations in Israel (Levy & Cohen, 1979) showed that in spite of a daily average of 11 BFH during the summer there was no epidemic outbreak of the disease. We assumed, therefore, that a more detailed model is needed to forecast the epidemics. The model presented in this work seems to provide a better tool for forecasting of the northern leaf blight of corn in Israel. Yet, a validation of this model in the field is needed.

REFERENCES

Berger R.D. (1970) Helminthosporium turcicum lesion numbers related to number of trapped spores and fungicides spray. *Phytopathology* 63, 930-933.

Cohen Y. & Rotem J. (1971) Field and growth chamber approach to epidemiology of Pseudoperonospora cubensis in cucumbers. *Phytopathology* 61, 736-737.

Cohen Y., Levy Y. & Eyal H. (1978) Sporogenesis of some fungal plant pathogens under intermittent light conditions. *Can. J. Bot.* 56, 2538-2543.

Levy Y. & Cohen Y. (1979) Seasonal fluctuations in epidemics of the Northern leaf blight of corn in Israel. *Phytoparasitica* 7, 59-60.

Levy Y. & Cohen Y. (1980a) Sporulation of Helminthosporium turcicum on sweet corn: Effects of temperature and dew period. *Can. J. Plant Pathol.* 2, 65-69.

Levy Y. & Cohen Y. (1980b) Sporulation of Helminthosporium túrcicum on sweet corn: Effects of light and sugars. *Physiol. Plant Pathol.* 18, (in press).

Waggoner P.E., Horsfall J.G. & Lukens R.Y. (1972) Epimay, a simulator of southern corn leaf blight. *Bull. Conn. Agr. Exp. Sta.* 729, 84 pp.

Levy Y. (1979) Epidemiology of the northern leaf blight of corn incited by Helminthosporium turcicum Pass. Ph.D. Thesis, submitted to the senate of Bar-Ilan University, 120 pp.

APPENDIX I. LISTING OF PROGRAM (CSMP)

```
TITLE EPID HEL TURC
DYNAMIC
*WEATHER
HOUR = AMOD (TIME, 24.)
TEMP = AFGEN (TEMPT, HOUR)
FUNCTION TEMPT = 0., 10., 12., 20., 24., 10.
WET = AFGEN (WETT, HOUR)
DRY = 1-WET
FUNCTION WETT = 0.,1.,6.99,1.,7.,12.,0.,16.99,0.,17.,1.,20.,1.,24.,1.
LIGHT = AFGEN (LIGHT, HOUR)
FUNCTION LIGHTT = 0.,-1.,5.99,-1.,6.,1.,20.,1.,20.01,-1.,24.,-1.
LEA = AFGEN (LEAT, TIME)
FUNCTION LEAT = 0.,7.0E+04,700.,9.7E + 04,1300.,10.0E + 04
*SPD SPORES DEPOSIT ON LEAVES
SPD = INTGRL (2000., RATE7 - ROKIL)
*RATE7=0
RATE2 = ROGE * SPD
ROGE = AFGEN (ROGET, TEMP) * WET
ROKIL = KILL * SPD/DELT
* ROKIL SPORES THAT DONT GERMINATE AT THE FIRST WET PERIOD
PROCEDURE KILL = DESS (WET)
KILL = 0
IF ( (WETP - WET) . GT.0.) KILL = 1
WETP = WET
ENDPRO
FUNCTION ROGET = 10.,0.01,15.,0.02,20.,0.04,25.,0.07,30.,0.06
SPGE = INTGRL (0., RATE2-RATE4-RKSGE)
RATE4 = RAFOR*SPGE
RAFOR = AFGEN (RAFORT, TEMP) * WET
FUNCTION RAFORT = 10.,0.01,15.,0.03,20.,0.06,25.,0.05,30.,0.02
*RKSGE GERMINATING SPORES THAT NOT PRODUCE APPRESSORIA
RKSGE = KILL *SPGE/DELT
SPAP = INTGRL (0.,RATE4- (SPAP* PUSHD)
*SPAP SPORES WITH APPRESSORIA
```

```
ISITI = INTGRL (0., SPAP*PUSHD) - ISIT1 * PUSHD)
*ISIT (2,9) INCUBATION PERIOD
ISIT (2,9) = INTGRL (0., (ISIT (1,8) * PUSHD) - (ISIT (2,9) * PUSHD) )
VLESIO = INTGRL (0., (ISIT9*PUSHD) - (VLES 10 * PUSHD) )
VLES (11,30) = INTGRL (0., (VLES (10,29) *PUSHD) - (VLES (11,30)*PUSHD) )
*S THE LENGTH OF AN HYPOTHETIC LESION
S = INTGRL (0., ROEL)
SLES (1,30) = VLES (1,30) *P
P = DELAY  (240., LPER,S)
LPER = 240.
PUSHD = PUSH/DELT
PUSH = IMPULS (24.,24.)
ROEL = AFGEN (ROELT, TEMP)
G = MODINT (0, HOUR, 1., ROEL)
FUNCTION ROELT = 10.,0.05,15.,0.16,20.,0.19,25.,0.2,30.,0.1
*IAR IS THE RELATIVE INFECTION AREA
TVLA = SUM1 (SLES (1,30) )
RCN1 = AFGEN (RCN1T, TEMP) * WET
RCN2 = AFGEN (RCN2T, TEMP) * WET
RCN3 = AFGEN (RCN3T, TEMP) * WET
* CON4 CON24 CON34 ARE DEAD CONIDIOPHORES
CON1 = INTGRL (0., RCN1*C*TVLES) *STEP (264)
CON (2,4) = INTGRL (0., (CON (1,3) *  PUSHD)- (CON (2,4) * PUSHD)
CON21 - INTGRL (0., RCN2*C*TVLES) *STEP (288)
CON(22,24) = INTGRL (0., (CON (21,23)*PUSHD) - (CON (22,24) * PUSHD) )
CON31 = INTGRL (0., (RCN3*C*TVLES) * STEP (312) )
CON (32,34) = INTGRL (0., (CON(31,33)*PUSHD) - (CON (32,34)*PUSHD) )
*PARAMETER C THE GROWTH OF A LESION WITHIN 24 HOURS
PARAMETER C = 3.7583
PARAMETER C = 4.115
PARAMETER C = 3.241
*RSP (1,3) RATE OF SPORULATION OF CON (1,3)
RSP1 = AFGEN (RSP1T, TEMP) * WET
RSP2 = AFGEN (RSP2T, TEMP) * WET
RSP3 = AFGEN (RSP3T, TEMP) * WET
```

```
* RSP (21,23) RATE OF SPORULATION OF CON (21,23)
RSP23 = AFGEN (RSP23T, TEMP) * WET
RSP22 = AFGEN (RSP22T, TEMP) * WET
RSP21 = AFGEN (RSP21T, TEMP) * WET
RSP33 = AFGEN (RSP33T, TEMP) * WET
RSP32 = AFGEN (RSP32T, TEMP) * WET
RSP31 = AFGEN (RSP31T, TEMP) * WET
FUNCTION RCN1T = 15.,2.4,20.,3.1,25.,4.1,30.,2.5
FUNCTION RCN2T = 15.,2.1,20.,2.8,25.,3.5,30.,2.0
FUNCTION RCN3T = 15.,1.8,20.,2.4,25.,3.0,30.,1.8
FUNCTION RSP1T = 15.,1.2,20.,4.2,25.,2.2,30.,1.2
FUNCTION RSP2T = 15.,1.2,20.,4.0,25.,1.9,30.,0.2
FUNCTION RSP3T = 15.,1.2,20.,2.3,25.,0.2,30.,0.0
FUNCTION RSP21T = 15.,1.0,20.,3.8,25.,1.9,30.,1.0
FUNCTION RSP22T = 15.,0.8,20.,3.2,25.,1.2,30.,0.1
FUNCTION RSP23T = 15.,0.8,20.,1.5,25.,0.1,30.,0.0
FUNCTION RSP31T = 15.,0.5,20.,3.6,25.,1.2,30.,0.5
FUNCTION RSP32T = 15.,0.4,20.,2.4,25.,0.5,30.,0.0
FUNCTION RSP33T = 15.,0.2,20.,0.1,25.,0.0,30.,0.0
* TVLES THE TOTAL NUMBER OF VISIBLE LESIONS
* TVLES = SUM1 (VLES (10,30) )
*TSP = SUMX (RSP (1,3), CON (1,3) + SUMX (RSP (21,23) CON(21,23) + SUMX...
(RSP (31,33), CON (31,33)
METHOD RECT
TIMER FINTIM = 1080, PRDEL=24,DELT=1.
PRINT TSP, CON (1,3) CON (21,23) (CON (31,33)
END
FUNCTION TEMPT = 0.,20.,12.,30.,24.,20.
END
FUNCTION TEMPT = 0.,15.,12.,25.,24.,15.
END
STOP
ENDJOB
```

Developments in Arid Zone Ecology and Environmental Quality
Edited by H. Shuval

9

THE ENVIRONMENTAL AND ECOLOGICAL IMPACTS OF THE ASWAN HIGH DAM

KHALIL H. MANCY

Professor of Environmental Chemistry
The University of Michigan, Ann Arbor,
Michigan, 48109, USA

ABSTRACT. Against an unprecedented amount of controversy concerning its potentially damaging side effects, the Aswan High Dam has been successfully operating for over eleven years, and it has significantly contributed to the economic development of Egypt. The Aswan High Dam was built for the purpose of (a) the storage of Nile flood water which used to be discharged unused into the sea, (b) river flow control, and (c) generation of hydropower. It is evident that these objectives have been fulfilled. Water storage and river flow control were accompanied by increased agricultural production and improved year-round river navigation. Egypt is no longer faced with the threat of high floods, in contrast it could be threatened only by an extended series of low floods.

Recent investigations on the environmental impacts of Aswan High Dam gave quantitative assessments of present ecological perturbations and their future trends. These studies included the socioeconomic and public health implications of these perturbations. A summary of research findings is presented. This reveals that early prognostications of the damaging side effects of the Dam have been greatly exaggerated. Taking into consideration the pressing economic needs of the country, the transformation of the Nile into a placid and fully controlled year-round source of water and energy has been a valuable asset for Egypt's development.

EGYPT: THE GIFT OF THE NILE

The fertile inhabitable land of Egypt is a riverine oasis in the vast North African Sahara Desert. On this narrow strip of arable land, about 2.5% of the country's total area, lives 98% of Egypt's 40 million people; the rest is barren desert. The Nile is Egypt's lifeline, the main highway, and a dominating factor in all aspects of life. The economic history of Egypt has been primarily one of a continuous readjustment to a variable water supply and repeated efforts to expand land use and increase agricultural production. Man's reverence to the Nile is as old as Egypt where it has been an object of worship and a symbol of life, prosperity and survival.

The Nile is the longest river in the world, extending for approximately 6,800 kilometers from its most remote source in the East African Plateau to the Mediterranean and thus traversing approximately half of the African continent. The Nile drainage basin, an area of nearly 3,100,000 square kilometers, occupies one-tenth of the African continent and is unique among major rivers in the world in that it includes three climatic zones; tropical, sub-tropical and desert. The Nile basin extends over ten riparian countries: Burundi, Buanda, Tanzania, Zaire, Uganda, Kenya, Central African Republic, Ethiopia, Sudan, and Egypt (Figure 1).

The Nile's annual yield has been subject to great variations from one year to the next. This is illustrated in Figure 2. During the 1971-75 period the annual discharge of the Nile at Donkola in Northern Sudan varied from a low, of 53,102 million cubic meters in 1972, to a high of almost double that volume, 100,045 million cubic meters in 1975. Seasonal variations of the Nile flow are even more dramatic. More than 80% of the river's total discharge occurs during the flood period (from August to October) and used to be directly discharged into the sea. During the remaining nine months, the river carried only 20% of the annual discharge.

Early in the ninteenth century, Egypt started large-scale irrigation schemes. This significantly expanded the total farmed area, permitted large-scale cultivation of cotton, and made it possible to harvest two to three crops per year. From 1821 to 1907, the success of the perennial irrigation schemes resulted in a 76% expansion in cultivated land and a 2.5-fold rise in crop production (Hamdan, 1961).

Early in the twentieth century, under British rule, Egypt and Sudan began a systematic river control program which resulted in the construction of six dams, including the Old Aswan Dam (Table 1).

TABLE 1: WATER STORAGE FACILITIES ON THE NILE

FACILITY	COMPLETED	CAPACITY (billion m^3)	LOCATION
Sennar	1925	0.9	Blue Nile, Sudan
Old Aswan Dam	1933	5.3	Aswan, Egypt
Jebel Auliya	1937	3.5	White Nile, Sudan
Owen Falls Dam	1954	200	Lake Victoria, Uganda
Khashm Al-Girba	1966	1.2	Atbara, Sudan
Roseires	1966	3.0	Blue Nile, Sudan
Aswan High Dam	1970	167	Aswan, Egypt

These dams did not provide sufficient protection against excessive floods or droughts and about 40% of the total annual river flow was still being discharged unused into the sea. In the meantime, Egypt's need for water was rapidly growing. This prompted the Egyptian authorities to build the Aswan High Dam to provide over-year storage of Nile floods.

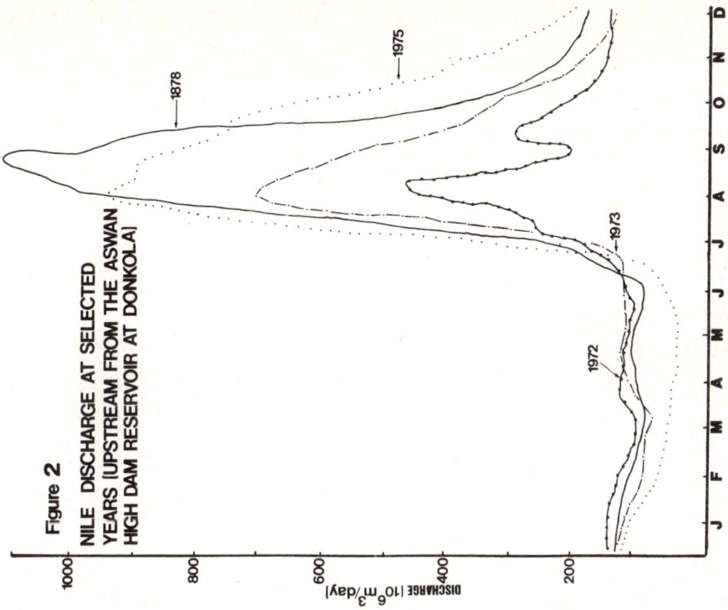

Figure 2
NILE DISCHARGE AT SELECTED YEARS [UPSTREAM FROM THE ASWAN HIGH DAM RESERVOIR AT DONKOLA]

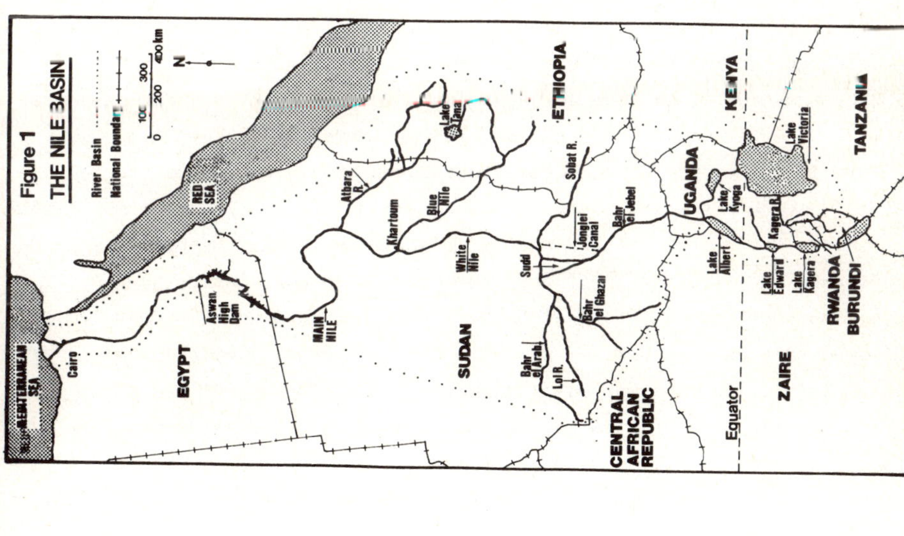

Figure 1
THE NILE BASIN

THE HIGH DAM AT ASWAN

The construction of the Aswan High Dam started in 1964 and was finally completed in mid 1970. The general features of the dam are listed in Table 2.

TABLE 2: THE ASWAN HIGH DAM IN FIGURES*

Measurement of the Dam	
Length of Crest	3920 m
Length of Part in River	520 m
Width at Crest	40 m
Width at Base	980 m
Crest Level	196 m (a.s.l.)
Measurement of the Reservoir	
Maximum Water Level	183 m (a.s.l.)
Total Capacity	167 billion m^3
Dead Storage Capacity (at 147 m a.s.l.)	31.6 billion m^3
Working Storage Capacity (at 175 m a.s.l.)	90.2 billion m^3
Flood Control Storage Capacity (a 183 m a.s.l.)	46.1 billion m^3
Length of Reservoir	500 kilometers
Area of Reservoir (at 175 m a.s.l.)	5000 square kilometers
Power Station Capacities	
Total Installed Capacity	2,100,000 kw
Maximum Available Annual Power	10 billion kwh

* High Dam Authority, Aswan, Egypt, 1970.

By any standards, the Aswan High Dam is an impressive engineering achievement. This enormous undertaking was neither frivolously conceived, nor expediently developed. In fact, it was studied by reputed international experts from Europe, USA and USSR. That it turned out to be a Soviet financed rather than an Anglo-American World Bank project was a result of Cold War politics of the mid 1950's (Meyer, 1980).

In 1955, the World Bank prepared a favorable report about the technical and economic feasibility of the Aswan High Dam. On December 15, 1955, the United States and Britain made a joint offer to lend Egypt $70 million to finance the first stage of the project. Six months later, on July 19, 1956, the American Secretary of State abruptly cancelled the loan. Britain followed suit and the World Bank announced that in view of the Anglo-American decision, its loan was also cancelled.

Historians and poltical scientists who wrote about the Aswan High Dam and the Suez Canal crisis agree that the withdrawal of the offers to finance the Dam by the west was politically motivated (Meyer, 1980). This included: the failure of the proposed U.S. settlement plans of the Arab-Israeli conflict, Britain's opposition to the newly emerging Egyptian-Arab nationalism which was thought to be determined to destroy the British influence in the area, and last but not least, Egypt's diplomatic recognition of the People's Republic of China.

Furthermore, the Aswan High Dam has long been the environmentalist's favorite example of development gone wrong (Economist, 1975). It has been an irresistable target to sensational expose' in the news media and the environmental sciences-conservation type literature. Facts were either unavailable or they were used from this or that perspective to speak for this or that party of interest.

For all its political controversies, the Aswan High Dam has stood as a symbol of Egyptian nationalism and struggle for survival and prosperity. Instead of being entirely dependent on river control projects in upstream riparian countries, the Aswan High Dam has provided effective over-year water storage and hydropower within the political boundaries of Egypt.

A COMPREHENSIVE ASSESSMENT PROGRAM

Eleven years after its completion, it is evident that the Aswan High Dam has fulfilled its objectives. Against early prognostications that huge losses by evaporation and seepage (Economist, 1975) could prevent the reservoir from filling, water storage has proceeded rather satisfactorily. The reservoir reached its operating level of 175 m above sea level in 1976. Following the 1975 flood, the Aswan High Dam Authority became concerned with the possibility of a series of high floods. This prompted the construction of a relief canal (Toshka Canal) for water release into a depression west of the reservoir, known as Toshka Valley (Abu El Atta, 1978).

Egypt is no longer faced with the threat of high floods, in contrast it could be threatened only by an extended series of low floods. River flow controls, afforded by the Aswan High Dam, significantly improved river navigation which extends throughout the year.

One of the main contributions of the Aswan High Dam is hydropower, an inexpensive, clean form of energy. This contribution has been in the order of 5 billion kwh in 1975, 6 billion kwh in 1976, and 7.15 billion kwh in 1977, which constitutes about half of the total electric power used in Egypt (U.S. Dept. of Energy, 1979). The availability of power from the Aswan High Dam (as well as the improved performance of power generation from the Old

Aswan Dam) was accompanied by the establishment of an impressive electrification network in Egypt. The main beneficiaries have been rural communities as well as various industrial plants (U.S. Dept. of Energy, 1979).

Agricultural expansion through land reclamation in the order of one million acres, has been achieved. In the meantime, the conversion of about 840,000 acres in Upper Egypt from basin irrigation to perennial irrigation has been completed. Summer water demands have been satisfied and rice production has exceeded expected goals (Abu El Atta, 1978).

With all its assets, it is quite apparent that the construction of the Aswan High Dam has caused major physical and ecological perturbations, which in turn exerted certain social, economic and public health implications. Changes in the Nile River system and their associated impacts have been the subject of a comprehensive project conducted by a US-Egyptian research team (Mancy, 1974).

SEDIMENTATION, EROSION AND SOIL FERTILITY

During the flood period, the Blue Nile brings large amounts of sediment loads, in the range of 80 to 100 million tons annually (Smaika, 1940). Since the construction of the Aswan High Dam, all river sediment load has been deposited in the reservoir and the water released from the reservoir is silt free. The main concerns have been (a) the effects of sedimentation on the capacity of morphology of the reservoir and (b) erosion of the riverbed and the Mediterranean coastline of the Nile Delta.

Sedimentation

Our studies utilized both field surveys and satellite remote sensing techniques with ground truthing observation during satellite overpass, to study the distribution of sedimentation in the reservoir. Remote sensing measurement of surface turbidity utilized Landsat digital imargery processed by the Multispectral Data Analysis System at the Remote Sensing Center in Cairo, and The University of Michigan in Ann Arbor, Michigan (Smith, 1980). Mosaics for several flood periods, showing the extend of penetration of the flood water in the reservoir, were correlated with the thickness of sediment deposits.

Typical results of sedimentation measurements are shown in Figure 3. The lower portion of the figure shows the accumulation of sediments along the old river canal (longitudinal section) in the reservoir. The upper portion is a remote sensing image of the whole reservoir showing the

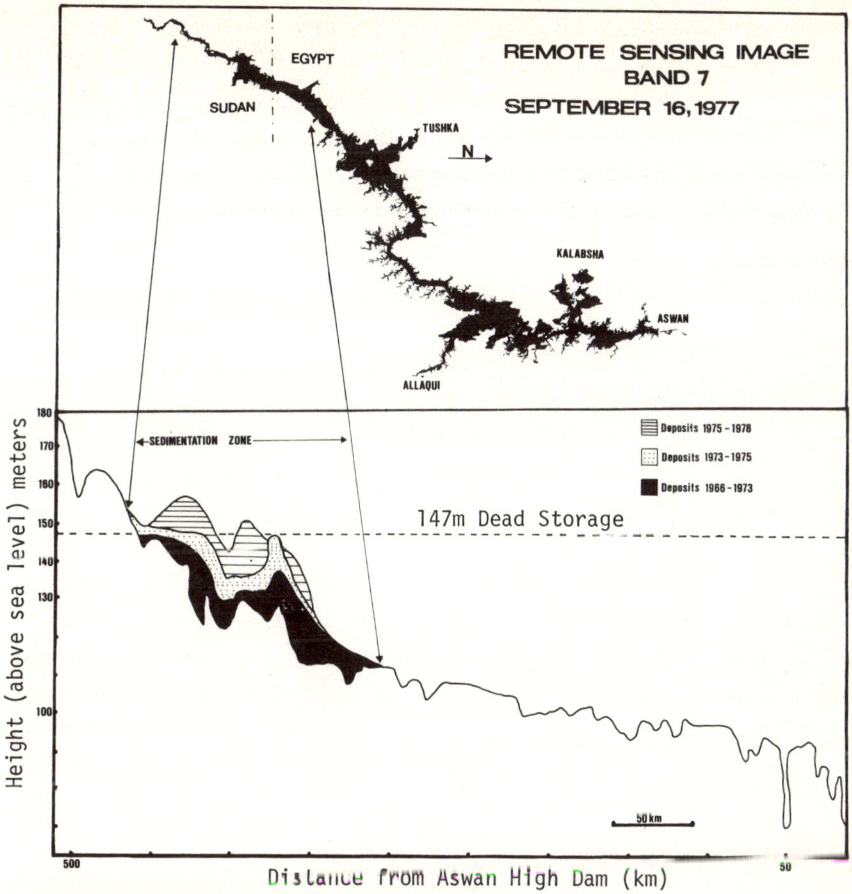

Figure 3 Sedimentation In Aswan High Dam Reservoir

sedimentation zone.

After the closure of the river in 1964, the first and second floods were accompanied with sedimentation throughout the reservoir. As the reservoir filled up, the advance of the sediment front in the reservoir retreated to the extreme southern portions. At the present time, the sedimentation zone extends between 260 km to 480 km south of the dam. In the meantime, a new delta is being formed in the area between 360 km and 450 km with noticeable morphological changes. The amount of distribution of sediment deposition are primarily dependent on the magnitude of the incoming flood, and stage and morphology of the reservoir.

The short term effects of sedimentation will be delta formation and aggradation in the backwater reach of the reservoir. This will be accompanied by marked changes in the shape of the reservoir in this area. The long term effects will be further accumulation of sediments in the deep dead storage pool (20 billion cubic meters). Preliminary calculations indicate that the lifespan of the reservoir is in the order of 300 years.

Riverbed Erosion

The scope of our investigations did not include riverbed erosion. Nevertheless, there seems to be a significant amount of erosion in the river stretch between Aswan and Cairo. The absence of suspended sediment is causing scouring and degradation of the riverbed and banks. This could eventually undermine the stability of bridges and barrages.

Riverbed erosion and the stability of hydraulic structures are being carefully monitored by the Egyptian authorities. At least one barrage, Esna Barrage, has been recently reinforced against erosion processes. In the meantime, water released from the reservoir is carefully controlled in an attempt to limit river scouring. The availability of the Toshka relief canal appears to offer the needed security against excessive water release from the dam (Abu El Atta, 1978).

Soil and Agricultural Productivity

Critics of the dam have repeatedly warned that the loss of the flood's annual sediment load would inevitably diminish the fertility of farmlands and render Egypt more dependent on chemical fertilizers. Over the millennia, this river sediment built the Delta and the fertile land of Egypt. In contrast, proponents of the dam took the position that the sedimentation has a minor nutritive value and could be easily replaced by chemical fertilizers.

Our findings show that nitrogen and phosphorus losses due to deprivation from Nile sediments is indeed insignificant when compared to current use of chemical fertilizers. Because of the low nitrogen content in the Nile sediment, Egypt has been heavily relying on chemical fertilizer. Furthermore, it was also noted that about 80% of the annual river sediment load used to be flushed out to the sea and only about 13 million tons actually deposited on the fields. The majority of this amount, about 9 million tons, were applied to the 840,000 acres in Upper Egypt under basin irrigation.

Coastal Erosion

The Mediterranean coastline of the Nile Delta has been subjected to a variety of changes throughout the years with alternate periods of accretion and erosion. Herodotus reported, around 450 B.C., that the Delta had seven outlets to the sea and around 1000 A.D. these were replaced by the present Rosetta and Damietta branches (Herodotus, 1866).

Since 1964, after the construction of the Aswan High Dam, large scale offshore erosion occurs in areas where winter erosion is no longer compensated by summer deposition. The main affected areas are the Rosetta and Damietta Peninsulas. In fact, the whole coastline is now subject to erosion and steepening (Nielson, 1973).

If erosion proceeds unchecked, the long term effect could be the breaching of the narrow strip of sandy barriers separating the northern brackish lakes from the sea. Sea intrusion would have a devastating effect on the northern delta cultivated land, which lie only three to four meters above sea level.

ECOSYSTEM AND WATER QUALITY ALTERATIONS

The major portion of our study has been devoted to river quality management for multipurpose use. The results clearly demonstrated the dependency of water quality and ecosystem characteristics on changes in river flow. A brief overview of this dependency can be categorized into two cases, each representing a different region with a unique hydrodynamic regime and the corresponding physicochemical and hydrobiological characteristics. The first is the Aswan High Dam Reservoir; the second is the river from Aswan to the Mediterrean (Upper Egypt and the Delta).

Reservoir Limnology

The Aswan High Dam Reservoir is located in one of the most arid regions of the world. A unique microclimate, created by the reservoir, prevails in this region. This is characterized by a higher probability of cloud cover than an equivalent area away from the reservoir. Cloud cover estimates were obtained from satellite remote sensing data (Smith, 1981).

The impoundment is monomictic and has a long season of stratification during the summer that parallels the air temperature regime. The breakdown of the stratification begins with the incoming floods in the southern region and the gradual cooling of the climate in the northern region. The northern two-thirds of the lake is lacustrine in nature, while the southern one-third varies from riverine to semi-riverine and lacustrine conditions depending on the magnitude of the annual floods (Abdel-Latif, 1979).

The distribution of phytoplankton and zooplankton communities in the reservoir are largely dependent on the (a) seasonal variations in upstream river conditions (i.e. in August flood waters originate primarily from the Blue Nile and the Atbara rivers while during March the main source is the White Nile and upstream impoundments), (b) hydraulic regime in the reservoir, and (c) reservoir seasonal climatic variations. Considering phytoplankton, the total counts have been in the range of 10^6 to 10^7 organisms/liters (Abdel-Latif, 1979).

Satellite remote sensing was used to monitor the spatial distribution of plants in the reservoir (Smith, 1980). This was based on chlorophyll a reflectance measurements in the near infrared region (0.7 to 0.8 μm) which were used to prepare maps showing plant distribution at different periods. Algae and aquatic plants seem to concentrate in the embayments, while the main channel is largely clear. Shoreline vegetation is predominate in the flat areas surrounding embayments. It is important to note, in contrast to other African lakes, the reservoir is devoid of water hyacinth.

Our studies emphasized reservoir fishes and fishery (Abdel-Latif, 1979). The primary impact of river impoundment is to preclude obligate riverine fishes, since they require flowing water for all their ecological requirements. Essentially this impact is due to an inadequate habitat for spawning and not due to changes in basic living conditions, such as feeding. Other minor factors include changes in water temperature and chemical properties.

The fishes of the reservoir are derived entirely from the Nile system. This includes 57 species which belong to 15 families. Different specie distributions exist bewteen the southern and northern parts of the reservoir which reflect the effect of different hydrodynamic conditions.

Reservoir's fisheries potential seems to be quite high (Abdel-Latif, 1979) Initial prediction of commercial fishing, made during the early stages of reservoir development, was estimated at 20,000 tons per year. The actual landings of fish increased from about 2662 tons in 1968 to 28,000 tons in 1979 (1). While the fishing landings have progressively increased, the rate of increase was much more pronounced in the first few years than in the last few years. One reason is that in the early stages of the reservoir the capacity of supporting growth was initially tremendous due to increased food sources and available inhabitable space. However, as the excess nutrients and energy sources were utilized and the ecological niches were filled there followed a gradual, but definite, leveling off of the overall production.

The increase in overall fish landings to the present level is primarily due to an increase in the number of fishermen. The predominant fresh species present in a typical catch include: Tilapia spp., Lates niloticus, Labeo spp., Bagrus spp., and Synodontus spp. The major salted fish species obtained include: Hydrocynus foshalu, Aletes spp., Labeo, Barbus and Europicus niloticus (Abdel-Latif, 1979).

There are a number of both social and industrial limitations in current fishing practices. The most prevalent social problem is the severe working conditions of the fishermen. They lack sufficient provisions, health care and free time, and have low incomes. The main limitation of the industry as a whole is the inadequate transportation of fish catches.

River Water Quality

Main emphasis in these investigations has been on the effect of flow controls on river water quality and its hydrobiological characteristics, and the assessment of the impact on waste discharge and agricultural return flows on river quality (El Gohory, 1979).

Obvious effects of river impoundment on water qulaity have been (a) decrease in suspended matter, (b) increase in total dissolved solids and (c) increase in algal counts. These changes can be directly related to sedimentation, evaporation and primary production in the reservoir. The water released from the dam is devoid of sediment and its suspended matter content is due to phytoplankton.

Changes in the total dissolved solids content has been assessed by a salt balance simulation model of the reservoir (Guariso, 1980) using a historical sequence of monthly inflows from 1871 to 1975. The increase in total dissolved solids in the outflow from the reservoir, as compared with inflows, is in the order of 10-15%. Total dissolved solids values of the inflows vary from 140 to 160 mg/l, while in the outflows, they vary from 155 to 180 mg/l, depending on the surface area of the reservoir and evaporation losses.

The total algal counts of water released from the reservoir are in the order of 10^5 to 10^6 organism per liter, as compared to 10^4 to 10^5 organisms per liter for pre-dam conditions.

Water released from the Aswan High Dam Reservoir exhibits the same seasonal variations and the same overall characteristics from one year to another. Downstream changes in river water quality and hydrobiology are primarily due to (a) different prevailing hydrodynamic regimes, (b) agricultural return flows, and (c) domestic and industrial waste discharges including oil and waste from passenger and cargo river boats. These changes are more pronounced as the river passes the densely populated urban and industrial centers at Cairo and the Delta region. These changes also occur in main canals, e.g. the Ismailia Canal which provides Nile water to the Suez Canal region (El Gohory, 1979).

Considerable amounts of agricultural drainage waters are being returned to the river and main irrigation canals. Drainage waters add nutrients to the river and promote algal and hydrophyte production. There is also the danger that pesticide residues, depending on their concentration levels and persistance, accumulate in river biota and fish, and may be ultimately harmful to man.

A detailed inventory of all major municipal and industrial waste discharges was made. Waste discharges into the river are being readily diluted and is only accompanied by localized effects of water quality deterioration, immediately downstream from the waste outfalls.

River transportation of both goods and passengers has dramatically increased in recent years. This includes tourist ships and floating hotels which are regularly making trips between Cairo and Aswan. The sharp increase in river traffic and the increased reliance on motor boats instead of sailing ships has resulted in noticeable pollution problems, largely due to oil spills, and in some cases due to human waste discharges. These problems are quite noticeable in the vicinity of river harbors and shipyards (El Gohory, 1979).

Changes in the phytoplankton population along the river are primarily due to (a) changes in the hydraulic regime at Aswan from lacustrine to riverine conditions, (b) growth stimulatory or inhibitory factors from point and nonpoint sources of pollution along the river, and (c) climatic conditions.

Following an initial decrease in total counts at Aswan, the total algal content generally increases as the river travels north due to the cummulative effects of agricultural and human waste inputs. Eutrophic conditions occur in the northern delta as the river velocity is greatly reduced. Total algal counts at Cairo range from 10^4 to 10^5 organisms per liter and this may reach 10^6 organisms per liter in the northern delta. The predominant species are diatoms, except at the ends of the Rosetta and Damietta Branches, where blue-green algae flourish occasionally. Bioassay experiments, conducted on twenty Nile water samples collected during the period of 1976-1977, indicated that algal growth was limited by nitrogen, followed by phosphorus and trace elements.

Riverine fishery has been adversely affected following the construction of the Aswan High Dam and subsequent downstream ecological perturbations (El Gohory, 1979). A reduction in the commercially desirable fish species has been recorded in places where river conditions deteriorated, e.g. in the northern delta.

Out of the 47 commercial Nile fish species recorded earlier in 1948 only 17 species now exist, as recorded from recent surveys (El Gohory, 1979). This does not mean that the other species have completely disappeared, but they were not found in sufficent quantity to appear in the commercial catch.

After five years of river investigations, it became apparent to us that there is a steady water quality deterioration, characterized by a number of localized pollution problems, especially in regions with high population density around Cairo and the Delta. Before the Aswan High Dam, the river cleansed itself annually. The Nile flood used to flush the accumulated waste into the sea. With increase river use after the dam, due to population increase and industrial activity, the impacts of waste discharges on river conditions are more pronounced.

SCHISTOSOMIASIS: AN ANCIENT DISEASE

Schistosomiasis is a chronic helminth infection in man caused by a trematode of the genus schistosoma (a parasite fluke). There is evidence that this disease occured in Egypt as far back as 1900 B.C. (Miller, 1978). Calcified ova of this parasite were found in Egyptian mummies dated 1250 to 1000 B.C. and a number of papyri contain remedies for it. In 1951, A.D., Theodore Bilharz identified the causative parasite in the Cairo Medical School. The intermediate host, the Bulinus snail was also identified in Cairo in 1915. Schistosomiasis poses a widespread problem and it has been estimated that over 200 million people are infected worldwide.

Schistosomiasis is a debilitating and not a fatal disease, although death can be caused from complications. The disease is essentially an occupational disease associated with farming. It has spread with the gradual conversion from basin to perennial irrigation since the last century. During basin irrigation, the snails could not survive the dry season during the low river period. Needless to say, the longer contact time associated with perennial irrigation enhances the chance for infection. Consequently, after the Aswan High Dam, the conversion of 800,000 acres in Upper Egypt from basin to perennial irrigation, was expected to be accompanied with a localized rise in disease prevalence.

One of the main activities of our study in Egypt has been devoted to find out the extent of the spread of that disease. Our investigations were based on conducting public health sample surveys of rural Egypt, which included 47 villages and a total of 20,000 individuals. Our findings, based on comparisons with historical data, indicate that there is a general declining trend in the prevalence of schistosomiasis (S. haematobium and S. mansoni) in the rural population Egypt which has not been reversed by the construction of the Aswan High Dam (Miller, 1978). This is primarily due to (a) a large shift of population out of rural high transmission areas to urban low transmission areas and (b) the effect of programs in environmental health including protected water supplies, chemotherapy, delivery of health services, and health education.

NUBIA: THE LAND THAT BECAME A LAKE

An immediate impact of the Aswan High Dam was the inundation of the Nubian Valley, previously extending from the city of Aswan to the town of Mareow in northern Sudan. While Egyptian and Sudanese sections of the Nubian Valley were separated administratively, they shared common traditions and customs. The reservoir inundated all Nubian villages along the banks of the river from Aswan to Wadi Halfa. That neccessitated the displacement of nearly 100,000 people; half of whom were from the Egyptian section and the other half were Sudanese Nubians. The Egyptians were resettled north in Kom Ombo while the Sudanese migrants were displaced at Khasm El Gerba east of Sudan.

A comprehensive investigation of the effects of the dam on the local population in southern Egypt and northern Sudan has been made (Fahim, 1979). This includes a study of the social, economic, and public health impacts of the displacemnt of Nubians and attempts for their resettlement. In spite of a noticeable improvement in their health status, the displaced Nubians are generally dissatisfied with their new way of life and are exerting pressure on the government to resettle them back on the shore of the reservoir.

Investigations of the desert nomads revealed significant changes in their way of life following the creation of the reservoir (Fahim, 1979). This included the establishment of a few settlements and the grazing of their goats and camels on vegetation along the shore of the reservoir. In certain places new pasture lands extend for several kilometers along valleys which receive Nile water during the flood period. It is interesting to note that these settlements are also being used as resting places for caravans of camel traders travelling between Egypt and Sudan.

PROTECTION OF ARCHEOLOGICAL SITES

Early Egyptian planners realized all along that the construction of the Aswan High Dam and rising water in the reservoir would drown some of the upstream monuments. This prompted unprescendented international rescue operations to save some of the threatened temples. The most spectacular of these efforts was the dismantling of the Abu Simble temple and reassembling it as higher elevations. Similarly, the Ptolemic temple of Kalabsha was moved five kilometers from its original place and reassembled on a small island facing the High Dam.

What the Egyptian planners did not expect was that the rise of the underground water table would damage the foundation of some of the downstream archeological sites. The rise of the underground water table was a direct effect of river flow controls and extensive irrigation, afforded by the Aswan High Dam. Furthermore, ground water becomes more corrosive if it is polluted with sewage, e.g. due to seepage from nearby cesspools or septic tanks.

The problem could be solved by artifically lowering the water table in areas of archeological significance. Constant pumping of underground water could be costly, but there are only a few close to the river, e.g. the Karnak Temple in Luxor.

CONCLUSION

Distant from the political controversies of the 1960's which seems to have once inhibited objective analysis, it is evident that the Aswan High Dam has not produced an ecological disaster. In contrast, the dam has significantly contributed towards the economic development of Egypt.

The environmental impacts of the dam have been thoroughly investigated. Research findings should provide the policy and decision makers in Egypt with sufficient information required to establish priorities for Nile resource development. Special attention needs to be given to river pollution control and the long term effects of coastal erosion.

Acknowledgement-These studies have been supported in part by grants from the Ford Foundation and the United States Environmental Protection Agency.

REFERENCES

Abdel Latif, A.F. (1979) "Limnology of The Aswan High Dam Reservoir", Nile Project Report, Academy of Scientific Research & Technology, Cairo, Egypt.

Abu El Atta, A. (1978) "Egypt and the Nile After the High Dam" Ministry of Irrigation and Land Reform, Cairo, Egypt.

The Economist ("What the Dam has Done", March 22-28, 1975); Washington Post (Jan. 6, 1971, p. 14; Feb. 17, 1971, p. 18; Feb. 20, 1971, p. 14; Feb. 24, 1971, p. 18; Oct. 16, 1973, p. 6); Los Angeles Times (March 9, 1975, sec. 6, p. 3; March 28, 1976, sec. 9, p. 2); New York Times (May 4, 1975, p. 30); Christian Science Monitor (March 7, 1975, p. 7; Aug. 4, 1976, p. 24); Wall Street Journal (Sept. 24, 1976, p. 1); National Parks and Conservation Magazine (Aug. 1971, p. 12). Articles in Scientific literature include: Claire Sterling's "The Aswan Disaster" in C. Giddings and M.B. Monroe (eds.), Our Chemical Environment p. 84, Canfield Press (1972), and in M.T. Farvar and J.P. Milton (eds.), The Careless Technology, New York, Natural History Press (1972), articles on schistosomias by Henry van der Schalie (p. 116), on fisheries by Carl J. George (p. 150), and on coastal erosion by M. Kassar (p. 179).

El Gohory, F. (1979) "Water Quality Characteristics of the River Nile", Nile Project Report, Academy of Scientific Research & Technology, Cairo, Egypt.

Fahim, H.M. (1979) "Dams, Environment and People: The Aswan High Dam Case" Nile Project Report, Egyptian Academy of Scientific Research & Technology, Cairo.

Guariso, G, Whittington, D., Abdel Samie, M.E., Kramer, C. (1980) "A Salt Balance Simulation Model of Lake Nasser", Water Supply and Management, 4, 73-80.

Hamdan, Gamal (1961) "Evolution de l'Agriculture Irrigue en Egypt" in A History of Land Use in Arid Regions, 183-61, UNESCO, Paris.

Herodotus (1866) "The Persian Wars. 450 B.C., The History of Herodotus", Translated by George Rawlinson, D. Appelton and Co., New York.

Mancy, K.H. and Hafez, M. (1974) "Water Quality Studies on The River Nile and Lake Nasser", Project Plan of Operation, Egyptian Academy of Scientific Research & Technology, Cairo, Egypt.

Meyer, G.E. (1980) "Egypt and the United States-The Formative Years" 130-150, Associated University Presses, Inc.

Miller, F.D., Hussein, M., Mancy, K.H. and Hilbert, M.S. (1978), "Schistosomiasis in Rural Egypt", U.S. Environmental Protection Agency Report EPA-600/1-78-070.

Nielsen, E. (1973) "Coastal Erosion in the Nile Delta", Nature & Resources, IX, 1, UNESCO, Paris.

Smaika, Y.M. (1940) "The Suspended Matter in the Nile", Paper No. 40, Physical Department, Ministry of Irrigation, Cairo Egypt.

Smith, S.E., Mancy, K.H. and Abdel Latif, A.F. (1980) "The Application of Remote Sensing Techniques Toward the Management of the Aswan High Dam Reservoir" Proc. 14th International Symposium on Remote Sensing of the Environment, San Jose, Costa Rica, pg. 1297-1307.

Smith, S.E. (1981) "Application of Remote Sensing Techniques to the Study of the Aswan High Dam Reservoir", Nile Project Progress Report (in preparation), Egyptian Academy of Scientific Research & Technology, Cairo, Egypt.

U.S. Department of Energy (1979) "Joint Egypt/United States Report on Egypt/United States Cooperative Energy Assessment" DOE/1A-0002/04, U.S. Government Printing Office.

Developments in Arid Zone Ecology and Environmental Quality
Edited by H. Shuval

10

THE IMPENDING WATER CRISIS IN ISRAEL*

Hillel I. Shuval

Director, Environmental Health Laboratory, Hebrew University - Hadassah Medical School, Jerusalem, Israel.

ABSTRACT

Israel is facing an impending water crisis of potentially major proportions due to severe water scarcity and increasing degradation of water quality. The gradual accumulation of chlorides and nitrates in the coastal aquifer may severely reduce the utility of this major water resource for both agricultural and domestic purposes by the year 2000, unless drastic steps to abate the process are initiated immediately. Lake Kinneret, the country's major surface watersource, is a highly sensitive ecosystem whose stability could be threatened unless organic pollution is strictly controlled. The relief in desalinization of sea water as some sort of technological "Messiah", providing solutions to both the problems of water quantity and quality is seriously questioned. The need to reorganize the country's water management system to meet the urgent problems of the impending crisis is proposed.

INTRODUCTION

While water is the most abundant natural resource on earth its distribution is not uniform and many areas are facing growing problems of water scarcity. As population increases, the demand for water for urban, industrial and agricultural purposes grows and leads to heavy withdrawals from the limited renewable freshwater resources.

Simultaneously these very same human activities generate wastes from urban and industrial centers and agricultural areas which are discharged directly or indirectly into the depleted water sources despoiling them. Thus the world in general and countries with a scarcity of water resources in particular are facing the dual problems of increasing demands on the

* Based in part on portions of "Water quality management under conditions of scarcity: Israel as a case study", Editor Hillel I. Shuval, Academic Press, New York (1980).

limited fresh water resources coupled with growing degradation of those very same water resources which may reduce their utility at the quality required.

Israel is a unique example of a country in a semi-arid zone facing this dual problem of severe water resources scarcity and creeping degradation of the quality of its water resources.

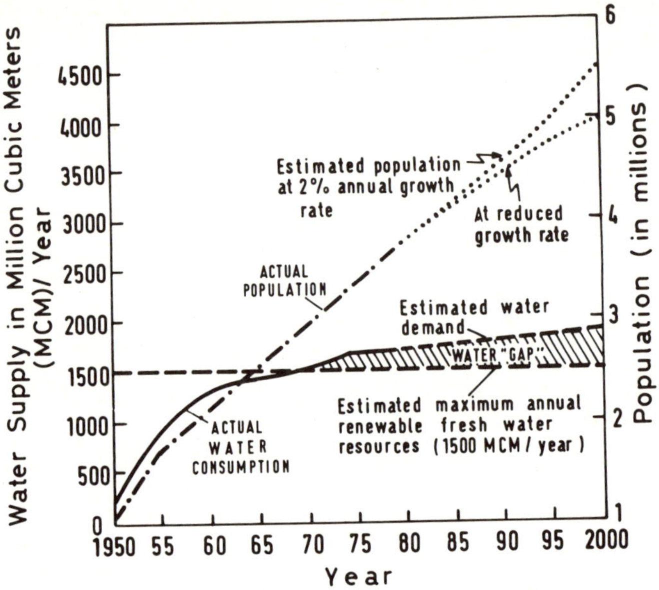

Figure 1: Estimated Population Growth and Water Consumption to the year 2000

As shown in Fig. 1 actual water consumption in 1980 was about 1700 million cubic meters, utilizing essentially all of the 1500 MCM/ year which are the known replenishable fresh water resources of the country. The additional 200 MCM/year was obtained by use of brackish water, wastewater recycling and the highly problematical "borrowing" of some 100 MCM/year from the one time reserve of ground water.

With an expected population growth of about 2% per year, including immigration, the estimated population for the year 2000 may reach 5,600,000 (Fig. 1).

Although official estimates of future water demand vary, a minimal estimate based on increased requirements of the growing urban and industrial sector and significant reductions in the agricultural sector indicates that the annual water shortage will grow to some 400 MCM/yr while Tahal (1979) estimates a gap of some 500 MCM/yr by the year 2000. Part of the gap will be closed by high priority wastewater recycling programs presently under intensive development.

Within this context of extreme scarcity one must consider the serious implications of degradation in water quality which has evolved and which threatens to seriously reduce the utility of the already severely limited water resources.

POLLUTION OF GROUND WATER FROM NATURAL SOURCES - ACCUMULATION OF CHLORIDES.

One of the most serious long term problems of water quality management under conditions of scarcity which has evolved from the Israel experience is the creeping accumulation of dissolved salts (mainly chlorides) from natural sources in the ground water aquifer of the coastal plains, the source of about 25% of Israel's water supply.

In a semi arid zone such as Israel with high evaporation rates and maximum exploitation of the available ground water resources, leaching of the soil by water and release of water to the oceans is minimized. Thus with intensive water recycling for agricultural irrigation, the ground water is continuously enriched by soluable minerals rejected by the normal physiological process of plants which selectively remove water of low mineral content.

Although providing the country with water at a suitable quality for vital agricultural expansion during the critical years of the country's development, the process of recharge and dilution of the higher salinity water from the Kinneret in the low salinity aquifer combined with the intensive ground water exploitation and irrigation activity in the central and southern areas has now begun to threaten the entire vital aquifer.

The model studies by Mercado (1980) indicate that in the areas studied of the heavily exploited coastal aquifer, chloride concentrations are increasing at the rapid rate of 1.5 mg/l per year. It has been estimated that if this process continues at the present rate, by the year 2000, the salinity of a major portion of the wells in the coastal aquifer will be above acceptable agricultural levels. In the final steady state which may be reached in one hundred years or more, the average chloride concentration will reach 900 mg/l meaning the effective total degradation of the aquifer.

This process presents severe problems both for agricultural and possibly for domestic use of the water. According to present policy in Israel the acceptable concentrations of chlorides in agricultural water are from 170-250 mg/l. The survival of economically vital agricultural branches such as citrus and avocado production may be threatened. Many other branches of agriculture may also be damaged by the detrimental effects of salt accumulation in the soil that results from irrigation with such brackish water.

The most effective and radical remedy to chloride pollution is to reduce present exploitation of ground water to less than half of its present rate. This solution has been rejected by the agricultural and water planning authorities since it would require immediate major cut backs in water allocations to agriculture.

The other alternatives are based on an optimal utilization of the natural buffering characteristics of the aquifer system. This could involve such measures as diversion of the somewhat more saline water from Lake Kinneret imported to the central and southern areas by the National Water Carrier, to the southeastern part of the coastal plain, where transit times in the aquifer are estimated to be several hundred years. Another proposal involves shifting the current centers for heavy exploitation of the ground waters to the coastal sand dune areas.

Both of these proposals could lead to a significant slowing down of the salination process but would require heavy investments in new wells and distribution systems. No decision has been taken on either of these options.

Considering the severe long term implications associated with the creeping salinization of the coastal aquifer on the future availability of water for both agricultural and domestic use it is urgent to reconsider all possible courses of action including a drastic reduction in the exploitation rate of the coastal aquifer.

Some sectors of the agricultural community have hopefully believed that such radical steps can be avoided by placing their faith in external technological solutions such as desalination of sea water which in theory could supply water low in salt content to dilute the deteriorating ground water as well as supply additional water for further agricultural development. A policy of inaction or a lack of policy will result in the socially questionable policy of handing costs of finding a solution to the next generation or "mortgaging the future". However this may, if delayed too long, lead to irreversible damage seriously prejudicing the future quality and availability of water.

Although the threat of salination of ground water resources has reached extremely serious proportions in Israel the problem is essentially common to all arid regions operating under growing conditions of water scarcity and high rates of ground water exploitation.

POLLUTION OF GROUND WATER FROM HUMAN SOURCES - ACCUMULATION OF NITRATES

While the problem of chloride pollution of the coastal aquifers discussed above relates to the accumulation of salts derived from natural sources, this second example will deal with the accumulation of nitrates derived primarily from human sources.

As estimated by Mercado (1980) the average background nitrate concentration in 1935-38 in the coastal aquifer was about 2 mg/l. This was essentially prior to the start of intensive human development and agricultural activity. This increased to an average of 30 mg/l in the period 1952-59 and to 53 mg/l in 1970. These increases indicate a constant pollution rate of 1.35 mg/l per year.

Nitrogen salts are rarely found in appreciable quantities in natural water and are derived almost exclusively from human activities. Studies have shown that in the coastal plain in Israel above the main aquifer, the intensive use of nitrogen fertilizer in agriculture contributes about 60% of the nitrate burden in the ground water with an additional 10% contributed by wastewater irrigation and refuse disposal. The other important source contributing 16% is the nitrates added by irrigation with local well water high in nitrates. This is essentially a matter of recycling the accumulating pollutants. The extremely rapid rate of nitrate accumulation that has developed in Israel is associated with the high rate of ground water exploitation as compared to the limited rate of replenishment, as well as the intensive agricultural and urban activities which take place on the overlying coastal plain. This is a typical water quality problem which in many areas can reach serious proportions under conditions of scarcity.

While nitrates within limits are beneficial to agriculture, as a fertilizer, an excess concentration in drinking water can be harmful to human health. Infant methemoglobinemia, a blood disorder effecting infants up to 12 months of age, has been associated with nitrate concentrations above 45 mg/l in water used to mix infant milk formula. The United States and some other countries have established the maximum allowable concentrations of nitrates in drinking water at 45 mg/l.

In 1974 the Israel Ministry of Health decided to set a Maximum Permissible Concentration of 90 mg/l for nitrates in drinking water while maintaining a Recommended Maximum Concentration at its previous level of 45 mg/l. This liberalization was influenced mainly by the fact that by 1974 a high proportion of the drinking water sources in coastal areas of Israel had surpassed the previous guidelines of 45 mg/l and little if any negative health effects were detectible (Shuval and Gruener, 1972).

The author did not support this relaxation despite the fact that few cases of overt methemoglobinemia had been associated with water in the high nitrate areas of Israel. This was mainly based on growing evidence that an excess of nitrates in water or food could be associated with other potentially harmful health effects in addition to the risk of methemoglobinemia. Among these potential risks, the direct carcinogenic effect of nitrites and the possible conversion of nitrates and secondary amines to highly carcinogenic nitrosomines within the human body should lead to a cautious policy in dealing with such increasing nitrate burdens in drinking water. The decision of the E.P.A. (1975) to set a mandatory nitrate standard in drinking water of 45 mg/l after many years of careful study and considerable pressure to relax the standard from the water industry is indicative of the present state of concern as to the potential negative health effects of nitrates.

An additional reason that lead me to oppose the relaxation of the standard from 45 mg/l to 90 mg/l in 1974 was the fear that the water planning and agricultural authorities would view this action by the health authorities as a "license" to further postpone action in initiating control measures which are both expensive and harsh in their agricultural implications. The need to take urgent remedial action in order to prevent irrevocable damage to the aquifer was already apparent at that time.

According to Mercado's model (1980) it is estimated that if no control procedures are initiated, by 1985 the average nitrate concentration in the aquifer will be 85 mg/l and 46% of the wells in the coastal plain will have to be taken out of service for municipal and domestic use. By the year 2000 the nitrate pollution problem will reach almost catastrophic proportions with the average nitrate concentration exceeding even the present more liberal maximum permissible nitrate standard resulting in some 80% of the wells being taken out of service for municipal use. This is a most severe threat to some 20% of Israel's most important water resources the effects of which will be felt dramatically as early as 1985 but which has already lead to the closing of a number of important wells.

Possible control strategies include: High level of reduction of nitrate concentrations in wastewater used for irrigation by tertiary treatment; reduction of the nitrogen fertilizer applications in agriculture; recharge and exchange of water in the coastal aquifer with low nitrate surface water leading to dilution of the nitrates in the aquifer; treatment of water for municipal water supply to remove nitrates to an acceptable level; abandon high nitrate wells for municipal supply purposes.

An effective combined strategy proposed by Mercado would require a 50% reduction in fertilizer application, tertiary treatment of all wastewater used for agricultural irrigation to remove 80% of the nitrogen and a temporary ground water recharge and exchange of 25 MCM for a 10 year period. This, if initiated immediately, could lead to a gradual reduction in the rate of nitrate pollution increase, levelling off in about 15 years. There is a lag time of five to 15 years before any control measure can take full effect. This combined strategy could keep down the percent of disconnected wells significantly. However, if its initiation is delayed until 1985, 45% of the wells will already have been disconnected and before the control measures would take effect another 10-20% of the wells might be out of action meaning that for all intents and purposes the usefulness of major portions of the coastal aquifer as a source of domestic quality water would be lost.

Water treatment technology for the removal of nitrates based on ion exchange resins or other processes is still in the developmental stage and will, at best be very expensive. It is hard at this time to evaluate the technical or economic feasibility of this strategy as a sole solution allowing the nitrate levels to increase in an uncontrolled manner.

The need for urgent action to salvage the threatened coastal aquifer is apparent: While some important educational activities among farmers to voluntarily reduce excess fertilizer application has been started and a program of ground water recharge in high nitrate aquifer areas has been undertaken, neither program is being carried out at a level sufficient to halt the creeping nitrate pollution of some 20% of the countries most accessible water resources.

LAKE KINNERET - A VITAL MULTIPURPOSE RESOURCE UNDER ENVIRONMENTAL STRESS.

Lake Kinneret plays a central role in the water economy of Israel both as the country's major long term storage reservoir and as the primary source of the National Water Carrier system which transports water to the central and southern areas. It also serves as the country's main recreational lake and as one of the prime inland tourist amenities. In addition it provides a limited although not insignificant source of fresh water eatable fish.

Lake Kinneret has many unique chemical, biological and hydrological features specific to its semi arid conditions. The most outstanding is the extremely large watershed area in relation to the lake volume. The ratio-watershed area: lake volume for the Kinneret is .69 m^2/m^3 compared to a ratio of 0.1 typical for sub-alpine lakes. This means that the lake is particularly sensitive to the pollution derived from human activity on the Jordan River watershed draining to the Lake.

While the trend of eutrophication as found in American and European lakes has not yet been detected in the Kinneret, due to the particular sensitivity of the lake a continuing increase in organic input would put the lake on the pathway of typical eutrophication with the many well known side effects such as algea blooms, development of unpleasant tastes and odors and possibly fish kills. Careful management of the Jordan River watershed, feeding Lake Kinneret, is essential in the protection and management of the Lake. Major sources of nutrients and organic pollution are derived from agricultural sources not only from urban and industrial activities. All of these must be controlled in an integrated watershed management program, including management to reduce the leaching of nutrients and the diversion of heavily contaminated effluents for irrigation or out of the watershed so as to reduce to a minimum the introduction of pollutants to the lake.

An independent environmental surveillance and control authority in the hands of an agency not directly involved in the economic activities of managing the Lake's water resources should be developed to assure long term protection of this vital resource.

WASTEWATER REUSE - A KEY WATER SOURCE TO BE UTILIZED WITH CAUTION.

Under such severe conditions of scarcity as exist in Israel it is unthinkable to allow wastewater from urban centers to be lost to the water economy by flowing to the sea.

However, this resource has to be utilized with care due to its burden of toxic chemicals and pathogenic microorganisms which cannot be completely removed by feasible wastewater treatment technology.

A higher percent of municipal wastewater is currently utilized in agriculture in Israel than in any other country in the world with plans now underway to ultimately tap 80% or more of the urban wastewater flows for such purposes.

The regulation of wastewater reuse will have to be tightened up to reflect our current understanding of the risks associated with such problems as long surviving viruses and aerosolized pathogens dispersed by spray irrigation. Israel is in the process of preparing such revised and upgraded regulations.

Another problem which was neglected in the early developmental stages of wastewater reuse was the disposal of effluent flows during the 6-8 months period when irrigation is not practiced. These excess effluent flows have lead to the serious pollution of all of Israel's natural flowing rivers and dry river beds which normally carry only flood flows. The only feasible solution, now official policy, is the construction of seasonal effluent storage reservoirs which will effectively serve both to solve the environmental problems and will significantly increase the utilization of wastewater effluent.

Although in the Dan Region Wastewater Reclamation Project plans were originally initiated with the intention of recycling treated wastewater for municipal use including drinking, these plans have wisely been modified to allow for more limited use for agricultural and industrial purposes only. This is a logical approach since some 75% of the national water demand is for agricultural purposes.

By allocating highly treated effluent for unrestricted agricultural use it will be possible to divert water of potable quality, presently used in agriculture, to meet the increasing urban and industrial requirements. This policy of resource exchange will assure the urban population with high quality water for domestic use and minimize the health risk associated with wastewater reuse.

ADEQUACY OF EXISTING ORGANIZATIONAL STRUCTURE.

The question that must be raised at this point is whether the existing administrative set-up is organically capable of meeting the challenges of the future which will require the highest level of integrated future oriented water quality management.

From the very beginning water policy was seen as an essential element of agricultural policy since the agricultural development of the country was integrally tied to water resource development. Thus the official government water management institution, the Office of the Water Commissioner is directl responsible to the Minister of Agriculture, Tahal - Water Planning for Israel and Mekoroth - the National Water Company also come under the wing of the Minister of Agriculture but are also directly under the powerful influence of the economic institutions of the agricultural sector.

This identity of purpose between agriculture and water policy has lead to the rapid and highly successful development of water resources which have made a major contribution to the economic growth of the country.

The future organization must emphasize managing the more or less fixed amounts now available, protecting its quality and making reallocations to meet growing urban and industrial needs.

Agriculture's unwillingness to accept the harsh realities of the impending water crisis is typified by those agricultural planners who still talk in terms of further agricultural expansion based on hopes of some yet unforeseen future technological "Messiah" which will supply plentiful cheap desalinated sea water. Most authorities agree, however, that there is little basis in reality in anticipating that the costs of desalination of sea water will ever allow for its use in agriculture where the return on water is relatively low.

While the long term dominance of the agricultural interests on water development has played a vital role in the country's economic development, the point has now been reached where a basic restructuring may be required in the water management administrative set-up to allow for the changes that must be instituted for proper future management of this scarce resource.

Now and in the years to come the major national goal in water management must be the provision of adequate water for urban and industrial needs resulting from immigration and normal population growth. While one farmer growing cotton uses some 125,000 cubic meters of water per year, that same amount of water could support 250 families in industry, based on 500 cubic meters per year. (Fig. 2).

Agriculture must be provided with those amounts of water that can be economically justified. This required policy of rational economic reevaluation of water allocations to agriculture must however give due consideration to the national policy and social commitment favoring the maintenance of a strong and healthy agricultural community. This social commitment to maintain a viable agriculture has loomed large in the development of Israel's national political and social thinking. However, the use of water as a pipeline for subsidies to farmers is open to serious question. All economic sectors should be charged the true price for water rather than the present highly subsidized rates for agriculture. Such artificially supported water prices leads to the misallocation of scarce resources and to improper economic evaluation of alternative costs.

The Quantity of Water Required to Provide a Livlihood for One Person (or Family) Working in Various Sectors in Israel

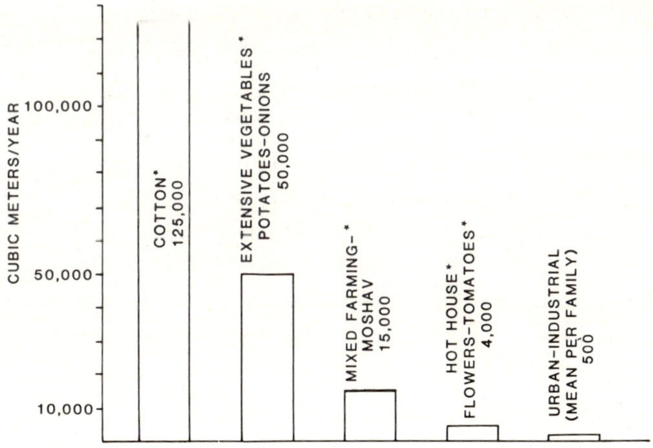

* Based on Data from: ***Advisory Service -Ministry of Agriculture***

Fig. 2

If charged the true price for water, agriculture would have to modify its practices to some extent so as to eliminate or reduce those branches not capable of giving a proper return on the investment. Since political and social policy favors the support of agriculture, the subsidies could, if required, be direct to the farmers themselves or to specific high priority communities rather than through one of the elements of production - water.

DESALINATION OF SEA WATER.

Most long term projections of water demand made by the official water authorities indicate that sea water desalination will be resorted to before the year 2000 to help ameliorate the salinity problem and to provide a net supplement to Israel's water resources, required to maintain certain levels of agricultural production.

In 1980, the best available estimated costs for producing desalinated sea water based on actual contract figures for a projected large scale dual purpose multi-stage flash distillation plant on the Arabian peninsula is $1.00-$1.50 per cubic meter (\$4.00-\$6.00/1000 U.S. gallons) operating at full design capacity.

Even assuming the lowest estimated cost, desalinated water would cost some ten times more than the current average prices paid for agricultural water in Israel.

The additional massive amounts of energy required for desalination plants of such magnitude would place inordinately heavy strains on the country's already precarious energy situation. The fact that almost all fuel is imported must lead to extreme caution in connection with any decision that would result in major long term commitments to significantly increase the demand for energy. This concern for energy conservation must remain a major factor in all future desalination decisions of non-fuel producing countries.

Based on the above considerations and unless some yet unforeseen and unanticipated break through leading to a drastic reduction in costs develops, I suggest that the prudent policy to follow at this time must be to rule out sea water desalination as an unfeasible option for the coming planning period of 20 years. The country should begin to manage its water resources and their quality control on the assumption that it will have to <u>make due</u> with its currently available resources and those additions from wastewater recycling and utilization of brackish water that are technically and economically feasible.

Such a declared policy decision will help to clearly establish the fact that a policy based on recognizing the existence of a total system shortage exists and that there is no point to postpone difficult decisions while waiting for some technological "Messiah" to provide a solution.

Plans should and can be made now to meet all the water demands for a population of 5,600,000 in the year 2000 on the premise of a rational reallocation of all the existing water resources based on their real cost. The operative results would undoubtedly be socially and ideologically painful cut backs in allocations in marginal agricultural uses now producing little return for the water used. As difficult as such a policy may be to carry out, it will be more easily accepted once the true cost of water is being paid and the reality of total system shortage is fully recognized. The unrealistic hope for a technical way out and the illusion of high agricultural profitability in some branches that has resulted from heavily subsidized water prices have helped to block any realistic appraisal to date of this serious problem by the agricultural planners and their clientele.

CONCLUSIONS

It should be apparent from the above discussion that the long term planning and policy development on such vital water issues of national concern transcend the scope of agricultural interests alone and should be allocated to an authority not directly committed by official mission to the protection of the interests of one economic sector.

In relation to the specific issues of water quality management it is clear that the scope of the problem now also goes beyond the area of legitimate interest and sole authority of agriculture.

There is no ideal solution to this problem and it is recognized that the local political reality will always play a major role in such reorganizational efforts but it appears logical and rational at this critical stage to transfer the entire water management structure out of Ministry of Agriculture into an appropriate strong organization dealing with all phases of natural resources management.

At present the Environmental Health Division of the Ministry of Health and the Environmental Protection Service of The Ministry of Interior have developed water quality control programs in addition to the program recently developed within the Office of Water Commissioner. All three programs have insufficient staff and resources alone to meet the growing demands for advanced environmental control of water quality. A unification of these three units within an agency having no direct vested interest in the water economy would provide a logical strengthening of the much-needed independent water quality control function.

Israel will have to mobilize its best scientific and management skills in an integrated program of future oriented water quality management if it is to quell the impending crisis of water quality degradation.

Other countries facing similar problems now or in the future can hopefully learn from the experience gained in Israel, both the successes and failures, in its pioneering effort to manage its severely limited water resources. No other country has yet faced such an extreme challenge effecting its very survival; for without water there is no life.

Knowing Israel's will for survival and ability to meet difficult challenges, I believe that the impending water quality crisis will be met in appropriate innovative modes, but we must act swiftly and clearly to assure the future of its vital water resources.

REFERENCES

Environmental Protection Agency (EPA) (1975) Interim Primary Drinking Water Regulations. Federal Register Vol. 40, No. 248, pp. 59566-59588.

Mercado A. (1980) The coastal aquifer in Israel: Some quality aspects of groundwater management in water quality management under conditions of scarcity, ed. H.I. Shuval, Academic Press, New York.

Shuval, H.I. and Gruener, N. (1972) Epidemiological and Toxicological Aspects of Nitrates and Nitrites in the Environment. Amer. Journ. of Public Health 62:1045-1051.

Tahal (1977) The Water Economy - Development and Management. Tel Aviv.

Developments in Arid Zone Ecology and Environmental Quality
Edited by H. Shuval

11

ENVIRONMENTAL EFFECTS OF COAL-FIRED POWER PLANTS

J.I. Bregman and R.B. Bogardus

WAPORA, Inc. 6900 Wisconsin Ave., Chevy Chase, MD, U.S.A.

ABSTRACT

Because of concurrent trends towards multi-unit power plant sites, increased use of coal as feedstock, and possibilities of regional as well as cumulative local environmental effects, environmental quality has received increased consideration in the siting of electric generating stations.

The significant environmental problems associated with the construction and operation of new coal-fired power plants include potential impacts due to location, raw material extraction or handling, power plant processes and attendant physical facilities, and pollution control processes. The two problem areas which most often are key decision factors in the siting of power plants are air emissions from coal combustion and the dissipation of heat from condenser cooling.

Air contaminants from coal combustion include sulphur dioxide, nitrogen oxides, particulates, and a variety of trace elements. After these pollutants have been emitted to the atmosphere, a number of chemical reactions may take place. Sulphur dioxide may be oxidized to sulfate in the presence of moisture, sunlight, certain metals, hydrocarbons, oxides of nitrogen, and ozone. Where sulfate particles are present, contact with moisture, whether in the atmosphere or in the lungs, can lead to formation of sulfuric acids and sulfate salts. The incorporation of sulfates and nitrates into cloud systems can result in acidic aerosols, mist, and acid rainfall on a regional scale. Acid rainfall may adversely affect fresh water aquatic communities, leach minerals from soil, damage crops and natural vegetation, and damage buildings and historical structure exteriors. Various approaches to the control of these pollutants, including both use of low pollutant coals and add-on decontaminant facilities, are discussed in this paper.

Several types of cooling systems are available for use in today's power plants. These include once-through cooling, wet cooling towers, dry cooling towers, cooling ponds, and spray ponds for canals. Each system has its own potential for producing adverse environmental impacts. The use of a once-through cooling system requires an assessment of the impacts of the thermal effluents and intake system on the ecology of the receiving waterbody. Wet cooling towers need to be examined carefully for blowdown disposal problems, potential fogging problems, and potential mixing of vapor and stack gases. Dry cooling towers are expensive, and thus far have only been used in small sizes. Spray canals are often impractical, and cooling ponds in dry climates suffer from excessive evaporation and accelerated eutrophication.

Developments in Arid Zone Ecology and Environmental Quality
Edited by H. Shuval

12

LEGAL ASPECTS IN THE SITING OF POWER PLANTS

Sanford Sagalkin

Ely, Guess & Rudd, Washington, D.C.

ABSTRACT

Energy production and the creation of electric power at large centralized power plants are necessary to achieve and maintain high standards of living in a modern society; however they also usually create substantial adverse environmental impacts -- impacts to air and water quality, impacts to endangered species of plants or wildlife, impacts to recreation areas or to the coastal zone, if located therein, and impacts to the perceived quality of life or psychological environment in the case of nuclear power plants.

In the United States, Congress has passed a variety of laws, some merely procedural, others of a substantive standard setting nature, in order to control and mitigate adverse effects of power plant siting decisions. The principal laws governing siting decisions (at the Federal level) are the Clean Air and Water Acts, the National Environmental Policy Act, the Endangered Species Act, the Coastal Zone Management Act (in the case of siting in the coastal zone) and the Atomic Energy Act in the case of nuclear power plants. Virtually all of these laws in the case of new facilities require the gathering of the available environmental information necessary to evaluate the adverse environmental impacts of the proposed facility.

Additionally, some of these laws, such as the Clean Air and Water Acts, and the Endangered Species Act are more than procedural laws. They are laws which either set very specific standards which must be met or which prohibit, outright, activities which endanger certain species. These standard setting laws can actually bar construction of projects failing to meet these acts' provisions.

The processes used in the siting of major power facilities provide a combination of substantive standards which must be met in all cases, and procedural requirements designed to highlight adverse environmental effects and mitigation measures. These laws require and encourage government and industry to behave responsibly toward the environment.

Developments in Arid Zone Ecology and Environmental Quality
Edited by H. Shuval

13

ECOLOGY IN TALMUDIC SOURCES

Samuel S. Kottek and Hanna Seligmann[1]

Division of the History of Medicine, The Hebrew University - Hadassah Medical School, Jerusalem, Israel.

ABSTRACT

The Talmud, which is a large corpus compiled from the 2nd century BCE till the 5th century ACE, is an interesting source for details concerning ecology in Jewish antiquity. Protection of nature : trees, waters, - is emphasized. Cemeteries, tanneries, quick-lime furnaces should be located at a fair distance outside town. Threshing-floors, sources of smoke, pigeon-houses - just the same. There are detailed rules concerning public water-closets, soiled waters, noisy trades - with special consideration towards individual sensitivity to noise. Water-supply is cared for in detail, and these waters must be kept clean. Paths and highways are submitted to planning and obstacles to traffic must be removed. Lights in the streets are supplied with oil from public funds. Towns should be girded by areas which remain free from habitations, for the aeration and the beauty of the town. These few examples show to what extent the Sages of the Talmud were aware of the importance of the environment for the well-being of the people.

INTRODUCTION

"The value of the air in town compared to the air of deserts or forests is similar to the value of thick and muddy waters compared to pure and refined waters. In towns the buildings are tall, the streets narrow and littered with waste products, rubbish and manure discarded by the inhabitants and their cattle ... So the air becomes spoiled, corrupted, thick and cloudy" ... But the individual is not aware of this influence of these factors on his health.

[Maimonides : Regimen Sanitatis, Part 4, chap. 2]

Maimonides lived in Spain, in the twelfth century. The Talmud is a vast compendium reflecting the ideas currently professed by the Jews in the Babylonian diaspora (Babylonian Talmud) or in Palestine (Palestinian Talmud) from the second century BCE till the fifth century ACE. It is thus obvious that it was influenced by Babylonian, Greek and Roman thought. However, the very fact that the Sages of the Talmud, mainly interested in legalistic discussions, had a lot to remark on ecological problems is certainly worth our attention.

PROTECTION OF NATURE

Destructions due to small cattle and birds

It has been ruled already from the earlier part of the Talmud (Mishna), that the raising of sheep and goats was prohibited in the land of Israel, the reason being that they ruin the fields and the gardens. But the raising of cattle and of domestic animals is permitted. [Baba Kama 79b] This ruling was abandoned during the middle-ages (Codes) and has not been renewed. A pigeon-house should also be built in open land, leaving at least 50 cubits empty in all directions (Mishna). The Talmud even wants it to be distant at least thirty Ris (ab. four miles) from any village. The reasons being evidently the depredation caused to seeds and crops. [Baba Bathra, fol. 23a]

Protection of trees

The Bible warns the Hebrews from cutting off fruit-trees, even during a war : "they are your food, do not destroy them". If you really need wood for the toils of the siege, you should cut trees which you know are not bearing edible fruit. [Deut. 20, 19-20]

Protection of waters

Twice in Scriptures, there are examples of Kings who decided, in the course of war, to obstruct springs or streams of water in order to deprive the enemy from finding water for his troops. [II Kings, 3, 19 and II Chronicles, 32, 4] Such actions are strongly opposed by the commentators and the Sages of the Talmud [Pesaḥim IV, 9].

AIR POLLUTION

Corpses : corpses (of animals) and cemeteries should be distant from town at least fifty cubits (because of the bad odours) [Baba Bathra, II, 9]

Trades : tanneries should be located outside town, at a minimal distance of fifty cubits. And they should be on the east side of the town. [ibid.] Maimonides explains that the wind coming from the east is hot and decreases the deleterious odours of the tannery.

Ritual offerings : according to Maimonides - vehemently countered by other authorities, - the fumigations[2] used at the Temple in Jerusalem twice a day were intended to remove the stench caused by the burned meat and the blood. [Guide of Perplexed, Part III, chap. 45]

Water-closets : the Bible rules that a special place should be chosen outside the army-camp and the stools should be covered (with earth). [Deut. 23, 13-14] The odours of the water-closet are considered as a serious damage which justifies removing the water-closet even if it has been there already for some time. [Baba Bathra, fol. 23 a]

Soiled waters : sewers and reservoirs of dirty waters should not be drained during the hot season, but it maybe done during the raining season. [Baba Kama, fol 6a] The bad odours are of course not the only reason for it, it is obvious that it spoils the street in warm weather and is washed off by the rain.

A pond used for steeping flax should be distant from a neighbor's vegetable garden (the distance is not stated in the Mishna but it is evaluated,as above,50 cubits.) Here the smell could possibly also damage the vegetables. [Baba Bathra, II, 10]

Air-borne agents : a threshing floor should not be permanently installed within 50 cubits from the town even in private property : there should be 50 cubits in all four directions free from houses, because it could cause damages. [Baba Bathra, II, 8] This law is extended by Maimonides to any industrial activity which brings forth dust or ashes. Not only could such agents harm the inhabitants of the village but it could also damage the vegetation.

Smoke : furnaces (for quick lime) should be sufficiently distant from town so that it could not harm people, especially their eyes. In the Talmud it is stated that there were no such furnaces in Jerusalem [Baba Kama, fol. 82 b], one of the reasons alleged being also that the smoke blackens the walls of the houses. The minimal distance, as always, is 50 cubits. Smoke and fumes are considered as a severe damage, just as deleterious odours, because they may harm the body - and not only the property [Nachmanides : gloses on Baba Bathra, fol. 59 b]

WATERS

Drinking water supply

A spring which sprouts forth in a town is the property of all the inhabitants - not of the landowner. One Sage claims that the owner gets a financial compensation, but the decision is that he gets no payment. [Baba Kama, fol. 81a] Other towns or villages may also use these waters without payment, but the first town has a right of precedency.

The Hebrews used of course springs and wells, but more frequently used rain-water which was kept in cisterns (dug into the earth) or reservoirs (built over the earth). The wells were usually covered with a heavy stone [Gen. 29, 2].

In Jerusalem, there was only one spring, that of the Giḥon, which was located outside the walls, but King Hezekiah built a 535 m. long canal which led the waters into the town. But there was also a long aquaduct which brought water from the reservoirs - still extant - called Solomon's cisterns.

Prevention from spoiling the waters

There was an annual inspection of the water supply, these justice officers devoting their attention to all necessary repairs of cisterns, wells or canals. [Toss. Shekalim I, 2]

In a public cistern, it was permitted to wash one's face, hands or feet - but not if really dirty (earth or dung). In wells or reservoirs of drinking-water this was altogether prohibited. [Toss. Baba Metsia, XI, 31]

Water which has been left during the night without cover

should not be drunk (fear from snake-venom). It is prohibited to drink water from a stream or a pool directly with the mouth or with one hand. [Avoda Zara, fol 30 a-b, and 12 b]

A dung-hole should not be dug close to a well, this is considered a risk of severe damage.

Protection of bathing waters

Somebody who has a skin disease should not bathe in a public bath, because of the danger of contagion. [Sefer Ḥassidim, ch. 161]

We shall not dwell in this paper on the importance of the public bath, which was well-developed in Hebrew antiquity. It is said that each one of the 12 tribes, once installed in the land of Israel, had 13 public baths. They had besides one bath in common, which was Tiberias with its warm water springs. [Lev. Rabba, 34, 3]

NOISE

There are in the Mishna several examples of noise being considered as causing damage to other citizens.

A grindstone should not be installed against the wall of a neighbor : there should be at least three tefaḥ (ab. 30 cm) from the bottom stone to the wall of the house. [Baba Bathra, II, 1] The Sages explain that there are two reasons : the shaking of the ground, and the noise of the mill.

If somebody opens a store in a courtyard, one of the neighbors may intervene, complaining that the noise of the people coming and going hinder his sleep,even if this store has been there already for a long time. But he may not complain that the sound of the hammer, or the mill, or the voice of children at school annoy him. [ibid, II, 3] In fact, a noisy activity may be objected by the neighbors at its beginning but not once it has been tacitly accepted for some time. The case of the noisy customers is an exception, as the store-keeper may resume his trade on the town-market. Noise, says Maimonides, is a definite nuisance just as smoke and dust. [Laws of Neighbors, 11, 5]

Moreover, noise may be an unbearable nuisance to sensitive individuals. Some barber-surgeons used to practice venesection under palmtrees and the blood attracted ravens, which spoiled the dates and were very noisy. There was an argument between the landowner who lived there and the surgeons. The final decision was that for him, who was very sensitive, this nuisance was as deleterious as smoke or a dung-hole (for others). [see Baba Bathra, fol. 22 b]

PATHS, ROADS AND TRAFFIC

The Talmud states that Joshua, as soon as he conquered the Land of Israel, made highways with stations. [Erubin fol. 22 b] Paths and short-cuts were also established, through private fields, to be used in case of need. The passage on private paths was authorized until the second rain of the season, then it was prohibited, except in cases of danger, [Baba Kama, fol. 81 b], or in case of somebody getting lost.

The breadth of a main-road (from town to town) should be eight cubits, to allow the passage of two carriages. [Baba Bathra, fol. 100 a] A main-road with much traffic on it should be twice as broad (18 cubits). The way leading to the "towns of shelter" even broader (32 cubits).

Repairs : one particular day was devoted to the inspection of public property and repairs were scheduled for streets and roads that had been damaged by the rains. [Tossefta, Shekalim, fol. 81 b]

Obstacles must be removed from the roads. For instance, a tree that bends towards a public way should be cut off in order to let free passage to a camel and its rider. [Baba Bathra, fol. 27 b] The same occurs to a tree that hinders the passage of the sailors who are dragging a boat along a stream or a canal. The duty to remove any obstacle is especially stressed as related to the roads of shelter (the roads leading to the town where the involuntary killer seeks refuge).

Rules for traffic are also dealt with in the Talmud. For instance if two boats cannot pass alongside, the boat that is farther from its goal has precedency. If one is laden, the other empty, the latter will let the other pass first. [Sanhedrin, fol. 32 b] The same applied also on roads or highways.

Lights and lamps were lit up in the streets with oil from priest-offerings which had become impure. Such oil was also used for public buildings such as synagogues. [Terumoth, 11, 10]

Accompanying strangers and travellers was considered a sacred duty and the inhabitants of a village used to go with them on a certain distance. If something happens to such a stranger, the town-officials are considered responsible. [Sota, fol. 45 b]

TOWN ENVIRONMENT

The Bible enjoins the children of Israel to build cities for the Levites within their territories, "towns to reside in and open spaces around them you will give to the Levites. The towns will be for them to dwell, and the outskirts will be for their cattle, for their possessions and all their well-being. The width of these outskirts will be 1000 cubits all around." [NUM. 35, 2-5] There was additionals surrounding land, dedicated to cultivation and vineyards,

The 11. century commentator Rashi explains that the outskirts are intended as land that remains uncultivated, without houses, for the beauty of the town and for its aeration.

Trees should not be planted closer to the walls of the town than 25 cubits, some kinds (as the carob-tree) even 50 cubits. The reason is that the latter are very leafy. [Baba Bathra, II, 7]

We have seen that tanneries are removed outside town. One of the reasons given in the Talmud is "for the beauty of the town".

The Levites' advice was sought by all other Israelites and it was considered important that such visits should take place in an atmosphere of calm and beauty. [Book of Education, ch. 342]

Our conclusion will be taken from the same passage of Maimonides as was our introduction. We read :

"It should be desirable to choose to dwell in a city which has an open horizon ..., situated on slopes of hills, with not too many trees and waters (or swamps). But if you have no choice, you should at least dwell near the walls on the north-eastern side, in a tall building with a large courtyard where the northern wind and the sun might freely enter." [Regim. Sanit., Part 4, chap. 2]

SHORT BIBLIOGRAPHY

Grünwald M. (1911) Die Hygiene der Juden. Dresden (325 pp.).

Kagan S.R. (1933) Hygiene among ancient Hebrews. Am. Med. J., 29 : 26-30.

Kotlar D. (1977) Human Ecology in the Ancient World. Jerusalem [Hebrew].

Preuss J. (1911) Biblisch Talmudische Medizin. Berlin (pp. 588-652).

Rakower N. (1972) Preservation of environment (Ecology) in Hebrew sources. Koroth, vol. VI, 1-2 : 100-118. [Hebrew]

Schachter M. (1932) Hygienic principles among ancient Hebrews. Rev. Hyg. et Med. Soc., vol. XI : 217-221.

Simon I. (1978) Etude des problèmes de l'environnement (écologie) et de la qualité de la vie dans les textes bibliques, talmudiques et rabbiniques ... Rev. Hist. Med. Hebr., 127 : 91-96.

Simon I. (1979) Id. (continued) Rev. Hist. Med. Hebr., 128 : 15 - 19.

NOTES

[1] Is completing a medical thesis on the topic : "Public Health in Ancient Jewish Sources" (Hebrew, English summary).

[2] These fumigations were made of several aromatic drugs mixed together.

Developments in Arid Zone Ecology and Environmental Quality
Edited by H. Shuval

14
THE BURNING ISSUE OF DISPOSING OF HAZARDOUS WASTES

Jack D. Lauber, P.E.

New York State Department of Environmental Conservation

ABSTRACT

Hazardous wastes in the environment represent one of the most serious problems facing us today. Our complex technological society has produced immense quantities of toxic wastes whose improper disposal has created serious air, water and land pollution problems. The tragedy of the Love Canal has shown that we cannot safely put toxic organic wastes in the ground, lest they become future toxic timebombs. One of the best means of converting hazardous organic chemical wastes to relatively harmless compounds is process incineration. Even chlorinated hydrocarbon wastes, including PCB's, the most difficult types to destory, can be completely destroyed by applying proper process incineration techniques.

Most organic chemical wastes are readily combustible and possess a significant heating value. Thus, such waste has the potential for being utilized as alternate fuels in some types of stationary combustion installations as well as various industrial processes. Utilizing such "synthetic" fuels, particularly in large high combustion efficiency boilers and in processes, such as cement kilns, stone and aggregate driers, can reduce the need for considerable quantities of fuel oil and other traditional fuels, if carried out in an environmentally acceptable manner.

Regulating the combustion of hazardous waste from the standpoint of air quality is not simple. Initially, concern was limited to the more common air contaminants associated with the combustion of industrial waste, such as sulfur and nitrogen oxides, hydrogen chloride and heavy metals. However, recent evidence suggests that complex chemical reactions can result in more exotic, potentially hazardous air contaminants, such as chlorinated dioxins, which can be present in the emissions, if the combustion of organic chemical waste is not complete.

We are now developing effective regulations for the proper incineration of hazardous organic wastes. New best available control technology (BACT) guidelines, will specify virtually complete degrees of waste conversion and control of toxic air contaminants; as well as state of the art environmental monitoring provisions for hazardous waste incinerators and processes that burn such materials as synthetic fuels.

We must convince the public that the proper incineration of hazardous wastes is a safe, major strategy for controlling our hazardous waste problems. This is the only way to remove these toxic timebombs from our environment forever, and to recover valuable energy from these hazardous wastes when properly used as synthetic fuels.

Developments in Arid Zone Ecology and Environmental Quality
Edited by H. Shuval

15

SOLAR ENERGY IN THE FUTURE ENERGY-MIX OF ISRAEL

Aharon S. Roy

Department of Chemical Engineering, Ben Gurion University of the Negev, Israel

ABSTRACT

The planned shift of Israel's energy economy from full dependence on imported liquid fuels to full dependence on imported coal is inadequate. The price of coal will not stay low and it will climb up following that of oil. From considerations of economy, national security and independence, and quality of life, solar energy should be given a fair chance to take a substantial part in Israel's future energy-mix. Recent technological progress has resulted in solar hardware capable of producing steam and electricity at low cost. Being environmentally acceptable, distributed solar systems near dwellings, industry and farms, may produce a substantial part of all forms of energy locally needed, while offering the economic advantages of cogeneration of electricity and process steam, air conditioning, or hot water. Prudent choice of technology may reduce the cost of solar energy to values below those of imported fuel. A huge local solar industry will offer many jobs and help abate depression trends due to the world energy crisis.

INTRODUCTION

In planning national energy strategies, several factors have to be considered; in particular, present and projected prices of various energy sources, pollution and hazards, and considerations of national security which calls for development of local energy resources. Very often it comes out that a mix of energy sources is most desirable. The proportions of the various ingredients in the "mix" vary by locality and they are also expected to change in time. Leading the direction of change of the mix, is one of the expressions of a national energy policy.

In Israel, 98% of our energy is imported as oil. Needless to say, this is a very unhappy situation. Of the remnant 2%, solar energy takes the prominent part. This comes in the form of warm water supply produced by conventional flat-plate solar collectors placed on roofs of buildings. The question whether the sun can supply a larger share of Israel's energy economy is a very legitimate one. Careful study of this question gives a positive answer. New technologies have recently emerged with which electricity, steam and air conditioning can be economically produced from solar energy. The percentage of energy supply from the sun can thus be made to increase yearly and meet a substantial part of our energy needs. This will be very benefitial for our national security and for decreasing pollution and hazards related to exaggerated dependence of fossil-fuel burning and nuclear plants. In addition, solar expansion entails substantial economic contributions which are discussed below.

Due to the late coming of the particularly valuable new solar technologies, they are not sufficiently known. Also, as the present cheap prices of coal in the world may cause many to believe that Israel's energy problem is solvable by a shift from imported-oil dependency to imported-coal dependency, other possible contributions to our energy mix are not adequately looked into. It is the purpose of this work to point out that Israel's energy policy should include efforts towards a major expansion in solar energy utilization, a task which is becoming feasible due to recent developments in solar energy technology.

THE CASE OF COAL

When planning a power plant based on a particular fuel, the projected prices of the fuel for a thirty-year span should be seriously considered. For the purpose of national energy policy, a longer time span is of interest. Today coal is cheap. Will it stay that cheap? The answer is: no. The reasons for this and an estimate of the projected price for international supply of coal follow.

Coal is a sedimentary rock containing more than 50% carbonaceous material (by weight), formed from the accumulation and burial of partially decomposed vegetation affected by varying biological, temperature and pressure changes over many years. Coal is ranked in terms pf carbon content. From the highest to the lowest carbon content the ranks are: antharcite, bituminous, subbituminous, and lignite. The term "coal resources" refers to potentially *recoverable* coal in the ground, implying a certain level of technology and a certain level of potentially acceptable costs. Coal reserves point to identified and *economically recoverable* coal in the ground. Resources are therefore not given by a unique number. Recoverability points to a figure of about 1000 billion (10^{12}) ton coal or roughly 1000 TW/a (10^{12} tera-Watt-year) of world coal resources (Hafele 1977). Global energy demands are increasing. It results from the continued economic growth of both industrialized and the less developed countries, and growth of world population. From the present energy consumption of the world of 9 TW, an increase to approximately 30 TW is expected by the year 2000, and to, say, 50 TW in post 2000 years (Hafele 1977). Being based on the use of coal only for global energy consumption, more than a third [20x0.5(9+30)/1000 = 39%] of the recoverable coal may be exhausted in 20 years and nearly three quarters [39%+(35x10)/1000 = 74%] in 30 years.

Needless to say, long before the above year-marks arrive, the cost of coal will rise sharply, in particular the coal in the international market. Energy experts' views are that the cost of production of coal will increase by at least a factor of three around the year 2000 (Hafele, 1980) (IIASA), while the selling price will escalate much faster. For internationally shipped coal, the price will most probably reach 80% of the price of oil (valorie vs. calorie at your boiler, including charges for shipping, handling, pollution control, etc.) (Starr, 1980) (U.S. DOE). This means even higher prices of coal in future than oil prices to-day. It is therefore quite probable that the average cost of electricity for Israel during the coming 30 years will not become lower than the nearly 10 cents per kWh of to-day, but, rather, *higher*, despite the shift to coal-burning powerplants. (True, coal will slow down the escallation of our energy bill but will not reduce it in the long run).

At such costs solar energy may provide a cost-competitive alternative, while being a non-polluting and a local, renewable energy resource. This involves contributions to both the quality of life and national security.

It should be pointed put that figures for geological coal resources without consideration of recoverability are larger than those for recoverable coal by a factor of, say, 8-10. Recent figures for U.S. (U.S. DOE, 1980) indicate 1700 billion tons of geological coal resources, while only about 220 billion tons coal can be recovered from demonstrated reserve base. Of this reserve, bituminous coal constitute 52%,and subbituminous coal, 38%. Starting with the present yearly production rate of 0.75 billion tons a year and assuming a 6% yearly increase (see for instance C&EN, 1978 and Cain et al., 1980), and using the equation of (Bartlett, 1978), the recoverable coal in U.S. will last for only 48 years. It has also been stated that oil companies in U.S. are purchasing coal mines.

All this is brought here to indicate that the present cheap price of coal is not here to stay. For the life span of planned new power plants in Israel, *coal cannot be considered as a proven, available cheap fuel.*

SOLAR ENERGY

The 1-2% of Israel's energy which is supplied from solar energy results from the use of simple solar flat-plate collectors (we see them on the roofs of our houses) for producing warm water. It may be expected that extended use of this conventional technology may increase the share of solar energy in our national energy bill to about, say, 2-3%, but probably not much beyond this, as there is a limit to the need for warm water. If steam and electricity could be economically raised by solar energy, several tens of percents of our national fuel bill could be replaced by solar. This is indeed the case. Though this cannot be done by the conventional flat-plate technology, however, the situation is rapidly changing as a result of new technologies which are just now emerging from the R&D

stage. Some of them have already reached commercialization stages.

Techno-economic assessments of various methods (Roy, 1979, Roy, 1978(I), Roy, 1978(II)) and other recent results indicate a substantial economical advantage of the solar-thermal high-temperature distributed-system concept. It is based on solar concentrators which are capable of producing high temperatures. The principle of their operation is the reflection of solar light from mirror-like curved surfaces onto a small area, where a small "receiver" absorbs the light which comes from a large area exposed to the sun. The focused sun-rays have an intensity of hundreds or thousands times of that of natural light. This concentrated energy heats up a small boiler located in the receiver, in which a working fluid or a heat-transfer fluid (water, oil, etc.) absorbs the energy at an elevated temperature such as 500°C. High-temperature high-pressure steam is thus obtainable which can produce electricity by a conventional turbine-generator. By-product process steam, hot water or air conditioning can also be generated with electricity and thus reduce the cost of the energy produced. Because the system is non-polluting, small and medium-sized systems may be dispersed near towns and villages and thus excercize the advantages of cogeneration and minimized transmission losses.

Experience is being assembled in this type of solar energy utilization. One of the interesting systems is being constructed these days at Shenandoah, Georgia, U.S.A., where 800 concentrators are going to supply 60% of the energy needs of a new knitting factory. It will supply electricity, steam, air conditioning and hot water. Calculations indicate that the cost of energy obtainable by this method is lower than that we have to pay to-day in Israel. For cumulative systems capacity of 100 MW (which corresponds to only 1-2% of Israel needs in the coming years), the produced energy cost would be 7 cents per kWh electricity and one third of this for steam (G.E. 1978). The economics are based on a discount-rate of 9% and fixed-charge rate of 19%.

It is worthy of note that in Israel the cost of produced solar energy should be substantially lower for two reasons:

(a) Solar insolation in Israel is greater than in Georgia by a factor of

1.2-1.3, and the energy cost is fairly inversely proportional to solar intensity.

(b) Labour costs is an important component of the solar system cost and labour is considerably cheaper in Israel. It is for that reason that at present installed solar collector systems (for hot water) are cheaper in Israel by a factor greater than 2 as compared to U.S.A.

One may also note that if we base the cost of money for the fossil-burning power plants in Israel by the same basis as noted above and without subsidies, the real present cost of electricity in Israel might be found to be already above 10 cents per kWh.

There are many advantages to using solar energy in replacement of imported fuel: clean air and environment, improvement of national security and political independence, and the provision of vast employment opportunities. This last point is of a very great significance. The energy world crisis will cause wide unemployment. It has started already. The supply of jobs which create energy alternative to imported fuel is of a major economic and political significance.

NATIONAL ENERGY MIX

Israel's national security and independence greatly depends on the possibility of establishing an energy economy that will not be exclusively reliant on imported duel. All conventional fuels, whether gas, oil, coal or nuclear, become more and more expensive continuously. In addition, if we continue to be dependent on imports of fuel from distant parts of the world, this also entails a security problem and a danger to our national independence.

While the prices of all conventional fuels climb upward in time, solar and some other renewable energies become cheaper. This is because of the new technological developments and the economy of production scale. By including renewable energy sources in the national energy-mix many advantages are expected. Solar energy is the most important option for Israel in this category. Hydropower, notably the projected Mediterranean-Dead

Sea Canal, is another very important renewable resource. As components in the national energy-mix, solar and hydropower are complimentary to each other in several ways. They lend themselves to mutual use of elevated lakes for energy storage, peak-shaving schedules, and schemes for minimizing the use of depleting and polluting fuels.

Studies show that after having brought to use the small limited amount of biomass exploitable in Israel, and the more significant, though with a defined upper limit, hydropower available; solar energy is the only viable renewable resource we have that is amenable to replace a very major part of our fuel imports.

REFERENCES

Bartlett, A.A. (1978), Coal: no superabundance for U.S., *Physics to Day*, Dec. 9-10.

Cain, W.C., Johning, C.E. & Shaw, H. (1980), Environmental R&D Needs for Advanced Power Plants, *Chem. Eng. Prog.* 76 (10), 62-69.

C & EN (Editor) (1980), Increased Use of Coal Deemed Safe Through 1985, *Chem. Eng. News,* Jan. 30, 22-23.

G.E. (1978), General Electric Space Division and U.S. DOE, Solar Energy - Large Scale Experiment at Shenandoah; Georgia, Final Report, *G.B. Doc. No. 78SD4243,* Sept, DOE Contract No. EG77-C-04-3985.

Hafele, W. (1977), Energy Options Open to Mankind Beyond the Turn of the Century, *International Conference on Nuclear Power and its Fuel Cycle,* Salzburg, Austria, 2-13 May, IAEA-CN-361538.

Hafele, W. (1980), World Energy Problems Effects on Small Countries (Plenary Lecture), *The Nuclear Societies of Israel Joint Annual Meeting at Ben Gurion University, Beer-Sheva, Israel,* Dec. 4-5.

Roy, A.S. (1978)(I)Economic Methodology for Solar Power-Generating Systems, *Proc. 13th IECEC (San Diego),* 20-24, Aug. 3, 2175-2179, (Published by Soc. Automotive Engineers, Inc., 400 Commonwealth Drive, Warrendale, PA. 15096, U.S.A.)

Roy, A.S. (1978)(II), Econometric Analysis of Concentrators for Solar Cells. *Solar Energy 21,* 371-375.

Roy, A.S. (1979), Concentrating Collectors, Chapter 8 (pp. 185-252), in the book *Solar Energy Conversion* (A.E. Dixon and J.D. Leslie, eds.), Pergamon Press.

Starr, C. (1980), International Aspects of U.S. Energy Policy (Plenary Lecture), *The Nuclear Societies of Israel, Joint Annual Meeting of Ben Gurion University, Beer-Sheva Israel,* Dec. 4-5.

U.S. Doe (1980), Coal: Energy Source Past and Future, *Energy Insider* 3 (25) (Dec. 8), 4.

Developments in Arid Zone Ecology and Environmental Quality
Edited by H. Shuval

16
A NOTE ON POLICIES OF ENVIRONMENTAL MANAGEMENT

Shmuel Amir

Environmental Protection Service, Ministry of the Interior, Jerusalem, Israel.

ABSTRACT

In this article, dynamic policies of renewable resource management are critically considered. It is shown that many policies which have been recommended in the literature cannot give appropriate answers to basic methodological questions. Hence their applicability in a closed ecological system is doubtful. A new dynamic policy is offered and its properties are analysed. It is proposed that a principle mechanism in systems undergoing rapid deterioration be at once interrupted so as to stabilize the system. Later policies are designed with their focus on the question of the resilience of the system. If the system is to be carefully managed, it is considered essential that programs be planned so that their ensuing dynamics trace none but the system's equilibrium states. Real policies are seen to deviate from carefully controlled policies to an extent that their execution requires both additional resources and deeper analytical tools. In particular it is shown that ambient and emission standards, when judiciously implemented may interact with any simultaneously imposed pollution charges to form a set of non redundant and even synergistic controls whose interplay makes a real managing policy environmentally sound and effective.

INTRODUCTION

The discharge of undesirable materials produced by ordinary human activities into the surrounding environment, as well as the unrestricted exploitation of renewable resources, are common phenomena. Viewing the environment as a huge storage tank whose content is unlimited has encouraged a rather shortsighted conduct, typical of the animate world as a whole.

Qualitatively speaking, the problem of environmental exploitation is an old one. It is therefore unfortunate that it has been granted due attention only recently. Nevertheless, as the deterioration of the environment becomes quantitatively significant, sound management policies have found their way into literature and their applicability has been tried out. Recent expositions (e.g. Baumol and Oates 1975; Fisher and Peterson 1976; Smith 1979) and references therein may leave one with the impression that any possible aspect of the problem both in its static context and its dynamic setup has been fully explored and expounded. This view seems to us superficial and not well substantiated. As a clarification it is intended here to examine some of the models' common biological and economic assumptions, so as to show that they allow narrower management policies than those offered and analysed in the literature. Consequently we are able to come up with new policy recommendations, whose applicability and suitability to the challenges of the real world are to be analysed and assessed in this article.

A SIMPLE MODEL OF ENVIRONMENTAL AND ECONOMIC INTERACTION

We shall look at a simple system composed of 'n' identical individuals. These individuals are used to working and consuming a product generated in a single known production activity. As the activity is going on some residual is formed and discharged into the environmental basin.

Let the flow of labor channeled into the production process be 'a' and the flow of output be $y=f(a,R)$, where R is some measure of the state or quality of the environment. The flow of residuals produced is assumed to be $r=g(a,R)$ and the relation $\frac{dR}{dt}=F(R,r)$ is supposed to summarize all environmental activities. Finally, it is proposed that individuals assess their own quality of life, or utility, with the help of a function $u=u(c,a,R)$, where 'c' is taken to denote the quantity of the product consumed by an individual.

This model is basically similar to many other models presented in the literature (Brown 1974; Clark 1973; Neher 1974; Plourde 1970 and 1979). It is flexible and general enough to accomodate many lines of research that are related to the characterization and evaluation of management policies in the fields of natural resources, exhaustible or renewable.

Many methods have been advocated in the literature to be suitable in deriving appropriate management policies. All share what appears to us a fundamental ambiguity: Resulting policies are necessarily time dependant whereas functional equations that are assumed to be in the background of the analytical process are so specified that they cannot but give rise to stationary behaviour.

Following the discussion launched by Georgescu-Roegen (1971), it should be clear that an undated production function describes a stationary process. It is the concept of stationarity which allows us to relate the flows of 'y' and 'a' without paying any regard to the physical details of the process, and without mentioning any stock variable which must exist so as to enable the smooth transformation of 'a' to 'y'. Moreover, Georgescu-Roegen has shown that modifying the production function by adding a time dimension, as for example in the equation $y(t)=f(a(t),R(t))$ or $y(t+\tau)=f(a(t),R(t))$ for a given τ, is also unacceptable. Claiming that these formulations are simplistic in their neglect of fundamental physical laws, he rejects them one by one in favour of a functional relation that degenerates into a simple function like that specified above, only under steady state conditions. Similar criticism applies to any other function mentioned in the current model, as all functions seek to delineate elementary physicl processes.

The arguement becomes more relevant when the environmental production function $F(R,r)$ is traced and assessed. This is so because the assumption that $F(R,r)$ exists amounts to an implicit proposition that $\frac{F(R,r)}{R}$ exists. The latter quantity is an intensive variable of the modeled system, that cannot emerge unless the system is in a state of equilibrium or in a steady state. This means that R can be kept constant only if positive or negative surplus denoted $\frac{dR}{dt}$ crosses the system's boundaries rather than remains in the system. Therefore either the steady state interpretation and consequently the simplistic notation used here, or the assumption that the system is controllable, should be rejected. As a consequence we cannot refrain from pointing out that as far as forces in a closed and equilibrated system are concerned, any realistic exposé of the system cannot disregard the condition $F(R,r)=0$. A similar approach has been mentioned rather incidentally by Smith (1968), but its principle proponent is Boulding, who has strongly asserted (Boulding 1966) that discussions concerning policies of environmental management have to be restricted to closed systems.

Otherwise, models have to explicitly specify the origins and destinations of all net flows. This is, practically speaking, an impossible task since the number of interconnections and interdependancies among different flows very quickly approaches infinity.

POLICIES FOR RESOURCE MANAGEMENT

In view of the criticism presented above, it is proposed to tackle the problem of selecting appropriate policies for resource management in stages. One factor which may affect our choice of effective policy tremendously is the type of state the system experiences. When a system is equilibrated, but for some reason it is desired to change its state to another equilibrium state, the kind of problems faced by the manager are entirely different from those which must be overcome when the system is not equlibrated and heads toward uncontrolled environmental disruption.

In the first case, some kind of optimization procedure should be adopted as the system is stabilized and costs and benefits can be evaluated and balanced. An appropriate way to accomplish this evaluation may be to consider the maximization of the individualistic and stationary utility function $u(c,a,R)$, subject to the constraints $c=y=f(a,R)$, and $r=g(a,R)$. Solving this problem enables us to derive the optimal actions taken by individuals when they face any environmental state R. Denoting these functions by $a=\bar{a}(R)$, $c=\bar{c}(R)$, and $r=\bar{r}(R)$, we realize that if individuals really optimize their actions and the system stays in equilibrium as assumed, this equilibrium state $\bar{R}$ must be determined from the equation $F(R,\bar{r}(R))=0$.

It is well known that not all social costs and benefits have been introduced into the calculation as performed above. A social and environmental planner will, due to external costs and benefits, find a more rewarding state R^* that is derived from the maximization of $u(c,a,R)$, given the constraints $c=y=f(a,R)$, $r=g(a,R)$ and $F(R,r)=0$. A sound management policy should instruct us on how to accomplish a move that is designed to transfer the system from $\bar{R}$ to R^*.

The proposal expounded here cannot over-emphasize how important it is to trace a completely defined path which links the two states. This path contains equilibrium states and nothing but equilibrium states. Tracing them carefully means that the system is undergoing a quasistatic process during which the system is always equilibrated with its environment. Besides unlimited time, a quasistatic process that is carried out successfully requires that any intensive variables of the state of the system will be effectively controlled by its external referrent.

Some policies may employ simple systems of taxes so as to accomplish the move from $\bar{R}$ to R^*. Two tax schemes immediately emerge: The first solves the problem by addressing the principle of income taxation, whereas the second chooses to introduce pollution taxes. Unfortunately the current model is incapable of singling out a favourable scheme, though it is believed that pollution taxes, in pointing right at the heart of the problem, better clarify the original intention of the planner to enhance environmental quality.

To derive appropriate management programs we return to the set of equations of the model, and maximize the function $u(c,a,R)$ subject to $c=(1-\alpha)\ f(a,R)\ +s$.
The actions chosen by individuals, $a=\hat{a}(R,\alpha,s)$, $c=\hat{c}\ (R,\alpha,s)$ and $y=\hat{y}(R,\alpha,s)$ and $r=\hat{r}(R,\alpha,s)$, are then balanced so that the right controls can be obtained. In other words, the equations $\hat{y}(R,\alpha,s)=s$ and $F(R,\hat{r}(R,\alpha,s))=0$ are solved to produce a tax $\alpha=\hat{\alpha}(R)$ and transfer $s=\hat{s}(R)$ schemes that should prevail when the system experiences the state R and in order that this state be really experienced.

A similar procedure is employed in deriving the second scheme. In that case $u(c,a,R)$ is maximised subject to $c=f(a,R)-\beta g(a,R)+q$. Then optimal actions are determined and introduced into the model equations to balance the system. Finally tax $\beta=\beta'(R)$ and transfer $q=q'(R)$ schedules are identified. Using these schedules, the planner, in principle, can direct the system to R^*, as he should be confident in his ability to completely control the system. That is of course not the case when taxes and transfers are statically fixed at some level such as $\hat{\alpha}(R^*)$, $\hat{s}(R^*)$, or $\beta'(R^*)$, $q'(R^*)$. In case the system is open, the net surplus $\frac{dR}{dt}$ may differ from zero, but it then has to cross the system's boundaries quasistatically. Hence it is not considered good economics to keep the assumption that closed and open systems do not differ in their qualities and that both share the same social costs, measured here in terms of labor requirements and the change in the state of the environment.

Before concluding the discussion on quasistatic management policies, the reader is reminded that these policies are applicable when the move from one equilibrium state to another is of one's own concern. It is frequently argued that we should not be bothered by the exact attributes of equilibrium states and that we should not compare them. Rather we have to be concerned with the fact that the system is not equilibrated and that no state of equilibrium is foreseen. That the system is not stable means more often than not that the environment is disrupted, and that reversing this trend is not economically possible.

There is no doubt that a quasistatic process cannot be started when the system is not stabilized. More than that, since an unstable system is surrounded with uncertainty and costs and benefits cannot be unequivocally measured, as is typical in any disequilibrium state, a crucial step towards stabilization of the system may have to be taken. This is achieved by setting somewhat stringent standards and regulations whose economic justification and reason cannot be clearly displayed at the time of their implementation, but whose economic vitality can be inspected later, when the system reaches some degree of stability.

It is clear that, practically speaking, quasistatic processes cannot be realistic since they cannot be completed in a finite time. Though the argument cannot be used to undervalue their importance, we are urged thereby to pay some attention to their possible modifications so as to be able to improve the chances of their implementation. Trade offs are quite obvious already at a first glance. The more rapidly we change any external referrent of an intensive variable, the more the system is disturbed, and the less its path is amenable to rigorous description and economic assessment. Actually the system is disequilibrated most of the time and costs and benefits cannot be estimated, other than in a very general and crude manner. In addition to this, the distribution of costs, benefits, risks, etc. changes continuously and hence constitutes a cause for social unrest.

Sometimes it is said that foresaking the quasistatic process altogether may benefit the whole population in that it approaches its highest obtainable utility in a finite time. This argument is wrong on two grounds. Firstly, assuming that benefits and costs could be measured along a realistic process, it would still not mean that the loss experienced along a realistic process is more than compensated later. This argument gains actuality when it is recalled that along a disequilibrated path costs cannot be evaluated in the same way as they are evaluated in a state of equilibrium or along a quasistatic path. Nevertheless it is instructive to launch programs to estimate costs and benefits in a crude and approximate way using methods that have demonstrated their suitability in assessing uncertain and risky events. Doing so we may come up with figures that are made commensurate and with two distinct paths that are made comparable. Secondly, it is doubtful whether in real systems the finally desired equilibrium is at all achievable. If the system is fully defined and constrained it may reach its equilibrium state; however it is well known that real systems tend to be underdefined and hence underconstrained so that they do not reach any equilibrium state once they are compelled to change.

Consequently, it is concluded that policies recommended in this article can easily stand possible criticisms as have been stated above. This is due to the fact that realistic environmental policies cannot be successful in achieving their goals if, during the period of change, they concentrate on using only one type of means. Our analysis shows that economic reasoning should be a principle tool in directing a change of stabilized systems. The same reasoning is however groundless when the system is not equilibrated, and costs and benefits as system functions do not exist. We have reasoned that in the first instance best results will be obtained by employing measures like taxes and charges, that have been advocated in the economic literature; but that in the second instance ad hoc changing of constraints and standards better serves our purpose. It is not surprising that policies should be devised so that, in real world situations, economic and non-economic measures together coexist and share the role of leading a partially controlled system to its final state.

REFERENCES

Baumol W.J. and W.E. Oates,(1975) The Theory of Environmental Policy. Prentice-Hall, Englewood Cliffs, New Jersey.

Boulding, K. (1966) The economics of Coming Spaceship Earth (p.3-14) in Environmental Quality in a Growing Economy, Jarrett H. (Ed.) The Johns Hopkins University Press, Baltimore.

Brown G. Jr. (1974) An Optional program for managing common property resources with congestion externalities. J.Pol. Econ. 82,163-173.

Clark C.W. (1973). Profit maximization and the extinction of animal species. J. Pol. Econ. 81,950-961.

Fisher A.C. and F.M. Peterson (1976). The environment in economics: a survey. J.Econ. Lit. 14,1-33.

Georgescu-Roegen, N. (1971). The Entropy Law and the Economic Process. Harvard University Press. Cambridge, Mass.

Neher P.A. (1974) Notes on the Volter Quadratic Fishery. J. Econ. Theory 8,39-49.

Plourde,C.G. (1970) A simple model of replenishible natural resource exploitation. Amer. Econ. Rev. 60,518-522.

Plourde C. (1979) Diagramatic representation of the exploitation of replenishable natural resources; dynamic iterations. J. Env. Econ. and Manag. 6,119-126.

Smith V.L. (1968). Economics of production from natural resources. Amer. Econ. Rev. 58,409-431.

Smith V.L. (1977) Control theory applied to natural and environmental resources: an exposition. J. Env. Econ. and Manag. 4,1-24.

Developments in Arid Zone Ecology and Environmental Quality
Edited by H. Shuval

17

A DESIGN APPROACH TO THE SOLUTION OF ECOLOGICAL PROBLEMS

I. Nemtsov

Design Research Department, Israel Centre for Packaging and Industrial Design, Tel-Aviv, Israel

ABSTRACT

Design activities in recent years have grown beyond the narrow confines of "industrial product design" and have penetrated deeply into many spheres of the technical and intellectual life of man.

This is so because of the basic principles of design, which can be defined as "esthetic pragmatism" - a term reflecting the organic integrity of the fundamental laws of technology, esthetics, and the procedures of scientific analysis.

These factors have converted design into an effective tool for the improvement of the environment of man into a medium free from the main causes of man-made ecologic hazards - namely, the pollution of the atmosphere and water resources, the destruction of animal and plant life.

Design can be used in a number of ways for the solution of major ecologic problems.

One of these is the establishment of an institute for the design solution of ecologic problems; alternatively, this task could be entrusted to an existing institute operating in a related field.

Standard samples of natural materials (soil, air, water, plants) from various ecologic regions of the country can be kept in the laboratories of such institutes. Geochemical, biochemical, hydrochemical, petrographical, spectral and other analytical procedures can be used for detecting any disturbances in these materials and determining their causes, dynamics, etc.

Side by side with this work the institute can develop ideas and recommendations for the prevention and elimination of such disturbances.

The design approach to the solution of ecologic problems will provide a totally new foundation for the development of agriculture and contribute a great deal to industry and education. All this will amount to a major advance toward the improvement of the environment of man in Israel and our geographic region as a whole.

Developments in Arid Zone Ecology and Environmental Quality
Edited by H. Shuval

18

ENVIRONMENTAL VARIABILITY IN A SALINE DESERT STREAM, AND SOME IMPLICATIONS FOR AQUATIC LIFE

Elizabeth Anne Colburn[1]

Department of Zoology, University of Wisconsin, Madison, WI 53706 USA

Introduction

Salt Creek, Death Valley, California, USA, is a small desert stream that provides a severe physical environment for aquatic life. The stream lies entirely below sea level and, as its name implies, has a high concentration of dissolved solids. Previous studies (R.R. Miller 1943, LaBounty 1968, G.A. Miller 1975) reported the presence of a salinity gradient in the stream, with low salinity of 10‰ at the source gradually increasing with distance downstream. Data on the extent of the gradient and the range of salinities present are variable. Information provided by the U.S. National Park Service, Death Valley National Monument, shows that Salt Creek is subject to flash flooding that can greatly alter stream flow patterns, usually experiences seasonal desiccation along much of its length, and undergoes large fluctuations in temperature and salt content throughout the course of a year. Clearly, organisms living in the stream must possess adaptations that allow them to deal with a wide variety of physical conditions not ordinarily encountered by their freshwater relatives.

In this paper I examine some physical and chemical characteristics of Salt Creek and discuss the implications of some of these for the survival of aquatic organisms in the stream.

Methods

Field studies were conducted at Salt Creek in January and June, 1976, January and April, 1977, January through May, 1978, October, 1978, through May, 1979, and January and April 1980. Physical and chemical factors measured included air and water temperature, water flow, specific conductance, dissolved oxygen, hardness, alkalinity, pH, chloride, sulfate, sodium, potassium, calcium, and magnesium. I determined the composition of the aquatic community at Salt Creek by dip netting and light trapping. Details of field procedures and equipment are presented in Colburn (1980).

Results

1. Source. Salt Creek lies 55 to 76 m below sea level in Death Valley, Inyo County, California, USA (SW 1/4 of Sec. 31, T. 15.S., R.46.E. through Sec. 5, T.16.S., R.46.E.) (Fig. 1). It

[1]Present Address: Biology Department, Williams College, Williamstown, MA 01267 USA.

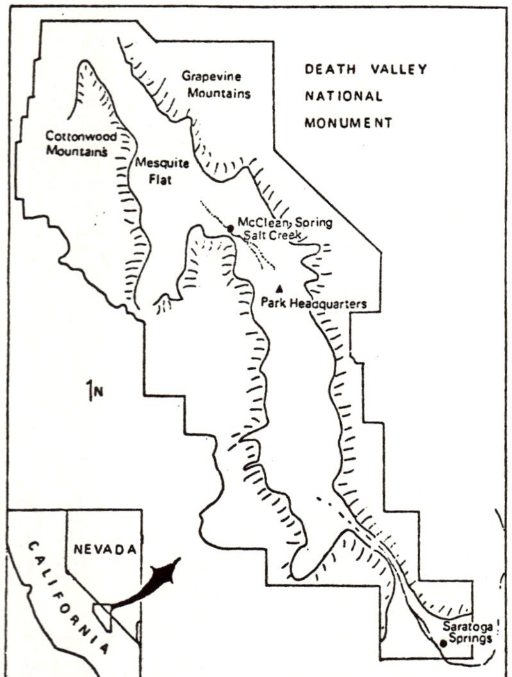

Figure 1. Map of Death Valley National Monument showing location of Salt Creek.

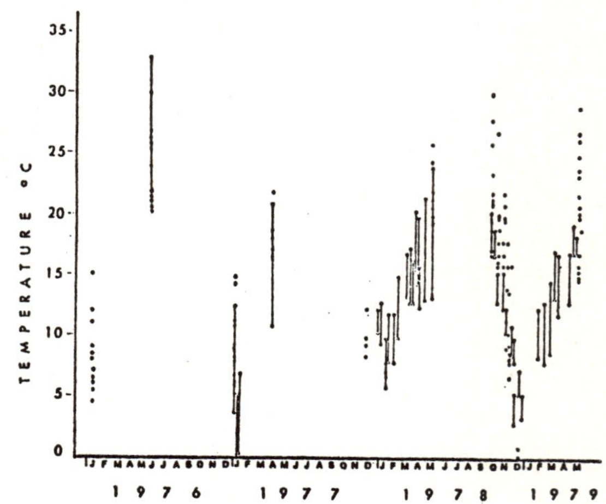

Figure 2. Temperature variations in Salt Creek. • — measurements in individual pools; •—• – 24 hours in a single pool.

receives groundwater from Mesquite Flat, a 260 km^2 expanse of alluvial plain between the Cottonwood and Grapevine Mountains. Permanent water reaches the surface at an uplifted shale bed at McClean Spring and flows SE for approximately 3 km before disappearing into the Death Valley salt pan (Miller 1975). Base discharge at McClean Spring varies seasonally from a summer low of 0.002-0.006 m^3/sec to a winter high of 0.012-0.018 m^3/sec (Lamb and Downing 1979). Mean daily discharge increases for one or two days after rains and can exceed 1 m^3/sec; it then returns to the normal seasonal level (Lamb and Downing 1979). Seasonal variations in flow at the downstream end of Salt Creek are more pronounced than at the source, with peak flows in midwinter and with flow ceasing entirely in sections in summer.

2. Temperature. Salt Creek experiences both daily and seasonal fluctuations in temperature (Fig. 2). Daily variations in fall and winter average 2.4 ± 0.6° C while summer variations exceed 10° C. Seasonally temperatures range from 0° C in winter to over 30° C in summer.

3. Salinity. Vast electrolyte deposits in the aquifer contribute high levels of dissolved solids to Salt Creek. In late spring and early fall, salinities in the stream range from 10-11‰ at McClean Spring to 22-23‰ in lower Salt Creek. This salinity gradient increases in summer, with salinities in the downstream end of the stream increasing to above 40‰ (Fig. 3a). However, in winter the salinity gradient disappears gradually as flows increase and evaporation and transpiration decrease (Fig. 3b).

As the gradient lessens in early winter, the salinity of lower Salt Creek approaches that of McClean Spring, 10-12‰ (Fig. 3b). If winter rains contribute to further increases in discharge, salinities in the stream rise. For example, the salinity of Salt Creek increased steadily from January through April, 1978, as a result of heavy rains throughout the winter (90.7 mm rain November 1977 through March 1978) (Fig. 3d). In April, 1979, after less rain (31.2 mm, November 1978 through March 1979), salinities were 15‰ as opposed to 18‰ in April 1978. Lamb and Downing (1979) report similar increases in total dissolved solids of McClean Spring at periods of high flow: in February, 1976, two weeks after rains produced a daily mean discharge of 1.19 m^3/sec (maximum discharge 10.3 m^3/sec within this period), Salt Creek at McClean Spring had a total dissolved solid content of 21.4 ‰.

The mechanism by which increased water flow leads to a rise in salinity is obscure. A possible explanation is that as water levels rise in the aquifer, the flow passes through electrolyte beds that lie above deeper deposits containing lower levels of salts.

As discharge decreases and temperatures rise in early summer, the salinity of McClean Spring again falls. For example, in June, 1975, with flow at minimum discharge of 0.002 m^3/sec, the total salt content at the spring was 11.3‰ (Lamb and Downing 1979). Concurrent with this decrease in salinity at the spring is an increase in dissolved solids at the downstream end of Salt Creek. Thus, while salinities change uniformly throughout the stream in winter, they change in opposite directions in the upper and lower stretches of Salt Creek in summer (Fig. 4).

Figure 3. Seasonal variations in salinity, measured as specific conductance, at Salt Creek.

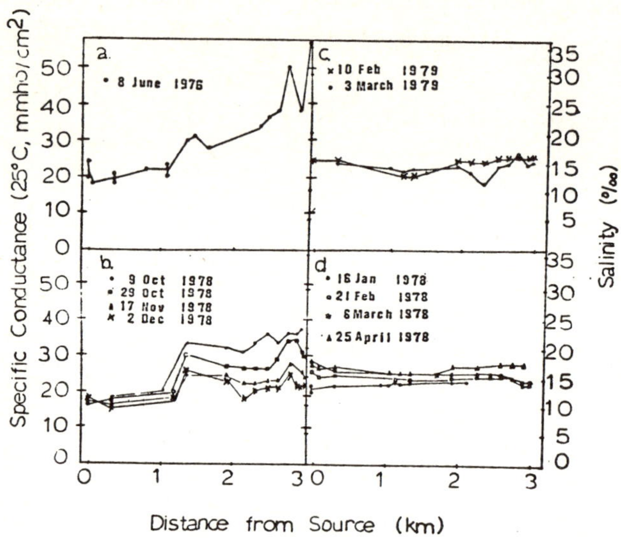

Figure 4. Patterns of salinity variation in upper and lower Salt Creek in years of high and low rainfall.

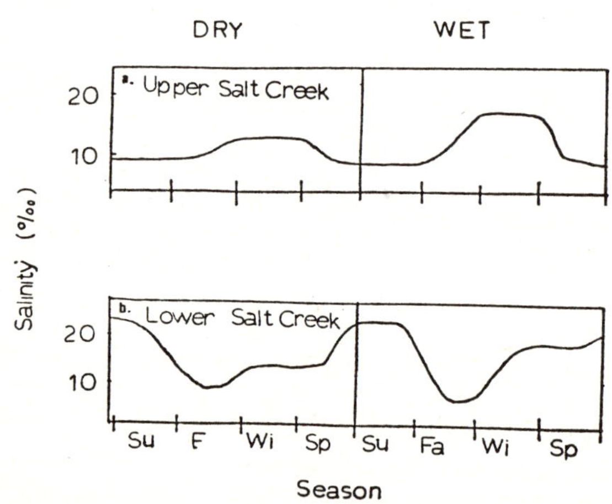

While salinities in most of Salt Creek range between 10 and 25°/oo, isolated pools exist in which salinities exceed 45°/oo in summer and do not fall below 30°/oo even in mid-winter. These pools are deep and commonly are stratified, with the deep saline water overlain by a layer of relatively dilute surface water. Often the dilute surface water connects to the main body of Salt Creek via a shallow channel.

Salt Creek, then, while saline, is not uniformly so in either time or space. At low flow, dissolved solid levels at McClean Spring are also low, 10°/oo. The salinity of the stream ranges between this base level and about 22°/oo. The extent of seasonal variation depends both on rainfall and on the section of the stream. Furthermore, deep pools of highly saline water exist in close proximity to the main stream year round. This spatial and temporal instability of the salinity levels in Salt Creek requires that organisms living in the stream either possess a broad physiological tolerance of such changes or have other mechanisms with which to counteract them.

4. Water chemistry. Sodium and chloride are the major ions in Salt Creek, contributing over 75% of the dissooved solids by weight (Table 1). High levels of magnesium, a metal toxic to most aquatic organisms and a major factor influencing animal distributions in inland salt waters (Bradley and Phillips 1977, Scudder 1969), occur in the stream.

5. Current. With the exception of a few cascades near McClean Spring and some short sections of stream channel along the lower third of Salt Creek, water currents in the stream are slow, 1-2 cm/sec or less. Surface currents in channels in lower Salt Creek average 11-20 cm/sec. The aquatic habitat of Salt Creek is thus, for the most part, more like that of a marsh or pond than that of a typical small stream.

6. Biological community. Salt Creek supports a diverse community of terrestrial and aquatic organisms. The plant community is the least diverse, with plant distributions restricted by groundwater salinity. The vascular plants are represented by only six species in the stream and surrounding marsh: in the upper kilometer of the stream, reed (*Phragmites communis*) and bullrush (*Scirpus olneyi*), both of which tolerate groundwater salinities up to 11°/oo (Hunt 1966), are common; they are accompanied by rush (*Juncus cooperi*) and salt grass (*Distichlis spicata*), which, tolerating salinities over 30°/oo (Hunt 1966) extend more than two kilometers downstream from McClean Spring; pickleweed (*Allenrolfea occidentalis*) can grow in solutions as saline as 60°/oo (Hunt 1966) and extends the length of the stream; and *Ruppia maritima* occurs in some of the larger pools. *Ceratophyllum demersum* was found in Salt Creek by Miller (1943) and Hunt (1966) but was not observed during my study.

The only aquatic vertebrate inhabiting Salt Creek is the pupfish *Cyprinodon salinus*. This small fish, which is endemic to Salt Creek, is found in large numbers in pools the length of Salt Creek, but is particularly abundant along the upper third of the stream.

There is no plankton in Salt Creek, but benthic copepods and ostracods form dense populations on some pool bottoms. Myriad

Table 1. Chemical composition of water from Salt Creek.

	Miller (1943)	LaBounty (1968)	Lamb and Downing (1979)		This study (1976-1979)
			McClean Spring	Lower Creek	
pH	8.6				8.0 - 9.6
Salinity (°/..)	23.6	11 - 22.7	11.3	21.4	15 - 18
Conductivity (mmho/cm^2, 25° C)	33.7		16.5	31.3	22 - 25
Sodium (mEq/1)	341.5	143.5 - 293	160.9	291.3	204 - 267
Potassium (mEq/1)	12.8	7.4 - 17	7.7	13.3	13 - 17
Calcium (mEq/1)	5.7	3.8 - 11.5	4.9	7	8 - 11
Magnesium (mEq/1)	30.4	8.5 - 35.2	17.3	34.6	37 - 58
Chloride (mEq/1)	290.5	39.5 - 273.4		257	206 - 345
Sulfate (mEq/1)	77.3	45. - 80.	42	84	16 - 20
Carbonate/bicarbonate (mEq/1)	12.9	10.3 - 14.4	11	10.9	
Nitrate (mEq/1)		0.3 - 0.6			
Arsenate (ppm)	0.4				
Boron (ppm)	38.8		22	3.5	
Fluoride (ppm)		1			
Silica (ppm)	29				
Phenylphthalein Alkalinity (ppm as $CaCO_3$)					0 - 70
Methyl Orange Alkalinity (ppm as $CaCO_3$)					502 - 572
Dissolved Oxygen (ppm)					4.6 - 11.4
Hardness (ppm as $CaCO_3$)					1686 -2331

dipteran larvae live in the mud of the pools, as well. Four species of corixids, numerous beetles, several odonates, a notonectid, and a limnephilid caddisfly, are common (Table 2).

Discussion

From an ecological perspective, the variability of Salt Creek is of interest in terms of its implications for physiological tolerances and life history strategies of the organisms inhabiting the stream. At its least extreme, in winter, Salt Creek is a severe physical environment, yet, a variety of organisms occur there. These organisms must, at a minimum, be able to tolerate relatively high salinities and high concentrations of potentially toxic ions such as magnesium and sulfate. In summer, pronounced spatial and short-term temporal variations within the stream increase the stresses imposed upon aquatic organisms. Summer water temperatures remain far below the ground temperatures, which commonly exceed 90° C, but they fluctuate over 10° C in a 24 hour period and can exacerbate physiological stresses of increasing salinities. Desiccation of much of the lower two thirds of the stream habitat in summer poses further problems for strictly aquatic organisms that lack mechanisms for dispersal into pools that retain water year round.

Some of the fluctuations in Salt Creek are predictable (e.g. seasonal variations in temperature, salinity, flow), while others are not (flash flooding, sudden salinity changes after rains or from one location to another). Some of the stream's inhabitants are able to tolerate the entire range of conditions and are widely distributed within the stream throughout the year. Others have more restricted distributions in space and/or time and often have life history patterns that allow avoidance of some of the predictable seasonal extremes.

The corixids *Trichocorixa reticulata* and *T. verticalis saltoni* are typical of the widely distributed, eurytolerant animals in Salt Creek. These waterboatmen occur in both the mainstream and hypersaline pools in the stream. This distribution implies well-developed mechanisms for physiological regulation. *T. reticulata* has been shown to regulate its osmotic and ionic composition over a salinity range from distilled water to 100°/₀₀ (Jang and Tullis 1980), and *T. verticalis interiores* likewise osmoregulates in hypersaline waters (Tones and Hammer 1975). At Salt Creek, gravid adult females and nymphs of all instars can be found at all times of year, so it appears that these animals tolerate not only extremes of salinity but also high temperatures. I have found large numbers of corixids in small, shallow pools in early June and have observed desiccated remains of nymphs in dried mud of former pool bottoms. It seems likely that as pools dry up in the course of the summer the winged adults leave and migrate to deeper water. Indeed, I have collected adults at light traps in late spring and early fall.

The pupfish *Cyprinodon salinus* is another full-time resident of Salt Creek. It also osmoregulates over a wide range of salinities (Stuenkel and Hillyard 1980), but unlike the corixids it has no

Table 2. Aquatic insects identified at Salt Creek, 1976 - 1979.

- Odonata
  - Aeshnidae
    - *Aeshna interrupta*
    - *Anax junius*
  - Libellulidae
    - *Tarnetrum corruptum*
  - Coenagrionidae
    - *Coenagrion resolutum*
    - *Ischnura* sp.
    - *I. perparva*
    - *I. demorsa*
    - *I. ramburii credula*
- Coleoptera
  - Dytiscidae
    - *Eretes sticticus*
    - *Hygrotus* sp.
    - *Deronectes panaminti*
  - Hydrophilidae
    - *Enochrus hamiltoni pyretus*
    - *Paracymus elegans*
    - *P. ellipsis*
    - *Berosus californicus*
  - Hydraenidae
    - *Ochthebius rectus*
  - Ptilidae
- Hemiptera
  - Corixidae
    - *Trichocorixa reticulata*
    - *T. verticalis saltoni*
    - *Corisella inscripta*
  - Notonectidae
    - *Notonecta unifasciata*
- Diptera
  - Chironomidae
  - Ceratopogonidae
  - Tabanidae
  - Culicidae
  - Ephydridae
  - Stratiomyiidae
- Trichoptera
  - Limnephilidae
    - *Limnephilus assimilis*

mechanisms by which to avoid entrapment in drying pools and exposure to high salinities and temperatures. The fish start to reproduce in early spring and by mid-April dense schools can be seen in most pools in Salt Creek. Many fish are trapped in shallow pools as the lower stream dries up in summer, and mortality is high (Miller 1943). In the pupfish, physiological tolerance combines with a very high reproductive potential to permit survival in Salt Creek.

In contrast to the above, caddisfly larvae (Limnephilus assimilis) occur only in the mainstream pools and marshes of Salt Creek. The range of salinities over which this insect regulates osmotic and ionic concentrations is narrower, 0-25°/₀₀, and the temperatures at which larvae can complete development to the adult stage are lower (<20° C) than for the more tolerant species. The caddisfly is able to survive in Salt Creek because its larval development and pupation are completed during the winter months and adults emerge in spring before salinities and temperatures reach lethal levels. The adults spend the summer in a state of reproductive diapause, with sexual maturation delayed until late fall, at which time mating and oviposition take place (Colburn 1980).

Adaptations of other organisms in Salt Creek involve other combinations of distributions, physiological capabilities, dispersal mechanisms, and life history strategies. The patterns of variability of the physical environment of Salt Creek, as well as the absolute extremes, are major factors around which adaptations have developed.

Acknowledgements

This research was supported by the University of Wisconsin Department of Zoology, the University of Wisconsin Graduate School Research Committee grant #100654 to J.J. Magnuson, and National Science Foundation Doctoral Dissertation Improvement Grant #DEB78-01288. I am grateful to the National Park Service, Death Valley National Monument, and to my friends and colleagues at the Laboratory of Limnology, University of Wisconsin, Madison, for their help and encouragement. Special thanks to P.G. Sanchez, N.P.S., Death Valley, and to S. Sharr and J.J. Magnuson, University of Wisconsin.

Literature Cited

Bradley, T.J., and J.R. Phillips. 1977. Regulation of rectal secretion in saline-water mosquito larvae living in waters of diverse ionic composition. *J. Exp. Biol.* 66: 83-96.

Colburn, E.A. 1980. Factors influencing the distribution and abundance of the salt-tolerant caddisfly Limnephilus assimilis in Death Valley. Ph.D. thesis. University of Wisconsin, Madison.

Hunt, C.B. 1966. Plant ecology of Death Valley, California. U.S.G.S. Prof. Paper 509. Washington, D.C.

Jang, E.B., and R.E. Tullis. 1980. Hydromineral regulation in the saline water corixid Trichocorixa reticulata (Hemiptera: Corixidae). *J. Insect Physiol.* 26: 241-244.

LaBounty, J.F. 1968. Some ecological and taxonomic considerations of Death Valley cyprinodonts. M.S. thesis. University of Nevada, Las Vegas.

Lamb, C.E., and D.J. Downing. 1979. Hydrologic data, 1974-1977. Stovepipe Wells area, Death Valley National Monument, Inyo Co., California. U.S.G.S. Open File Report 79-205. Washington, D.C.

Miller, G.A. 1975. Reconnaissance appraisal of the water resources of Death Valley, California and Nevada. U.S.G.S. Water Supply Paper XXXX. Washington, D.C.

Miller, R.R. 1943. Cyprinodon salinus, a new species of fish from Death VAlley, California. *Copeia* 1943(2): 69-78.

Scudder, G.G.E. 1969. The fauna of saline lakes on the Fraser Plateau in British Columbia. *Verh. Intern. Verein. Limnol.* 17: 430-439.

Stuenkel, E.L., and S.D. Hillyard. 1980. Effects of temperature and salinity on gill Na^{+}-K^{+} ATPase activity in the pupfish, Cyprinodon salinus. *Comp. Biochem. Physiol.* 67A: 179-182.

Tones, P.I., and U.T. Hammer. 1975. Osmoregulation in Trichocorixa verticalis interiores (Sailer) (Hemiptera-Corixidae), an inhabitant of Saskatchewan saline lakes, Canada. *Can. J. Zool.* 53: 1207-1212.

Developments in Arid Zone Ecology and Environmental Quality
Edited by H. Shuval

19

A PRELIMINARY STUDY OF THE ALGAL FOOD SELECTIVITY BY THE RABBITFISH SIGANUS RIVULATUS IN THE GULF OF ELAT AND AT THE MEDITERRANEAN COAST OF ISRAEL

Barbro Lundberg

Department of Botany, the Hebrew University of Jerusalem, Israel.

ABSTRACT

By migrating through the Suez Canal and settling in the eastern Mediterranean the herbivorous rabbitfish *Siganus rivulatus* had to adapt to new food resources. The fish feed on benthic algae and seagrasses. The composition of the algal and seagrass vegetation was studied at 3 stations in the Gulf of Elat and at Mikhmoret, the Mediterranean coast of Israel. The composition of the vegetation differed in the two seas. By analyses of contents of stomachs and intestines the composition of the plant diet of fishspecimens captured in the two seas was determined. It could be concluded that the fish diet changed considerably when the fish settled in the new environment. Likewise selection and avoidance of plant species changed. The considerable capacity to adapt to new food resources as well as to high salinities (when migrating through the Suez Canal) and lower winter temperatures indicate that *Siganus rivulatus* is very suitable for mariculture.

In view of the constantly increasing shortage of food resources the interest in large scale mariculture of fish has been intensified. Fishes belonging to the family Siganidae, rabbitfish, are among those suggested for mariculture (Hawaii, 1972). Rabbitfish dwell in shallow waters (Popper and Gundermann, 1975; Tobias, 1976) where they feed on algae and seagrasses (Tsuda and Bryan, 1973; Westernhagen, 1973a, b, 1974; Bryan, 1975; Tobias, 1976; Lundberg and Lipkin, 1979; Lundberg, 1980). Their capability to utilize the resource of marine plants make them valuable objects for mariculture. It is, for instance, conceivable to combine fish culture with algal culture which in the future ought to be the most economical maricultural method.

Siganids are mainly distributed in the tropical parts of the Pacific and Indian oceans. Four of about 20 known species are found in the Red Sea and two of those species, *Siganus rivulatus* and *S. luridus*, have successfully migrated to the Mediterranean Sea through the Suez

Canal. The natural food of siganids has been studied in Guam (Tsuda and Bryan,1973;Bryan,1975;Tobias,1976) in the Philippines (Westernhagen 1973 a,b,1974) and in Israel at the Gulf of Elat,the Red Sea (Lundberg and Lipkin,1979) and at Mikhmoret,the Mediterranean coast, (Lundberg,1980). These studies show that rabbitfish are herbivores. They feed mainly on algae and seagrasses even though the composition of the plant diet varies in different regions. Likewise preferences and avoidances of species differ in different regions.

In the present study the plant food resources available for Siganus rivulatus in its original Red Sea environment will be compared to those found in the colonized Mediterranean environment. The diets grazed by the fish in the two seas will also be compared and related to the food resources available . By comparing the composition of the diets with the composition of the vegetations it will be possible to calculate degree of selection and avoidance of species.

MATERIALS AND METHODS

The diet descriptions are based on stomach-content data from 45 Siganus rivulatus specimens captured at Dahab,El Kura and Shorat el Manqata in the southern basin of the Gulf of Elat (Fig.1). The fish were captured during different seasons during 1972 -1979. These data will be compared to stomach-content data from 307 S. rivulatus specimens of which 5-10 were captured monthly at Mikhmoret during 1972-1974. The total length of the Red Sea specimens varied between 83 and 337 mm whereas the total length of the Mediterranean species varied between 81 and 265 mm. The fish were captured with traps,gillnets,hook and line and by divers using handnets. All the individuals studied had well filled stomachs. Contents of stomachs and intestines were suspended in large Petri dishes. Plant fragments were identified at least to genus by the aid of stereoscope and microscope. Quantitative estimates were made by examining ten random fields and estimating the percentage of each alga per field. Mean-percentage for ten fields was calculated and designated the relative abundance (RA),i.e. the amount of the respective alga in the food biomass of the said fish. Mean relative abundance (MRA) values for the different plant species were calculated for the fish samples from the both seas. Vegetation samples were taken with a metal frame (28 x 22 cm).

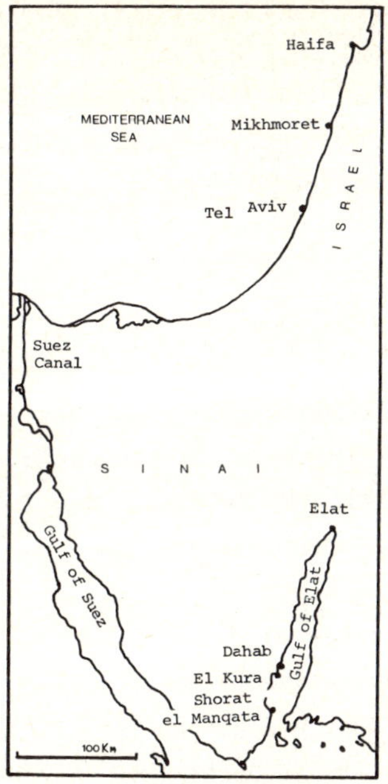

FIG. 1. Sinai and Israel, locations mentioned in text where algae were sampled and fish captured.

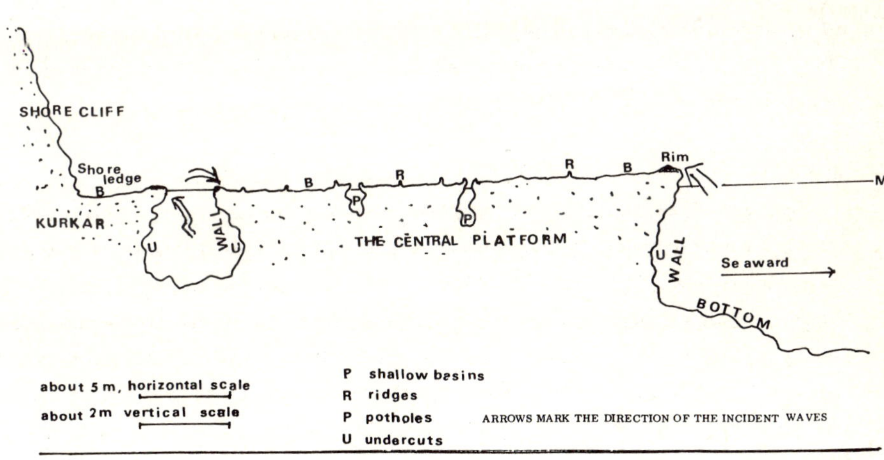

FIG. 2. Schematic transect of the platform site at Mikhmoret, the Mediterranean coast of Israel (Freely after Safriel, 1975)

In the Gulf of Elat stations the frame was placed at intervals along transects perpendicular to the coastline. The transects ran from the sea level down to depths of 6 meter on reef flats and fore reefs and on sandflats at Dahab and El Kura (92 samples) and on a shallow reef flat in Shorat el Manqata (15 samples). Vegetation samples were taken during different seasons during 1972 - 1979. In Mikhmoret monthly samples were taken during 1972-1974 (585 samples). About one third of these were taken on the horizontal platforms at about sea level ,on third from the vertical walls and one third from the bottoms at depths down to 6 meters (Fig. 2 ,cf Lundberg,1980). The relative cover of each plant species in the rectangle was visually estimated in the water as a percentage of the total plant cover of the rectangle. The plants were then collected by scraping the substrate and taken to the laboratory for analysis. Field estimates of the relative cover had to be corrected to include tiny epiphytes and other small algae overlooked in the sea. The corrected values are referred to as RC's. Mean relative covers (MRC's) for the plants were calculated. Hence MRA's are close estimates of the percentages of the different algae in the fish diet and MRC's of those in the vegetation. Ivlev (1961) determined the electivity index (E) of a certain food species as

$$E = \frac{MRA - MRC}{MRA + MRC} \quad .$$

The electivity index reflects the relative feeding preference of a given fish for a given plant species in nature. Electivity index of -1, 0 and + 1 respectively indicate complete avoidance,consumtion in proportion to occurrence and maximal selection. Selections and avoidances of Siganus rivulatus in the Gulf of Elat and in Mikhmoret were determined on the basis of the electivity index.

RESULTS

The reef flats which were sampled in the southern Gulf of Elat sloped gently seawards towards the vertically steepening fore reefs.The depth at the seaward rims reach about 1,5 meter at high tide being only about half a meter at low tide at which large areas close to shore are almost dry. The intertidal roch surface is rough with corals and rocks sticking up between deeper holes and crevices. The vegetation on the reef flats changed gradually from the shallow to the deeper areas . Ulva sp., Enteromorpha clathrata and Laurencia papillosa were for instance most abundant close to shore whereas Sphacelaria spp. covered stones close to the fore reef and tiny species,like Leveillea jungermannoides, were found as epiphytes on algae growing in the deeper areas. During high tide schools of *S. rivulatus* invade the intertidal rocks and feed (Fishelson, 1977). They may reach as close to shore as the tide allows.

The vegetation was dominated by *Laurencia* spp., including several species,like the soft snd tiny *L.obtusa* mostly growing at the deeper areas. The coarse,brown *Cystoseira myrica* and the large,fleshy *Caulerpa racemosa* are also very abundant. These three genera make up more than 40 % of the algal vegetation . Other high cover species include the red calcareous *Jania adherens*, the green *Dictyosphaeria cavernosa* and the brown *Sargassum dentifolium*,*Hydroclathrus clathratus* and *Padina* spp. In addition a large number of soft fleshy or filamentous algae were found in small amounts,e.g. *Hypnea* spp., *Polysiphonia* spp.,*Champia irregularis* and *Cladophora* spp. The mean vegetation cover was about 65 % of the area.

The fish diet was also dominated by *Laurencia* spp. ,which even was slightly selected (Fig.3). The remaining high cover algae were less abundant in the diet than in the vegetation. Some of them,however, like *Caulerpa racemosa*,*Sargassum dentifolium*,*Cystoseira myrica* and *Hydroclathrus clathratus* contributed considerably to the diet although avoided by the fish. They had electivity indices between - 0.11 and - 0.53 (Fig.3). Others like the calcareous *Jania adherens*,*Dictyosphaeria cavernosa* and *Padina* spp. were strongly avoided (electivity indices between -0.59 and - 0 .99). On the other hand, the selected algae were rare in the vegetation. *Hypnea* spp. with E + 0. 81 contributed considerably to the diet but also blue green algae,*Champia irregularis* ,*Digenia simplex* and the seagrass *Halodule uninervis*,which all were selected,were important components of the diet. Selected ,though less important diet components, were also *Sphacelaria* spp.,*Leveillea jungermannoides*,*Enteromorpha clathrata* and *Cladophora* spp. (Fig.3). Most of the selected species were soft and fleshy or filamentous. The filamentous *Polysiphonia* spp.were avoided as were the leaflike *Ulva* sp. and *Dictyota* spp.

In Mikhmoret the habitat differs greatly from those in the Gulf of Elat. Instead of the gently sloping reef flats with its gradually changing vegetation which can be grazed by fish during high tides the Mediterranean locality is characterized of three well distinguished habitats with different ,however,overlapping algal vegetation which never is covered by high tide but which may be exposed to relatively high water levels only when the tide coincides with high waves (Fig.2). At about sea level subhorizontal platforms extend. Species,like *Laurencia papillosa* and *Ulva lactuca* were abundant on platforms and scarce in

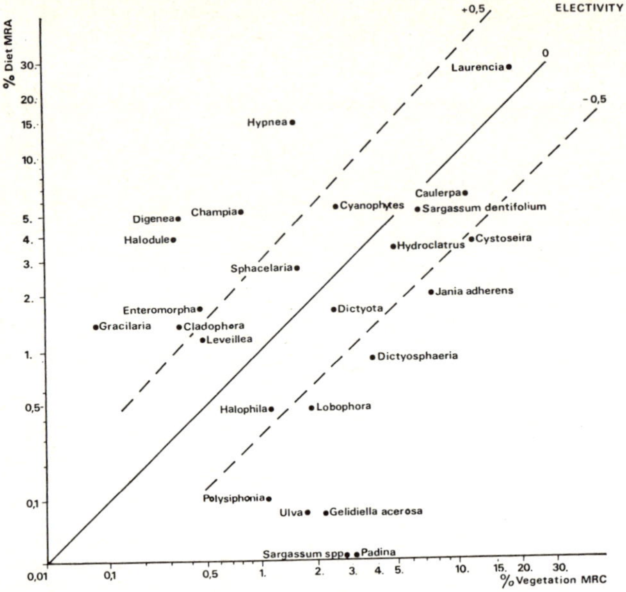

FIG. 3. Electivity for some species common in the vegetation and in the diet of Siganus rivulatus in the Gulf of Elat. Points to the left of the solid diagonal line represent selected species while points to the right represent avoided species.

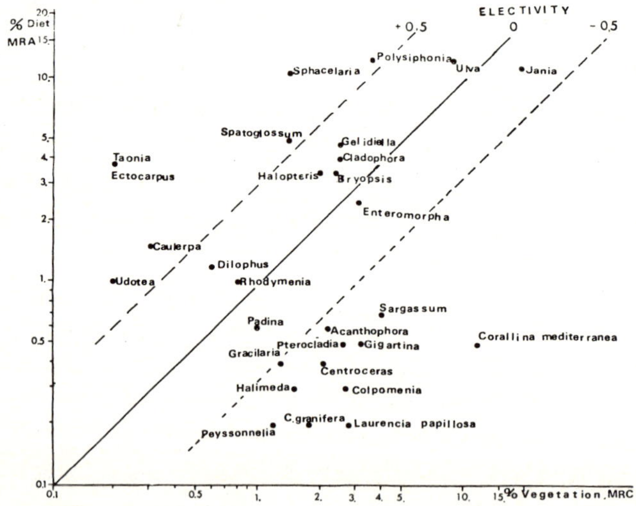

FIG. 4. Electivity for some species common in the vegetation and in the diet of Siganus rivulatus at Mikhmoret, the Mediterranean coast of Israel. Points to the left of the solid diagonal line represent selected species while points to the right represent avoided species.

the other habitats. From the rims of the platforms down to about 2 meter depth subvertical walls steepen. Characteristic wall algae were Pterocladia capillacea and Rhodymenia pseudopalmata. The platforms are surrounded by deeper areas depth between 2 and 6 meter :the bottoms. Several small filamentous algae,like Sphacelaria spp.,Polysiphonia spp. and Gelidiella tenuissima were more abundant on bottoms than in the other habitats The vegetation in Mikhmoret was strongly dominated by calcareous red algae like Jania rubens,common in all the three habitats and Corallina mediterranea common at walls and bottoms. Ulva lactuca had also high cover as had Sargassum vulgare,Polysiphonia spp. and Laurencia pappilosa ,however,less so. Most species were found only in small amounts. Most common among them were Pterocladia capillacea,Gigartina acicularis,Colpomenia sinuosa,Bryopsis spp, and Gelidiella tenuissima (Fig.4).

The diet grazed by Siganus rivulatus in Mikhmoret was greatly dominated by the strongly selected Polysiphonia spp. and Sphacelaria spp.,the selected Ulva lactuca and the common Jania rubens, which ,however ,was grazed in much smaller proportion than its occurrence and therefore was avoided(Fig.4). Important diet algae were also the filamentous Ectocarpus spp.,Cladophora spp.,Bryopsis spp,Halopteris scoparia and Enteromorpha spp.which except for Enteromorpha were selected , and the leaflike Spatoglossum asperum and Taonia atomaria, which both were strongly selected.

In the Gulf of Elat most of the reef flat algal habitat was accessible for the rabbitfish:atleast during high tide. From the vegetation where Laurencia spp. ,Caulerpa racemosa and large mostly coarse brown algae were the main components the fish chose to graze a diet where Laurencia spp. and Hypnea spp. where most abundant and the rare Champia irregularis and other soft species were selected whereas the brown algae were avoided. In Mikhmoret where the calcareous red Jania rubens and Corallina mediterranea dominated thin filamentous species like Sphacelaria and Polysiphonia were most abundant in the diet together with Ulva lactuca which was abundant only on the platforms.(Fig 3 and 4).

DISCUSSION

Siganus rivulatus migrated early to the Mediterranean. It was recorded as early as 1902 in the Suez Canal by Tillier (in Por,1978). The early migration implied resisting high salinities since the salinity in the Canal was greater than 5.5.% at that time (Por,1978). The fish was recorded in the eastern Mediterranean at Haifa as early as 1927 (Steinitz,

1927). In terms of area extension in the Mediterranean the rabbitfish is now regarded as one of the most successful migrants from the Red Sea (Por,1978). The settling in the Mediterranean implied also adaption to lower winter temperatures which might have slowed down the growth of the fish during the winter. This may explain the smaller size of the Mediterranean specimens (Ben Tuvia pers.comm.) When colonizing platform shores like the one in Mikhmoret which are common along the coasts of Israel and Lebanon (George,1972) Siganus rivulatus found a habitat where fleshy algae like Laurencia were scarcer than in the Gulf of Elat and where L.papillosa which is fairly coarse was the most common Laurencia species and that it and most other likeable species were difficult to reach because they grow on the platforms at about sea level. At the easily accessible bottoms and the fairly accessible walls hard calcareous algae dominated. Also the brown algae which were so abundant in the Gulf of Elat habitats were scarcer in the Mediterranean locality. In addition the species occurring in Mikhmoret were harder and probably more difficult to ingest. Sargassum vulgare of Mikhmoret is harder than Sargassum dentifolium and Colpomenia sinuosa is harder than the related Hydroclathrus sinuosa. Lacking the bulk of its diet from the Gulf of Elat S.rivulatus consumed the small filamentous species which grow intertwined with Jania rubens particularly on the bottoms and by taking them also Jania rubens was ingested. However,since these species are relatively scarce,the fish probably had toswim over large areas in order to find them. It is known that S. rivulatus swim fast over large areas (Popper and Gundermann,1975).. The filamentous algae like Polysiphonia and Sphacelaria which also were greatly selected were mostly grazed during the colder seasons (Lundberg,1980). During spring when Ulva lactuca became very abundant though mainly on the platforms the fish consumed large amounts of that alga. This may be interpreted as "switching" as described by Murdoch and Oaten(1975). It is common in nature that animals feed on species when abundant and ignore them when scarce. The authors found that switching has a tendency to occur as a change of habitat when food in the first habitat becomes scarce while food becomes abundant in another habitat. Here it wouldmean that the filamentous algae were overgrazed so that "jumping"towards the platforms at least the rims was a better c choice. It would be possible at least during days with high waves. During the late summers and the autumns the diet included much of the big brown leaflike Spatoglossum asperum which grows at the lower parts of the walls and adjacent parts of the bottoms . The grazing of that alga was interpreted as another example of switching. The consumtion of the scarce filamentous algae and the effort to switch to other habitats may be understood as efforts necessary in order to survive in the new environment.

The great adaptability shown by Siganus rivulatus when adapting to the new habitat and the food resources available there as well as the capacity to adapt to high salinities and low winter temperatures indicate that Siganus rivulatus is a species which would suceed in large scale mariculture.

REFERENCES

Bryan P.G.(1975) Food habits,functional digestive morphology,and assimilation efficiency of the rabbitfish Siganus spinus(Pisces, Siganidae)on Guam. Pac.Sci. 29, 269-277.

Fishelson L. (1977) Sociobiology of feeding behavior of coral fish along the coral reef of the Gulf of Elat (=Gulf of Aqaba),Red Sea. Isr.J. Zool. 26, 114-134.

George C.J. (1972) Notes on breeding and movements of the rabbitfishes Siganus rivulatus (Forsskal) and S. luridus (Ruppel), in the coastal waters of the Lebanon. Ann. Mus. Stn. Nat. Genova. 79, 32-44.

Hawaii University (1972) The Siganid Fish Farming Program; the Siganid Mariculture Group.

Ivlev V. S. (1961) Experimental ecology of the feeding of fishes. Yale Univ. Press, New Haven, Conn. 302pp.

Lundberg B. (1980) Selectivity of food algae by the herbivorous fish Siganus rivulatus in the marine vegetation at Mikhmoret (the Mediterranean coast of Israel). Ph. D. thesis, The Hebrew University of Jerusalem.

Lundberg B. & Lipkin Y. (1979) Natural food of the herbivorous rabbitfish (Siganus spp.) in the northern Red Sea. Bot.Mar. 22, 173-181.

Murdoch W. W. & Oaten A. (1975) Predation and Population stability. In: Advances in Ecological Research (MacFadyen A. ed.) Academic Press, London, N.Y. and San Francisco.

Popper D. & Gundermann N. (1975) Some ecological and behavioural aspects of siganid populations in the Red Sea and Mediterranean coasts of Israel in relation to their suitability for aquaculture. Aquaculture 6, 127-141.

Por F. D. (1978) Lessepsian migration: The influx of Red Sea biota into the Mediterranean by way of the Suez Canal. Springer Verlag, Berlin,Heidelberg and N.Y. 228 pp.

Steinitz W. (1927) Beitrage zur Kenntnis der Kustenfauna Palastinas. I. Publ. Staz. Zool. Napoli 8, 311-353.

Tobias W. J. (1976) Ecology of Siganus argenteus (Pisces:Siganidae) in relation to its mariculture potential on Guam. Techn. Rep. Univ. of Guam Marine Laboratory 29, 58-93.

Tsuda R.T. & Bryan P.G. (1973) Food preferenc of juvenile Siganus rostratus and S. spinus in Guam . Copeia 3; 604-606.

Westernhagen H. von (1973 a) The natural food of the rabbitfish Siganus oramin and S. striolata . Mar.Biol. 22, 367-370.

Westernhagen H. von (1973 b) A preliminary study on the food preferences of Siganus concatenata (Cuvier and Valenciennes). Philipp. Scient. 10, 61-73.

Westernhagen H. von (1974) Food preferences in cultured rabbitfishes (Siganidae). Aquaculture 3, 109-117.

Developments in Arid Zone Ecology and Environmental Quality
Edited by H. Shuval

20
EUTROPHICATION OF THE NAIROBI-ATHI RIVER IN KENYA

Stephen G. Njuguna

Dept. of Botany, University of Nairobi, P.O. Box 30197, Nairobi, Kenya

ABSTRACT

The rapid rate of industrial, agricultural and urban growth in Africa has led to eutrophication and increased deterioration of the water quality in its tropical lotic systems. A study of the physico-chemical and biological conditions of a tropical river receiving agricultural and urban effluents was undertaken between 1976 and 1978 to evaluate the effects and extent of the pollutants.

Conductivity from the source to the mouth ranged from 44-870 μmhos cm^{-1}. The river constituted a highly buffered system with a pH range of 6.9 to 7.6. Other parameters varied as follows: dissolved oxygen (DO) 0.00 - 8.24 mg l^{-1}, biochemical demand (BOD) 0.10-72.0 mg l^{-1}, suspended solids 11-4827 mg l^{-1}, ammonia-nitrogen 0.06-1.70 mg l^{-1}, nitrite-nitrogen 0.02-0.62 mg l^{-1}, total organic nitrogen (kjeldahl) 0.6-14.1 mg l^{-1}, orthophosphate (SRP) 0.06-1.80 mg l^{-1} and total phosphorus 0.07-7.29 mg l^{-1}.

Results of physico-chemical analyses and the examination of the macrofauna and riverine flora describe a river unable to cope with the heavy eutrophication and pollution load from Nairobi City in its upper reaches. Floods and heavy sediment load appear to be important factors affecting the ecology of the middle and lower reaches.

Developments in Arid Zone Ecology and Environmental Quality
Edited by H. Shuval

21
SKELETON SHEDDING IN PERIDINIUM CINCTUM FROM LAKE KINNERET - A UNIQUE PHYTOPLANKTON RESPONSE TO NUTRIENT IMBALANCE

C.M. Criscuolo*, Z. Dubinsky+ and S. Aaronson*

**Biology Department, Queens College, City University of New York, Flushing, New York 11367, and +Department of Life Sciences, Bar-Ilan University, Ramat-Gan, Israël*

ABSTRACT

Many cases of slowing down growth and its subsequent complete cessation in nutrient deficient phytoplankton are known. In Peridinium cinctum from Lake Kinneret (Israel) we found a divergent course of events. The dinoflagellate, Peridinium cinctum fa. westii, grown in a nutritionally balanced medium, showed a parallel increase in cells and thecae with increasing incubation time. However, when grown in the above medium with nitrogen (N) omitted, cells and thecae increased in parallel to the seventh week of incubation after which thecae, but not cells, continued to increase in number to the ninth week. When phosphorus (P) was omitted, cells multiplied to the third week, while the thecae increased steadily; after the third week the cell number declined, although the thecae continued to increase. Omission of both P and N gave results similar to results with P omission alone. We infer from these data that thecal (carbohydrate) synthesis can be uncoupled from cell division during major nutrient deprivation in the presence of adequate light and inorganic carbon. The implications of this phenomenon for organic carbon cycling in natural waters are discussed.

INTRODUCTION

In the course of growing axenic mass cultures of the dinoflagellate, Peridinium cinctum fa. westii, we observed the ongoing accumulation of a white pile of cell walls (thecae) for months after the population had reached the stationary phase. This seemed contrary to our expectation that the two thecae would be shed prior to cell division. Assuming "perfect" growth conditions, the number of accumulated thecae should be twice the number of living cells in the culture less the two for each cell in the original inoculum. Furthermore, any cell wall degrading

[1]Dedicated to Prof. Luigi Provasoli for his many contributions to phycology.

enzymes, released prior to cell division, and involved in the removal of the dinoflagellate skeleton and the production of two thecae might lead to the destruction of the accumulated thecae. Nevo and Sharon (1960), studying the carbohydrate chemistry of the cell wall of P. cinctum, noted, however that the cell wall was unusually resistant to β-glucanases of the snail which usually hydrolyze polyglucans and presumably to other polyglucan-hydrolyzing enzymes. This accumulation of cell walls in our cultures had no simple explanation, and it does not take place in nature or in non-axenic cultures (Hertzig, Dubinsky and Berman, this volume). P. cinctum multiplies in enormous numbers during the spring in Lake Kinneret (Sea of Galilee) in Israel; its numbers increase so greatly that they color the waters brown (Berman Rodhe, 1971; Pollingher and Berman, 1975; Dubinsky Berman, 1979). A similar accumulation of large amounts of photosynthetically-fixed carbon due to cell wall accumulation might also be occurring in Lake Kinneret towards the end of the bloom. Nitrate but not total nitrogen and perhaps more importantly, phosphorus become limiting as the Peridinium bloom peaked and declined (Serruya et al., 1975). Does N-and/or P-depletion affect cell wall productionµ

MATERIALS AND METHODS

Peridinium cinctum fa. westii was obtained as an axenic culture from Dr. T. Berman, Kinneret Limnological Laboratory, Tiberias, Israel. Peridinium was maintained and grown in 300 ml of the medium of Lindstrom and Rodhe (1978) in 1 l Pyrex Fleakers at 23°C under continuous fluorescent light at $20\text{-}40 \times 10^7$ ergs $cm^{-2} sec^{-1}$. Experimental cultures were inoculated with aliquots of an exponential phase culture of Peridinium containing ca. 50 active cells ml^{-1}; depleted culture inocula were washed once with N- and P-free medium. Samples of 1 ml were taken aseptically at two-week intervals from control, N-free, P-free and N- and P-free cultures, filtered through Millipore type RA filters and with a pore size of 1.2 µm and the filters were dried and examined under a Bausch a Lomb dissecting microscope. The cells and empty thecae were counted in a random sample of 5 boxes, the average per box calculated and multiplied by the number of boxes (26) equivalent to the area over which 1 ml was filtered. The experiment was done in five Fleakers for each experimental value and three samples (of 1 ml) were counted for each

Fleaker at each time interval for a total of 15 data bits for each experimental value. The experimental values are presented as the means $\pm$ standard errors for one experiment. The experiments were done twice.

RESULTS AND DISCUSSION

Cultures of P. cinctum grown in Lindstrohm-Rodhe medium showed an appreciable lag period (about 5-6 weeks) before the cells began to multiply (Fig. 1). For the first eight weeks, thecae number paralleled cell number at about twice the cell number, in agreement with expectations and only diverged significantly after the 8th week. Nitrogen-depleted cultures showed a similar lag followed by an increase in cell number, from week 5-7, and then by an arrest of further multiplication (Fig. 2); thecae formation increased slightly during the first 5 weeks, paralleled cell multiplication from week 5 to 7 and then increased dramatically relative to cell number. Cells grown in P-free medium showed a slight increase in cell number and then declined (Fig. 3); thecae number paralleled cell number for three weeks and then increased steadily with a large increase after the 7th week. Cells grown in N- and P-free media resembled cells grown in P-free media (Fig. 4). When the ratio of skeletons to cells was plotted (Fig. 5), we see that in all cultures there was no marked increase in skeleton producton over cell production up to the 7th or 8th week of incubation but this was followed by a marked increase in P-free media where skeleton production seemed to be completely uncoupled from cell production and relatively few surviving cells continued to produce large numbers of skeletons without further cell multiplication. From this we infer that even if cell growth and division stop, photosynthesis and anabolism continue and the cells can produce relatively large amounts of extracellular organic substances, i.e., thecae. This accumulation of organic carbon (thecae) could occur only as long as light and inorganic carbon are abundant and phosphorus or nitrogen become limiting. In such a case organic compounds like Peridinium thecae (Nevo & Sharon, 1960) lacking significant amounts of phosphorus and/or nitrogen, could continue to be synthesized. Indeed, such cases of secretion of a large variety of organic molecules have been reported (see Hellebust, 1974 for a recent review). The shedding of carbohydrate thecae is an analogous case where primary productivity (as

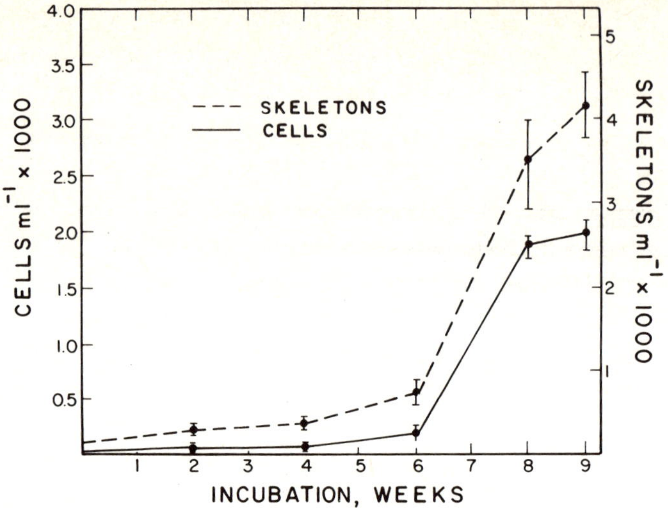

FIG. 1. Production of Peridinium cells and skeletons in nutritionally adequate medium.

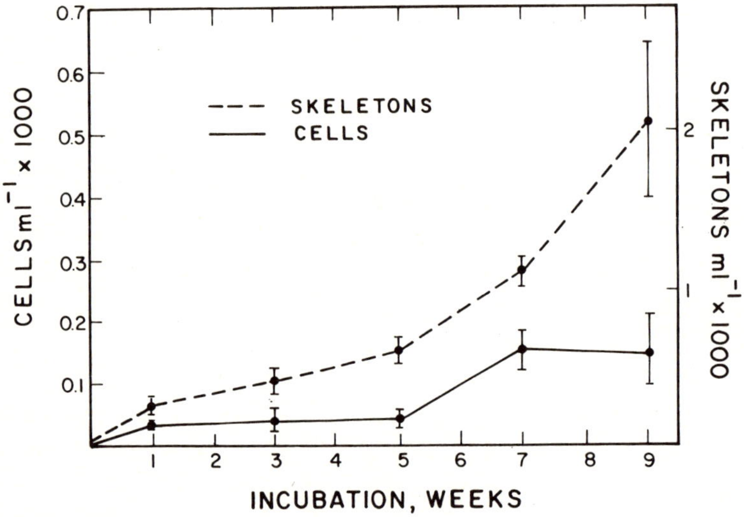

Fig. 2. Production of Peridinium cells and skeletons in nitrogen-free medium.

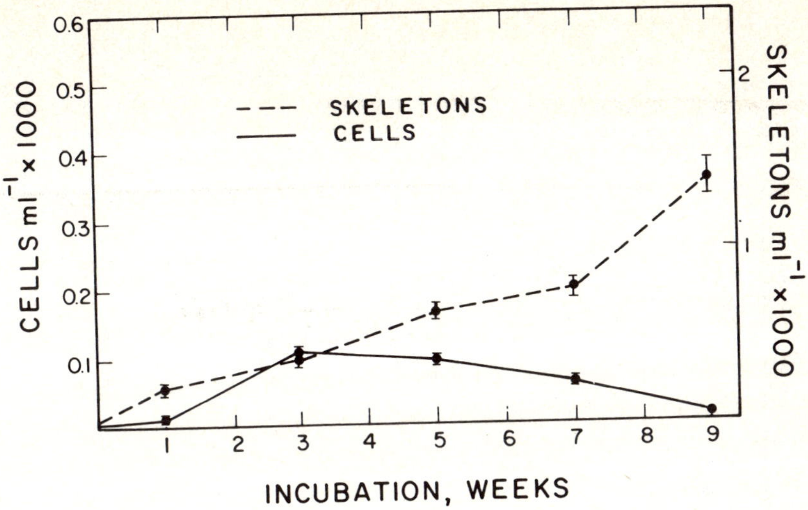

FIG. 3. Production of Peridinium cells and skeletons in phosphorus-free medium.

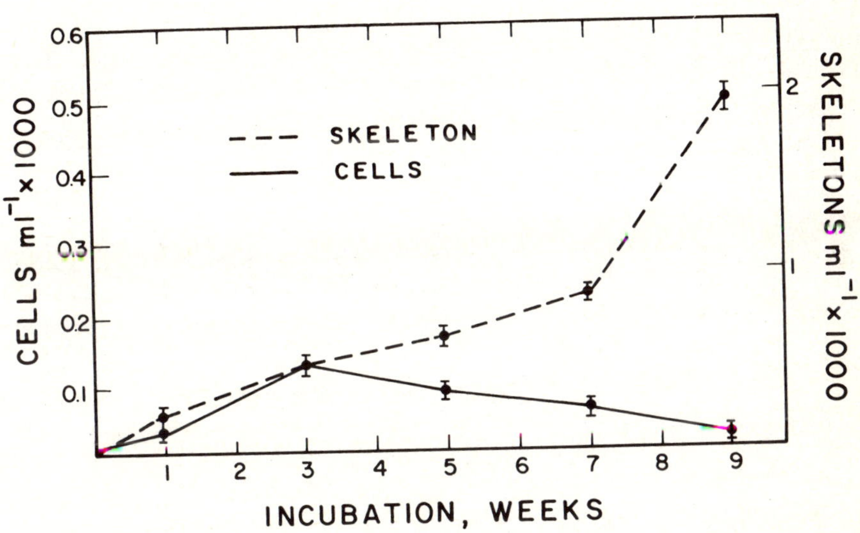

FIG. 4. Production of Peridinium cells and skeletons in phosphorus and nitrogen-free medium.

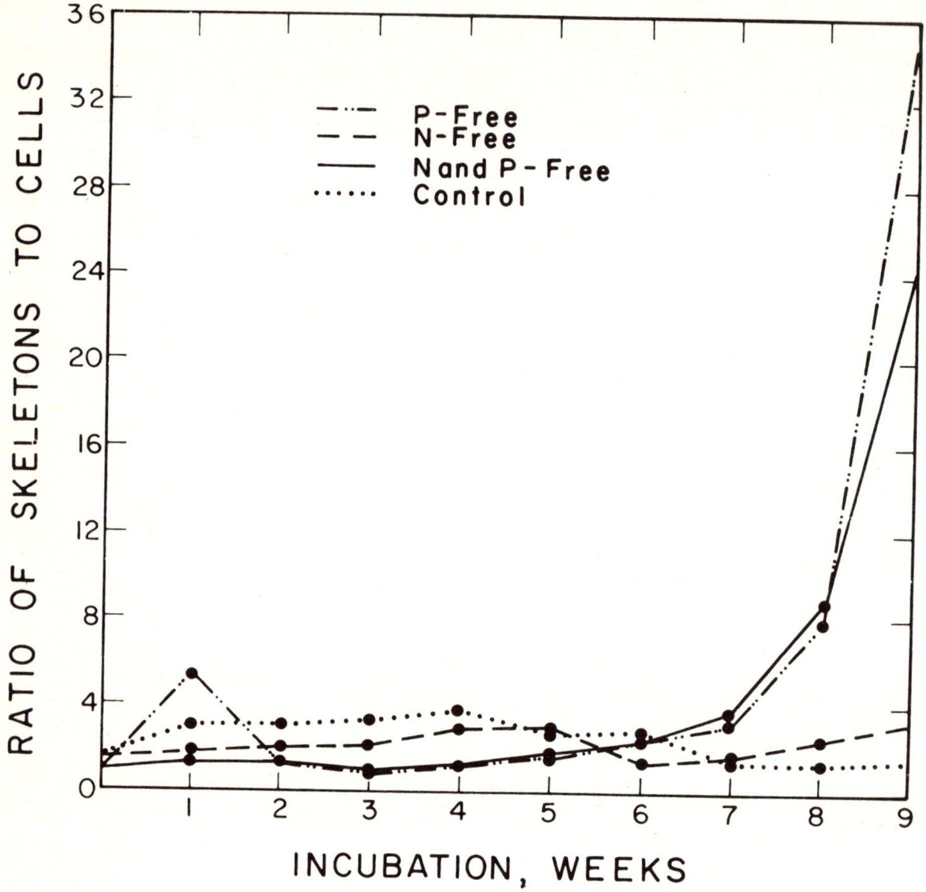

FIG. 5. Ratio of skeletons to cells grown in various media.

carbon fixation) becomes separated from cell multiplication. Similar observations on the loss of thecae due to drastic changes in environment, i.e., increased temperature and oxygen deprivation have been noted in Peridinium from Lake Kinneret (U. Pollingher, personal communication). Entze (1926) noted a shedding of the skeleton of Peridinium Borgei when the concentration of salts in the medium increased. Similar (in vitro) observations have been noted in this laboratory with other algae; acid phosphatase secretion in Chlamydomonas reinhardi (Patni et al., 1977) and acid phosphatase (Aaronson & Patni, 1974) and β-glucosidase (Meyer, 1976) and vitamins, amino acids, and macromolecules (Aaronson, 1971; Aaronson et al., 1971) in Ochromonas danica. Similar increases in extracellular carbohydrate were observed when a diatom was cultured in N- or P-depleted media; P-depletion was more effective (Myklestad & Haug, 1972). The uncoupling of the synthesis of organic molecules can occur in vitro, albeit slightly, under nutritionally adequate conditions (Fig. 1) and does occur in significant amounts under nutritionally inadequate conditions (N- and/or P-depletion; Figs. 2-4); it is likely that it may occur in nature, as well, where algae blooms peak and decline in response to the presence of major nutrients, i.e., P and N sources. If similar types of uncoupling of biosynthesis from multiplication occur in nature, they may well help to account for at least part of the organic compounds found in natural waters and perhaps the collapse of the bloom as well.

ACKNOWLEDGMENTS. This work was supported in part by a grant No. 5-S05-RR-07064-09 to Queens College from the National Institutes of Health, and a grant No. 665 from the U.S.-Israel Binational Science Foundation.

REFERENCES

Aaronson S. (1971) The synthesis of extracellular macromolecules and membranes by a population of the phytoplagellate Ochromonas danica. *Limnol. Oceanogr.* 16, 1-9.

Aaronson S., De Angelis B., Frank O. & Baker H. (1971) · Secretion of vitamins and amino acids into the environment by Ochromonas danica. *J. Phycol.* 7, 215-218.

Aaronson S. & Patni N.J. (1974) The secretion and cell content of phosphatases by Ochromonas danica during the population cycle and in resting suspensions under some conditions of chemical and physical stress. *Colloques Internat. du C.N.R.S. No.* 240, 137-144.

Berman T. & Rodhe W. (1971) Distribution and migration of Peridinium in Lake Kinneret. *Mitt. Internat. Verein. Limnol.* 19, 266-276.

Dubinsky Z.& Berman T. (1979) Seasonal changes in the spectral composition of downwelling irradiance in Lake Kinneret (Israel). *Limnol. Oceanogr.* 24, 652-663.

Entz G. (1926) Beitrage zur Kenntnis der Peridineen. I. Zur Morphologie und Biologie von Peridinium Borgei Lemmerman. *Arch. Protistenk* 56, 397-446.

Hellebust J.A. (1974) Extracellular products. In: *Algal Physiology and Biochemistry* (Edited by Stewart W.D.P.), University of California Press, Berkeley, pp. 838-863.

Lindstrom K. & Rodhe W. (1978) Selenium as a micronutrient for the dinoflagellate Peridinium cinctum fa. westii. *Mitt. Internat. Verein. Limnol.* 21, 168-173.

Meyer D.H. (1976) Secretion of β-glucosidase by Ochromonas danica. *Arch. Microbiol.* 109, 263-270.

Myklestad S. & Haug A. (1972) Production of carbohydrates by the marine diatom Chaetoceros affinis var. willei (Gran) Hustedt. I. Effect of the concentration of nutrients in the culture medium. *J. Exp. Mar. Biol. Ecol.* 9, 125-136.

Nevo Z. & Sharon N. (1969) The cell wall of Peridinium westii, a non cellulosic glucan. *Biochim. Biophys. Acta* 173, 161-175.

Patni N.J., Dhawale S.W. & Aaronson S. (1977). Extracellular phosphatases of Chlamydomonas reinhardi and their regulation. *J. Bacteriol.* 130, 205-211.

Pollingher U. & Berman T. (1975) Temporal and spatial patterns of dinoflagellate blooms in Lake Kinneret, Israel (1969-1974). *Verh. Internat. Verein. Limnol.* 19, 1370-1382.

Serruya C., Pollingher U. & Gophen M. (1975) N and P distribution in Lake Kinneret (Israel) with emphasis on dissolved organic nitrogen. *Oikos* 26, 1-8.

Developments in Arid Zone Ecology and Environmental Quality
Edited by H. Shuval

22
LOTIC ECOSYSTEM DYNAMICS IN XERIC AND MESIC REGIONS

Robert J. Naiman

Woods Hole Oceanographic Institution, Woods Hole, Massachusetts 02543, U.S.A.

ABSTRACT

The predominant trophic base, level of productivity, detrital characteristics and food web diversity of streams in mesic and desert regions are summarized and compared at the ecosystem level. Desert streams in the American southwest are shown to differ from their mesic counterparts in the boreal forests of Canada by having a predominantly autotrophic base, a lack of permanent headwater channels, an absence of significant amounts of woody debris, and intermittent and variable flow patterns. The main departure of desert streams from the River Continuim Concept, which is developed from lotic ecosystems in mesic regions, is that most semi-permanent desert streams would be reanked as about 4th order under mesic conditions. The structural and funtional properties of spring fed desert streams are compared to desert streams subject to flash floods, and the role of catastrophic flooding is shown to be reflected in community development and life history strategies. However, the basic nature of aquatic ecosystems in the American southwest has been drastically altered since 1850 by grazing and concomitant erosion and these changes are briefly reviewed to add perspective to our understanding of lotic ecosystem dynamics in desert regions and its application to current ecological thought.

Developments in Arid Zone Ecology and Environmental Quality
Edited by H. Shuval

23
BREAKDOWN OF *PERIDINIUM* BIOMASS IN LAKE KINNERET

Ronnie Hertzig* Zvy Dubinsky* and Thomas Berman+

**Department of Life Sciences, Bar-Ilan University, Ramat-Gan, Israel, and +Kinneret Limnological Laboratory, P.O. Box 345, Tiberias, Israel.*

ABSTRACT

Degradation of *Peridinium* in Lake Kinneret was studied by incubating lyophilized cells under various conditions *in situ* or in the laboratory. The rate of breakdown was most rapid at the end of the *Peridinium* bloom season, when >90% of the cell material disintegrated to particles (10μm) or soluble compounds within 24 hr. Degradation was dependent on specialized bacterial populations and accelerated in the presence of protozoans and by turbulence.

INTRODUCTION

The annual bloom of the dinoflagellate, *Peridinium cinctum*, in Lake Kinneret has been described in great detail (e.g., Berman & Pollingher, 1974; Pollingher & Serruya, 1975; Pollingher & Berman, 1975). Large standing crops of this alga appear in the lake from February through mid-June, generally with maximum concentrations from late March through May. The amount of *Peridinium* has been estimated as between 70 to 80% of the total phytoplankton biomass and standing crops in the lake can exceed 20000 tons wet weight with maximum concentrations of 1000 cells ml^{-1} (Berman & Rodhe, 1971). In early to mid-June, the population of *Peridinium* drops rapidly and during the rest of the year only low numbers of organisms (20 cells/ml^{-1}) are observed.

The sediments of Lake Kinneret have a low organic matter content (about 3% dry weight, Serruya et al., 1974) and although there is increased microbial activity in the sediment-water interface in May-June and July (Cavari, personal communication), it appears that relatively little of the *Peridinium* biomass reaches the lake bottom. Also, *Peridinium* is not grazed extensively by the zooplankton or fish, with the exception of Sarotheredon (Tilapia) galilaeum (Serruya et al., 1980).

This background suggested that most of the Peridinium biomass is degraded within the water column. The present report is the first of a series which will describe various aspects of the detrital pathway (Serruya et al., 1980) of Peridinium breakdown.

MATERIALS AND METHODS

Most experiments were carried out using lyophilized Peridinium cells which had been collected during the bloom period from the lake (for details, see Wynne & Berman, 1980). Preliminary trials showed that there were no significant differences between this and freshly collected and dried material.

Portions (100 mg) of lyophilized Peridinium were placed in specially constructed glass holders which were closed at both ends with Nitex netting (usually 10 μm mesh, see below). Experiments were begun by placing these holders either in situ in the lake or in a controlled temperature, water bath, shaker in the laboratory. At various times the

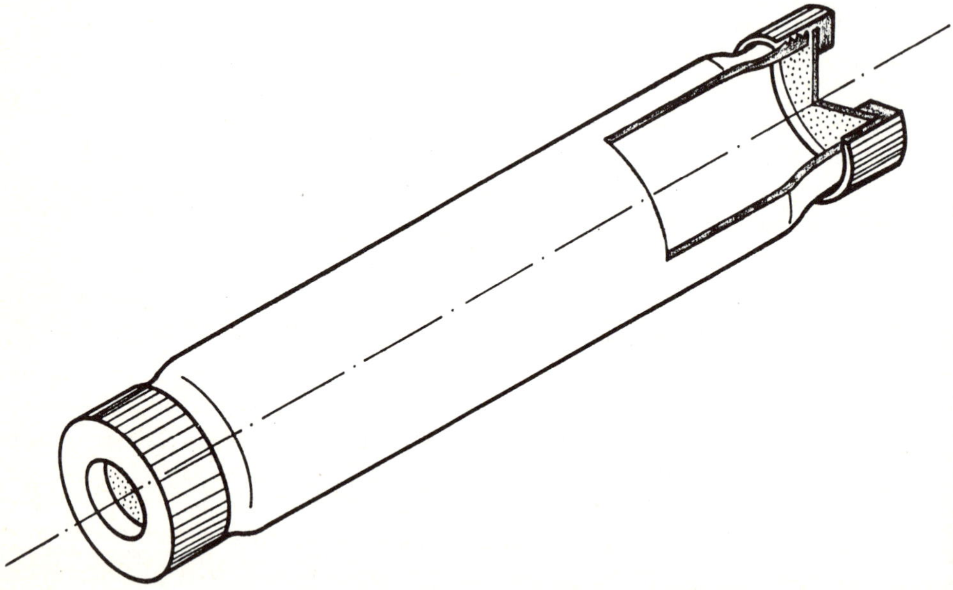

FIG. 1. Schematic of glass holders for Peridinium breakdown experiments.

holders were removed and the remaining particulate material was carefully collected on preweighed GF/C filters, dried and weighed. The results, generally from triplicate samples, were expressed as the percentage difference of dry weight particulate material from the initial 100 mg cell material.

RESULTS AND DISCUSSION

It is important to note that in all these experiments we define 'breakdown' operationally as any process by which the Peridinium cells (diameter range 40-80 μ) are degraded into soluble substances or into particles capable of passing through a 10 μm mesh net. Such particles should provide excellent substrates for heterotrophic bacteria and should also be available as a food source for fish, zooplankton or protozoans grazing on detritus. This fact has very significant implications for the entire Kinneret food web (Serruya et al., 1980), but a discussion of these matters is beyond the scope of the present paper.

Degradation in situ. The results of a series of experiments in which holders with lyophilized Peridinium were suspended in the lake for 24 hours at 3 depths (5m; or epilimnion; 14 m, or thermocline; and 20 m, or anaerobic hypolimnion) are shown in Fig. 2. The fastest rates of

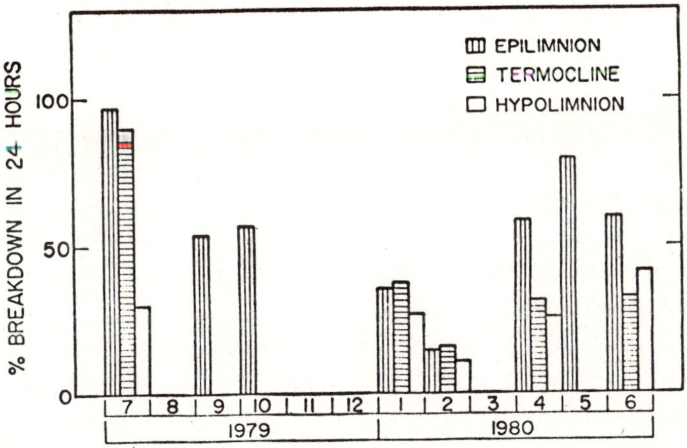

FIG. 2. Seasonal variation of Peridinium degradation rates measured at 3 depths in Lake Kinneret. Breakdown expressed as % of initial dry weight material degraded within 24 hours.

breakdown were observed in the epilimnion at the end of the bloom period. Rates were generally lower in the 14 m, and the lowest rates were in the 20 m samples. Not surprisingly, these results indicated that a specific microbial population capable of acting on Peridinium cell material, develops especially towards the end of the bloom season and that most degradative activity takes place in the more oxygenated upper water layers. Microscopic observations showed that the hard polyglucan (Nevo % Sharon, 1969) theca was the first part of the cells to be decomposed. The degree of cell 'breakdown' varied with the mesh size of the net closures, i.e., on the minimal size of cell fragments produced (Table 1) but, as noted above, for all other experiments we routinely used a 10 μm mesh closure.

TABLE 1. Effect of closure pore size on loss of Peridinium particulate matter in holders.[+]

Pore size	% Loss
10 μm	85.4
8 μm	81.7
3 μm	29.9

Factors affecting rates of breakdown. We assumed that cell degradation was affected by at least several environmental factors. The importance of mechanical agitation was shown by comparing breakdown in samples incubated with or without shaking in the laboratory (Table 2). During the summer months in the lake, the upper water layers are strongly mixed by the almost daily westerly winds (S. Serruya, 1975) and this undoubtedly accelerates Peridinium cell breakage.

Decomposing Peridinium in Erlenmeyer flasks was always accompanied by numerous bacteria and non-pigmented flagellates and ciliates (Fig. 3). In order to examine the influence of these microorganisms on the rate of Peridinium breakdown, we placed holders with lyophilized cells inside watertight plastic bags which were filled as follows: 1) Untreated lake water (1 l) - "Control". 2) Lake water (1 l) passed through sterile GF/C

TABLE 2. The effect of turbulence, temperature and biological inhibitors on Peridinium breakdown

Experimental Conditions	Date	% Degradation in 24 hr	
a. No shaking		21	(± 5.2)
Shaking		57	(±15.0)
b. Temperature 31°C		44	(± 7.0)
24°C		33	(± 0.2)
20°C		32	(± 1.4)
17°C		32	(± 4.3)
c+. Unfiltered lake water		57	(± 7.0)
"Bacteria only"		41	(± 1.2)
"Bacteria inhibited"		34.5	(± 5.0)
"Biologically inhibited		54	(± 3.0)

+See details of experimental treatments in text. Results of this experiment were from a 48-hr incubator.

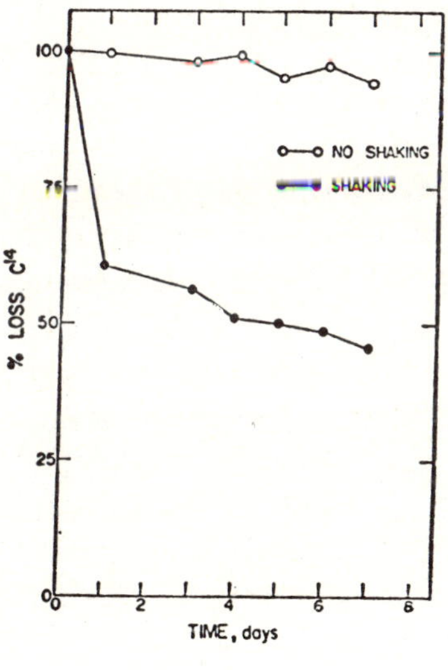

FIG. 3. Loss of ^{14}C particulate cell material in decomposing Peridinium.

filters - "Bacteria only". 3) Lake water (1 l) with added antibiotics, streptomycin (50 mg) and penicillin (100 mg) - "Bacteria inhibited", and 4) Lake water (1 l) with 40% formalin - "Biologically inhibited". The bags containing the holders were then suspended in the lake for 48 hrs. Water which contained "bacteria only" showed lower degradative activity than untreated water which also contained protozoan populations (Table 2). (Subsequently, Sherr, Sherr & Berman (in preparation) have demonstrated experimentally that the presence of a microflagellate will accelerate the bacterial degradation of Peridinium cells). Material suspended in "Bacteria inhibited" water broke down more slowly than the "Bacteria only" sample - presumably mainly due to mechanical agitation. We cannot fully explain the rapid degradation of the formalin-treated material, which was surprising, but may have resulted from chemical action. The results of this experiment (Table 2) indicate that over the 48-hr period about half of the observed breakdown could be attributed to mechanical turbulence. Also, although bacteria alone were capable of enhancing breakdown, other larger organisms, presumably protozoans, further accelerated the process.

As expected, the rates of breakdown were also affected by temperature (Table 2). We did not check temperatures greater than 31°C, which is slightly higher than the maximum found on the lake surface in summer.

Autolysis. We also attempted to examine the role of autolysis in Peridinium degradation. This was done by placing uniformly ^{14}C-labelled Peridinium cells (grown in axenic cultures with ^{14}C-bicarbonate over several generations) in autoclaved filtered lake water and following the drop in particulate radioactivity. As can be seen in Fig. 3, there was virtually no loss in particulate ^{14}C in static flasks. Only samples which were shaken showed a drop in particulate ^{14}C.

Therefore, it appears that autolytic processes are rather unimportant in transforming particulate Peridinium cell material into dissolved substances although they may be necessary to weaken some structures which are then detached by turbulence.

In summary, we have confirmed that the breakdown of Peridinium cells to small (10 μm) particles and dissolved compounds can be very rapid in the waters of Lake Kinneret, especially at the end of the bloom season.

This process appears to depend on a specialized population of bacteria, and is accelerated in the presence of protozoans and by turbulence.

REFERENCES

Berman T. & Rodhe W. (1971) Distribution and migration of Peridinium in Lake Kinneret. *Mitt. Int. Ver. Theor. Angew. Limnol.* 19, 266-276.

Berman T. & Pollingher U. (1974) Annual and seasonal variations of phytoplankton, chlorophyll and photosynthesis in Lake Kinneret. *Limnol. Oceanogr.* 19, 31-55.

Nevo Z. & Sharon N. (1969) The cell wall of Peridinium westii a non-cellulosic glucan. *Biochim. Biophys. Acta* 173, 161-175.

Pollingher U. & Berman T. (1975) Temporal and spatial patterns of dino-flagellate bloom in Lake Kinneret. *Int. Ver. Theor. Angew. Limnol. Verh.* 19, 1370-1382.

Pollinger U. & Serruya C. (1975) Phased division of Peridinium cinctum fa westii and the development of the bloom in Lake Kinneret. *J. Phycol.* 11(2), 155-162.

Serruya C., Edelstein M., Pollingher U. & Serruya S. (1974) Lake Kinneret sediments, nutrient composition of the pore water and mud water exchanges. *Limnol. Oceanogr.* 19(3), 489-508.

Serruya C., Gophen M. & Pollingher U. (1980) Carbon flow patterns in Lake Kinneret. *Archiv fur Hydrobiologie*(in press).

Serruya S. (1975) Wind, water temperature and motions in Lake Kinneret: General pattern. *Int. Ver. Theor. Angew. Limnol. Verh.* 19, 73-87.

Wynne D. & Berman T. (1980) Hot water extractable phosphores: An indicator of nutritional status of Peridinium cinctum in Lake Kinneret. *J. Phycol.* 16, 40-46.

Developments in Arid Zone Ecology and Environmental Quality
Edited by H. Shuval

24

DYNAMICS OF RAIN POOLS ALONG THE COASTAL ZONE OF ISRAEL, AN-OVERVIEW

Avital Gasith, Tel Aviv University, Faculty of Life Sciences, Institute for Nature Conservation Research

ABSTRACT

Rain pools along the coastal zone of Israel are shallow temporary aquatic systems. The pools vary in size and form from pond-like system in open fields to channel-like systems along the roads.

The water source to the pools are direct rainfall, runoff and/or ground water. The duration of the wet period vary from several weeks up to 8 months. The soil of the coastal zone is typically brown red sandy soil. Limnologically these pools can be characterized as heterotrophic (R >P), mesotrophic systems (D.O. > 50% saturation, BOD < 10 ppm). Conductivity usually ranges from 150 to 450 μmhos/cm at 20°C. The pH is slightly alkaline (7.5 - 8.5). Water turbidity varies among pools depending mainly on the principle source of water, the morphometry of the pool and the extent of macrophyte development (secchi depth 5 to > 50 cm).

From an energetic point of view rainpools of the coastal plain are detritus drived systems. The detritus is of autochthonous origin, that is organic matter which is produced within the rainpool basin during the wet and dry seasons and of allochthonous origin which is contributed to the pool from the drainage basin via runoff. The autochthonous source can be divided to:

a. Dead aquatic plant material remaining from the previous wet season.

b. Dead plant material of terrestrial origin which is produced in the pool's basin during the dry season.

c. Fresh plant material of terrestrial origin which sprout and grow in the pool's basin during the early phase of the wet season prior to the flooding of the pool.

The relative significance of the various sources of organic matter depend primarily on land use in the drainage and pool basins during the dry season. From this point of view rainpools are year-round systems consisting of dry and wet phases.

From an ecological point of view rainpools can be described as exploding systems in which species diversity and biomass develop during extremely short periods of time. Biologically the rainpools are characterized by two main animals components: The "specialist group", those organisms specially adapted to temporary water situations (i.e. Eubranchiopoda, Triturus vittatus) and the "opportunistic group", those organisms which inhabit temporary as well as permanent bodies of water (i.e. insects, other amphibians, turtles, birds). The richness in detrital matter and the lack of an efficient predator (such as fish) results in a system consisting of relatively high zooplankton biomass and large size organisms.

As a consequence of their consideration as a non-resource, the existence of rainpools in Israel is continuously threatened, placing their unique community on the verge of extinction.

Developments in Arid Zone Ecology and Environmental Quality
Edited by H. Shuval

25
SOME RESPONSES OF HERMATYPIC CORALS TO AMBIENT LIGHT

D. Sherf P. Falkowski* and Z. Dubinsky

Department of Life Sciences, Bar-Ilan University, Ramat-Gan, Israel
*Brookhaven National Laboratory, Upton, New York, U.S.A.

ABSTRACT

The widespread hermatypic coral Stylophora pistillata was found over two orders of magnitude range of light intensities. Shallow water, lagoon colonies were more profusely branched, with thinner individual branches than reef samples. This morphological difference resulted in higher area-to-weight ratios in these corals. The colonies adapted to high-light were markedly lighter in color than samples collected from low-light locations, like deep water and shaded crevices. The chlorophyll levels in the shade-adapted corals were up to eight times higher than in the high-light ones. This increase in chlorophyll was not caused by a change in numbers of the symbiotic zooxanthellae in the corals, but from a corresponding increase in cellular pigment concentration in the algae. These changes in pigmentation were also reflected on the ultrastructural level, as differences in the number of chloroplasts and thylakoids and the spacing among them which were observed by transmission electron microscopy.Transfer of colonies from shade to light resulted in reversal of the above-described pigment traits within four weeks

Developments in Arid Zone Ecology and Environmental Quality
Edited by H. Shuval

26
LAND APPLICATION OF SEWAGE SLUDGE WITH SPECIAL EMPHASIS ON COMPOSTING

Eliot Epstein and Joel E. Alpert

E&A Environmental Consultants, Inc., P.O. Box 529, Sharon, Massachusetts, 02067, U.S.A.

ABSTRACT

Legislative, environmental and economic constraints in the past decade have resulted in major changes in sewage sludge disposal and management practices in the United States. Land application and composting are emerging as the most promising and acceptable solutions. This paper discusses these alternatives with respect to current and proposed regulations and discusses in detail the recently developed Forced Aeration Composting System which is receiving wide attention.

INTRODUCTION

There are seven major systems for sludge disposal and utilization currently used in the United States. The U.S. Environmental Protection Agency (EPA), classifies the seven systems into (a) Treatment and Volume Reduction Methods which include incineration, composting, and surface impoundments and (b) Ultimate Utilization and Disposal Methods which include landfills, ocean dumping and discharge, land spreading and distribution and marketing.

Table 1 provides information on the percentage of sludge used in each of these categories by community size. In the past five years EPA has been strongly promoting composting and land application. This emphasis has been the result of legislative actions, energy cost and improved composting technology. The phasing out of ocean dumping of sewage sludge by December 31, 1981, as stated in the Marine Protection Research and Sanctuaries Act of 1977, and air emission regulations promulgated by the Clean Air Act Amendments of 1970 and 1977, will undoubtedly contribute to a major increase in land application and composting.

The objectives of this paper are to examine the regulations which affect sludge disposal decisions, present constraints to land application methodologies including composting, discuss land application and composting as related to present and projected regulatory requirements.

TABLE 1. Distribution of Sludge Managment Practices in the United States by Community Size

Method	Total Sludge Distribution %	Communities Small[2] %	Medium[3] %	Large[4] %
I. Treatment and Volume Reduction				
Incineration	21	0	9	21
Composting[1]	-	-	-	-
Surface Impoundment	6	3	8	6
II. Ultimate Utilization and Disposal				
Landfills	24	35	26	23
Ocean Dumping and Discharge	18	0	2	21
Landspreading	19	55	45	15
Distribution and Marketing	12	7	10	14

Ref. Prince et al 1980; Anonymous 1980.
[1]Included in distribution and marketing and landspreading
[2]Less than 3785 c.m. (1 mgd)
[3]3785 - 94700 c.m. (1-25 mgd)
[4]Greater than 94700 c.m. (25 mgd)

CURRENT AND PROPOSED REGULATIONS

The regulatory control over land application in the U.S. follows a multi-tiered approach. The first tier is comprised of the Applicable Federal Laws which provide the regulatory authority for different agencies to promulgate regulations and produce guidelines which can be considered the second and third tiers. The Applicable Federal Laws are shown on Table 2, while the major Federal Regulations Relating to Sewage Sludge Disposal, are shown in Table 3. In addition to these regulations the EPA has produced a series of guidelines against which the operation of a sludge disposal facility can be measured. Before a treatment authority sludge management plan can be approved for Federal funding it must meet all applicable Federal Laws, Regulations and Guidelines. The States and Regional Offices of the EPA however, do have the ability to provide variances to Federal Regulations if local ordinances or conditions warrant.

TABLE 2. Applicable Federal Laws

1. CWA-Clean Water Act of 1977, PL 95-217 and Federal Water Pollution Control Act, 1972, PL 92-500. Federal funding of 75%(85% for innovative & alternative projects) for POTW & sludge treatment facilities; NPDES (National Pollution Discharge Elimination System) for point source and non-point pollution; implementation of industrial pretreatment.
2. RCRA-Resource Conservation and Recovery Act of 1976, PL 94-580. Financial assistance to state and local governments for development of solid waste management plans; regulations for hazardous and non-hazardous waste disposal.
3. MPRSA-Marine Protection Research & Sanctuary Act of 1977, PL 92-532. Phasing out of ocean disposal by December 31, 1981; compliance schedule for land disposal.
4. CAA-Clean Air Act Amendments of 1970 & 1977, PL 91-604 and PL 95-95. State Implementation Plans(SIP's) for meeting Federal air quality standards; emission policies for incinerators.
5. SDWA-Safe Drinking Water Act of 1975, PL 93-523. Protection of drinking water.
6. NEPA-National Environmental Policy Act of 1969, PL 91-190. Requires Environmental Impact Statements(EIS) for facilities.
7. TSCA-Toxic Substances Control Act of 1976, PL 94-469. Regulates disposal of hazardous wastes. PCB's in sludges.

TABLE 3. Major Federal Regulations Relating to Sewage Sludge Disposal

Sludge Disposal	Regulation	Date of Promulgation	Applicable Federal Laws
1. Landspreading			
a. Food-chain application	40 CFR Part 257	9/79	RCRA/CWA
b. Non-food chain application	40 CFR Part 257	9/79	RCRA/CWA
c. Distribution and Marketing	40 CFR Part 258	12/81 (estimated)	CWA and others
2. Land Disposal			
a. Solid wastes (non-hazardous)	40 CFR Part 257	9/79	RCRA/CWA
b. Hazardous wastes	40 CFR Parts 260 et seq.	5/80	RCRA
c. PCBs; criteria modification	40 CFR Part 761	5/79	TSCA
3. Incineration			
a. New stationary sources of air emissions	40 CFR Part 60	10/75	CAA
b. Hazardous pollutants	40 CFR Part 61	10/75	CAA
c. Hazardous wastes	40 CFR Parts 260 et seq.	5/80	RCRA
d. PCBs; Criteria modification	40 CFR Part 761	5/79	TSCA
4. Ocean Dumping	40 CFR Parts 220-230	1/77	MPRSA

PUBLIC HEALTH ISSUES

The public health issues related to land application of sewage sludge or sludge based products, can be classified into the following categories:

- Pathogens
- Heavy Metals
- Organics
- Odors and Aesthetics
- Others - NO_3, etc.

Pathogens

A large variety of disease organisms is present in raw sewage. These pathogens include protozoa, parasitic worms, pathogenic bacteria and viruses. The amount present in any sludge depend primarily on the health of the community served and treatment processes used. The efficacy of removal of pathogens from wastewater varies with the type of treatment (Kabler, 1959, Wang and Dunlop, 1959, Foster and Engelbrecht, 1973, Wellings, et al., 1975; Wellings et al., 1976, Wright, et al., 1942).

The effect of land application of sludge on human health is a matter of great concern. Previous studies indicate that disease problems related to soil application of human wastes have been caused primarily by the use of raw sewage effluent, raw sludge, or night soil (Sepp, 1971). Parsons, et al. (1975) summarized various data on the survival of certain pathogens in soils and on plants. While most pathogens survive in soil for several days to a few months, the eggs of intestinal worms such as Ascaris Lumbricoides can survive for several years.

Soil moisture, pH, and temperature greatly influence the survival of pathogenic organisms. Absorption and movement of pathogens in soil are affected by the clay and organic matter content. Movement of bacteria through soils is generally restricted to the upper few centimeters (Romero, 1970). However, Bouwer, et al. (1974) showed that in porous soils subjected to high flow rates of sewage effluents, bacterial movement can occur to a depth of several meters.

Bitton (1975) cites several references regarding adsorption of viruses on soil particles and their movement through soils. Migration was generally limited to the upper 50 cm. However, in porous media or situations in which fissures, fractures, or cracks occurred in the substream, movement of viruses to ground water is possible (Hori, et al, 1970). Keswick and Gerba (1980) summarized the data currently available on isolation of viruses beneath land treatment sites.

The U.S. EPA, recognizing the potential dangers of applying untreated sewage to the land, requires that all sludge or sludge products be treated to a degree sufficient to destroy pathogens. The Agency has divided treatment processes into two levels of pathogen reduction: Processes to Significantly Reduce Pathogens (PSRP), which result in 90% reduction of bacteria and viruses, and Processes to Further Reduce Pathogens (PFRP) which result in reduction of all pathogens to negligible numbers. Treatment processes which will meet these criteria are listed below. Based on the final use of the sludge product, either PSRP or PFRP treatment processes are required as follows:

A. Processes to Significantly Reduce Pathogens (PSRP)
Aerobic digestion - 60 days @15 ^{0}C - 40 days @ 20^{0}C
Air drying - 3 months with 2 months above 0^{0}C
Composting - 40^{0}C for 5 days with 4 hours above 55^{0}C
Lime stabilization - pH 12 for 2 hours

B. Processes to Further Reduce Pathogens (PFRP)
Composting - static pile 55^{0}C for 3 days
windrow 55^{0}C for 15 days; minimum 5 turns
Heat drying - moisture reduced to 10%; 80^{0}C or greater
Thermophilic aerobic digestion - 10 days @55-60^{0}C

Heavy Metals

Sewage sludge applied to land either directly or after processing e.g. heat drying or composting, can result in heavy metal contamination of soils. Uptake of certain elements can cause phytotoxicity or accumulate in plants and be toxic to humans or animals. Sommers, (1976) showed that heavy metals in sewage sludge varied considerably among treatment plants and within a treatment plant. Although EPA is primarily concerned with those elements which can accumulate in food crops and be toxic to man, ecologists and agriculturalists are also concerned with those elements which can be phytotoxic or toxic to animals and wildlife.

The availability of heavy metals to plants, their uptake and their accumulation, depend on a number of soil, plant and other factors. The soil factors of concern include pH, organic matter, phosphorous, cation exchange capacity, moisture, temperature and aeration. The plant factors are species and variety, organ of the plant and plant age. Other factors to be considered include reversion of metals and forms and type of metal of concern.

Plant species and even varieties differ widely in tolerance to toxic metals. Vegetable crops that are very sensitive to toxic metals include members of the beet family (chard, spinach, redbeet, and sugarbeet), turnip, kale, mustard, lettuce, and tomatoes. Beans, cabbage, collards and other vegetables are less sensitive. Field crops such as corn, small grains, and soybeans are moderately tolerant. Most grasses (fescue, lovegrass, Bermudagrass, orchardgrass, perennial ryegrass) are tolerant to high amounts of metals. Unusually metal-tolerant ecotypes of the grasses are found on ore outcrops containing extremely high amounts of metals (Antonovics, et al. 1971). Crops differ in their susceptibility to injury by different metals. Plant age and different organs in the plant accumulate heavy metals differentially. Grain and fruits generally exclude metals to a greater extent than the foliage (4 to 70% of foliage metal concentration).

Heavy metals that accumulate in plant tissues can enter the food chain subsequently, reaching human being directly by ingestion or indirectly through animals (Page, 1974, Chaney and Giordano, 1977). The elements in sludge of greatest concern to human health are Cd, Pb, As, Se, and Hg. (Braude, et al. 1975). Industrial contamination of sludge with As, B, Ba, Cd, Co, Cr, Cu, Mn, Ni, Pb, V, and Zn can result in localized agronomic problems.

Cd is the element of greatest concern to human health when wastewater sludges and sludge composts are applied to land. When added to soils, Cd is readily absorbed by plants and can accumulate in edible parts and thereby enter the food chain. Most human exposure to Cd comes from food, principally grain products, vegetables, and fruits. Dugan and Corneliussen (1972) showed that 26.5, 26 and 10% of the calculated daily intake of Cd came from grains and cereals, vegetables, and fruits, respectively. Increased levels of Cd in foods can be toxic to human beings. (Sandstead,et al. 1974, Elinder,et al. 1976).Environmental pollution of soils with Cd and subsequent accumulation of Cd in rice resulted in the itai-itai ("ouch-ouch") disease, which occurred in the Jintsu River Basin of Japan (Yamagata and Shigematsu, 1970). The World Health Organization has established that the maximum permissible level of dietary Cd should not exceed 70 μg/person/day. Studies in the U.S. and Canada have shown that dietary intake of Cd is between 50 and 100 μ g/day; consequently, a further increase in our dietary intake of this element would not be acceptable. However, these past studies on dietary intake and the risk associated with cadmium are being reviewed.

The toxicity of nickel to humans is low. Application of wastes containing high concentrations of nickel probably will be phytotoxic before reaching levels dangerous to man.

Mercury is very hazardous to human health, but levels in sludge are generally low with little chance of increased levels in plants.

Lead is a heavy metal of concern to health authorities (Braude, et al. 1975) and hence its application to soils is regulated by EPA. High levels of lead

(>3000 ppm) have been found in urban soils which is probably the result of automobile, industrial or house paint sources. Soluble lead added to soils becomes insoluble due to reactions with clays, phosphates, carbonates, hydroxides sesquioxides and organic matter. This reduces its potential uptake by plants. Sabey and Hart (1975) showed that when sewage sludge was applied to land, the lead content of leaves and fruit or seeds were not significantly altered.

Organics

The problem of organics has historically been associated with chlorinated hydrocarbon pesticides and PCB's but the passage of regulations such as RCRA and increased environmental awareness of carcinogenic nature of many organic compounds, has led to growing concern about the fate of these compounds. Pahren, et al. (1977) and Jelinek and Braude (1977) have examined the impact of organic compounds on land application practices. The concerns with organic compounds fall into several categories, uptake of the compounds, adsorption of the compounds, translocation of the compounds, direct ingestion of the compounds by grazing and foraging animals, and the formation of breakdown products and subsequent movement of these compounds in the environment.

Very little information is available on land contamination by sludges or residues containing organic chemicals. Most municipal sludges contain relatively small amounts of organics. The concentrations of two polychlorinated biphenyls (PCB's) in sludge from several municipal wastewater treatment plants in Michigan ranged from less than 0.1 ppm to 352.0 ppm, (Anonymous, 1973). Furr, et al. (1976) reported PCB levels ranging from less than 0.01 to 23 ppm in sludges from 16 U.S. cities. Four cities had less than 1 ppm in the sludge and 10 cities had less than 5 ppm. Only two cities had in excess of 10 ppm, which is the limit suggested by the U.S. Food and Drug Administration (FDA) (Jelinek, 1976). Similarly, data on pesticides indicate that municipal sludges generally contain low levels of these chemicals. Analysis of raw sludge from the Blue Plain Wastewater Treatment Plant in Washington, D.C., showed BHC (lindane) and DDT at concentrations of 1.22 and 0.6 ppm respectively, with a PCB level of 0.24 ppm. Furr, et al. (1976) found a dieldrin concentration of under 2.2 ppm in only 1 city of the 16 examined. The remaining 15 municipalities had dieldrin concentrations of under 0.44 ppm. Taffel, et al. (1980) reported that sludges from several cities in Puerto Rico contained less than 1.0 ppm PCB's and 0.57 to 2.69 ppm chlorodane (Table 5). Dean (1975) indicated that the hazards from organic pesticides and chlorinated organics in sludges applied to land appear to be minimal. The FDA (Jelinek, 1977) indicated that sludges containing 10 ppm PCB's on a dry weight basis should not be used for growing food crops.

TABLE 5. Organic Chemicals in Several Puerto Rico Sludges

City	Dieldrin	Chlorodane	4,4'DDT	4,4'DDD	Endrin	PCB -1254
	Chemical ppm					
Ponce	.14	.57	.01	.01	.09	.21
Coamo	.02	1.12	.02	.09	.01	.08
Guayanilla	.05	2.69	-	-	-	.06
Guanica	-	1.20	-	-	-	.35

Odor and Aesthetics

Perhaps the single most important consideration in planning and implementing a land based sludge disposal program is odor control. Odor problems can begin at the point of initial sludge handling and can extend for a significant period of time after cessation of activity. Sludge odors have been attributed to mercaptans, skatoles, sulfides and amines (Henry and Gehr, 1980).

Other aesthetic considerations are control of insects (flies) and rodents which might be attracted to the land application site. Highly liquid unstabilized wastes are generally more susceptible to these type of problems than are drier stabilized products such as compost.

Others - Nitrates

A final public health concern involves nitrogen transfers in the soil and the movement of nitrate to the ground water table. Nitrate movement to ground water or other water resources depends on the rate of mineralization and plant uptake (Epstein, et al. 1978). For this reason, sludge application rates are based on the amount of nitrogen necessary for crop growth first and then the rates are checked to see that this will not cause an excessive load of heavy metals or a phosphorous accumulation (EPA, 1978). Generally, with sludges, phosphorous causes no problems, but in some unusual cases it may be problematical.

LAND APPLICATION

Advantages and Disadvantages

As shown in Table 1 approximatley 19% of the sludge generated in the U.S. is disposed of by some form of landspreading activity. The main advantages of land application are, 1) that it allows for the utilization of the nutritive and organic value of the sludge, 2) it is technologically quite simple to implement, 3) it has a long history as a successful operation, 4) it has certain inherent economic advantages over alternative options, 5) it is given preferred treatment under EPA Construction Grants Formulas as an alternative treatment. Details of these aspects are discussed below:

1. The USDA in 1978, estimated that nitrogen, phosphorous and potassium in sludge generated in the U.S. represents 0.5, 3.3 and 0.4% respectively, of the total fertilizer used in the U.S If all the sludge was to be used on land the value would be 21,69, and 3 million dollars for N, P, K respectively. (USDA, 1978). However, in addition to these macronutrients, sludges also contain many of the essential micronutrients which are necessary for plant growth such as Cu, Ni, Zn, B, Se, Fe, etc. The presence of these micronutrients can sometimes be as valuable as that of the macronutrients in terms of enhanced plant response. In many instances, perhaps the primary value of adding sludge to the soil is the increase in organic matter which increases soil crusting, improves soil tilth and generally improves the soil as a plant growth medium.

2. Sludge application can often be accomplished with traditional agricultural equipment, e.g., manure spreader. The mode of application is dependent on a number of variables including moisture content, distance from plant to field, type of crop grown, topography, etc.

3. As Jewell and Seabrook (1979) described in their review of sludge application, various communities have been utilizing this method of disposal since 1650 with a major surge in interest in the mid 1800's with the first U.S. applications occuring in 1872 in Augusta, Maine.

4. Walker (1979) pointed out that landspreading of sludges was much more economical than incineration. He compared the costs of Salem, Oregon which uses land application with those of Warren, Michigan which incinerated its sludge and landfilled the ash. The O&M costs were $55/dry ton for Salem, OR, vs $155/dry ton for Warren, MI. Cost savings in 1977 for land application vs incineration for the same size treatment plant (1.14×10^5 C.M.) was 1.423 million dollars. Walker did point out that landspreading was unfeasible at Warren, Michigan due to the high heavy metal loading. The City of East Fitchburg, Massachusetts switched from incineration to land spreading showing an actual annual savings of $120,000 or $109/dry ton.

5. On October 3, 1977, the EPA Administrator, Douglas Costle issued a memorandum to Assistant and Regional EPA Administrators directing that the Agency "will press vigorously for publicly owned treatment works to utilize land treatment processes to reclaim and recycle municipal wastewater." He further stated, "Because land treatment processes contribute to the reclamation and recycling requirements of P.L. 92-500 they should be preferentially considered as alternative wastewater management technology." The effect of this designation is that alternative systems can qualify for an additional 10% Federal Construction Grants funding.

The disadvantages or constraints to utilizing land application are shown in Table 7.

TABLE 7. Constraints to Land Application of Sludge

Health and Public Acceptance

- Concern that hazardous constituents[a] will enter food chain
- Asthetics - odors
- Concern for possible adverse impact on ground and surface water
- Cultural attitudes toward reuse of human waste

Technical and Economic

- Inflexible arrangements for utilization of private land
- Loss of tax revenues from municipality purchased land
- Poorly trained operators
- Insufficient design example data base

Legal and Institutional

- Inflexible or overly restrictive regulations
- Zoning and other local restrictions
- Liability/responsibility for potential damages
- Superficial consideration during planning

[a] Pathogens, toxic heavy metals, and persistent organics

Ref. Walker, 1979

Land Application Systems

The type of land application system chosen depends on such factors as site availability including loading limitations, acreage available, topography, permeability, drainage, geology, etc.; type of crop to be grown including agricultural, forested or non-agricultural; and finally the sludge limitations, including water content, nutrients, heavy metals and any other constraints. The EPA (EPA 1978) soil limitations criteria are: permeability of the most restricting layer above 152 cm, which includes both high $>$51 cm/hr and low permeable soils $<$0.5 cm/hr; soil drainage class; runoff; flooding; and available water capacity in the 0-152 cm or root layer.

States often become more restrictive as in the case of Wisconsin which includes depth to seasonal water table (60 to 140 cm), depth to bedrock (60-140 cm) and permeability of the most restricting layer to 91 cm (EPA 1978). New York (Olson 1976) published guidelines for soil suitability to sludge application which further includes slope, trafficability, erosion, stoniness, pH in B horizon and texture.

The geology underlying the site is important in determining the groundwater flow both to and from aquifers. Depths to groundwater and seasonal groundwater table fluctuations as well as present and potential use of ground and surface water bodies need to be considered.

The process design is a function of crops chosen, sludge application rates and method of application. The basic criteria can be sited as maximization of crop yield and quality with a minimization of environmental damage. With agricultural land, the application rates are generally designed to supply the optimal amount of nitrogen required for crop growth. Along with the application of the sludge, it is generally necessary to maintain the soil pH near or above neutral to minimize movement of hea y metals in the soil matrix. While there is a vast store of information on nutrient needs for agronomic crops, the same is not true for forested and reclaimed sites. Hart (1974) provided data on the removal of different elements from soil by various crops. For example, alfalfa hay removes 744 kg/ha of nitrogen as compared to 179 kg/ha, 305 kg/ha, 20 kg/ha for corn grain, corn silage and loblolly pine. Similar data was shown for P, K, Ca, Mg, Fe, Mn, Cu, and Zn.

Forested sites offer many advantages as sludge application sites since the product, trees are not consumed by humans, and the forests generally have low population densities, therefore minimizing the potential for human contact with the sludge. The trees, however, do not utilize as much plant nutrients as do crops , therefore application rates need be lower or NO_3^- leaching may occur.

The use of sludge in reclaiming soils may be the best use in many cases. This use takes advantage of the soil stabilization characteristics of the organic matter and generally takes place where no food will be grown, therefore lessening the concern with heavy metals content.

COMPOSTING

Advantages and Disadvantages

Aerobic composting is the rapid biodegradation of the organic matter in sewage sludge. A 10-15% reduction in volatile solids occurs over the composting period. Since sewage sludge is predominately organic matter, stabilization of the volatile organics and destruction of pathogens will result in a material that can be used as a soil conditioner. The presence of nitrogen, phosphorous, potassium, calcium, magnesium and minor elemnts, can provide necessary plant nutrients in sufficient quantities depending on application rates. Proper composting can achieve temperatures in excess of $55^{0}C$ for several days which will result in pathogen destruction (Epstein, et al., 1978).

Composting systems can be simple, utilizing standard public works equipment, or highly sophisticated equipment consisting of pugmills, conveyors, stackers, screens, etc. This wide range of flexibility makes the process very attractive to municipalities since it can be adapted and modified to existing systems and local conditions. The low technology aspect of the process makes it highly attractive for developing countries.

Sludges varying in solids content ranging from 4 to 80% can be composted. Thus, dewatering is not a prerequisite. However, the economics of composting low solid sludge depends on availability of bulking materials and the cost of recycling or materials handling. This aspect can be utilized in composting night soil (Shuval, 1977).

The major disadvantage to the process is the larger land requirements over high technology processes, e.g. incineration or landfills. Approximately 13 metric tons of sewage sludge solids can be composted per hectare. The process is often described as labor intensive. This aspect however, depends on the extent of mechanization. The high labor feature of unmechanized systems further makes this process attractive to developing countries.

Systems

Prior to 1972, virtually none of the sewage sludge produced in the U.S. was composted. In 1972 U.S. Department of Agriculture at Beltsville, Maryland, initiated studies on windrow composting (Willson and Walker, 1973) and in 1974 developed the Forced Aeration System (Epstein and Willson, 1975; Epstein, et al., 1!

Accelerated aerobic composting can be achieved by two basic processes: windrow and forced aeration. A third system often referred to in the literautre as "enclosed or within vessel systems" essentially consists of the principals of the previously mentioned processes.

The windrow method has been used predominantly in agriculture for stabilizing animal wastes or in the preparation of bedding material for the mushroom industry In this process the waste is placed in a long, narrow (3-5m wide by 1.5-2.5 m hig pile, termed a windrow, and periodically turned. A windrow turning machine aerate the organic matter by periodic mixing. Further aeration is achieved by convective air flow. Temperatures are generally lower with wide variations among different parts of the windrow which results in non-uniform biodegradation of the organic matter and incomplete pathogen destruction. The process is affected by climate and hence is generally more suited for dry climates where enclosed facilities are not needed.The largest windrow composting system in the U.S. is operated in Los Angeles County, California.

U.S. EPA in its regulations concerning pathogen destruction (40 CFR 257) indicates that as a "process to further reduce pathogens" that "the solid waste attains a temperature of 55^0C or greater for at least 15 days during the composting period. Also during the high temperature period there will be a minimum of five turnings of the windrow." Burge,et al. (1978) showed that total coliforms, fecal coliforms, salmonella and f2 bacterial virus survived during the windrow process. The indicator virus survived for over 70 days with digested sludge and 55 days in the raw sludge under cool humid conditions.

To overcome the difficulties in pathogen destruction in windrow and to negate the effect of climate on the process, the Forced Aeration or Beltsville Aerated Pile Method was developed (Epstein, et al. 1976). Figure 1 and 2 show process flow diagrams for sludge solids ranging from 4 to 35%. Taffel, et al. (1980) composted sludge from the Ponce, Puerto Rico treatment plant drying bed which contained 80 to 90% solids using bagasse as a bulking material. Bulking materials are needed for both moisture control and adequate porosity. Occasionally a bulking material is used to adjust the carbon/nitrogen ratio. Bulking materials which have been used include: wood chips, bark, shredded paper, peanut hulls, rice hulls, straw, bagasse, and shredded rubber tires.

The basic process consists of placing a mixture of sludge and bulking material over an aerated system (Figure 3). Aeration systems can be premanently imbeded in the working pad or consist of flexible perforated plastic pipe (Epstein, et al., 1978). The entire pile is covered with a 30-cm layer of screened compost. Fifty tons of sludge containing 20% solids and using a bulking material at a volume ratio of 3:1 bulking material:sludge, produces a pile with a triangular cross section and approximate dimensions of 16 m long, 7 m wide, and 2.5 m high. The aeration pipe system is connected to a blower. A 0.33 hp unit will handle the sludge generated by the daily production from a city of approximately 250,000 persons. Aerobic composting conditions are maintained by drawing air through the pile. The effluent air stream is conducted into a small pile of screened, cured compost, where odorous gases are effectively absorbed. The blower is operated intermittently to provide oxygen levels between 5 and 15% during the composting period. Reversing the blower accelerates drying which can assist in processing. The sludge-bulking material is usually composted for 21 days or until the sludge is stabilized by the rapid decomposition of volatile organics and odors are abated.

A modification of the aerated pile is the aerated extended pile in which the sludge-bulking material mixture is placed in a pile which uses the shoulder (lengthwise dimension) of the previous day's pile, forming a continuous or extended pile. The extended pile reduces the operating area and cover by about 50% as well as base material. A pile base is essentially eliminated when the aeration system is imbedded into the pad.

After the composting period and depending on the composition of the waste it may be necessary to further stabilize it by additional curing. Curing is an ex-

FIGURE 1. Process Flow-Forced Aeration Pile Method With Sludge Solids Content of 4 to 15 Percent

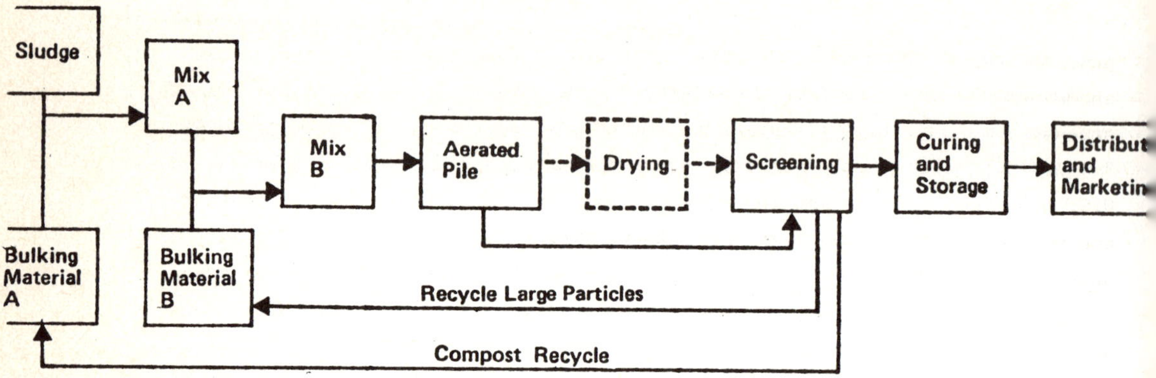

FIGURE 2. Process Flow-Forced Aeration Pile Method With Sludge Solids Content of 15 to 35 Percent

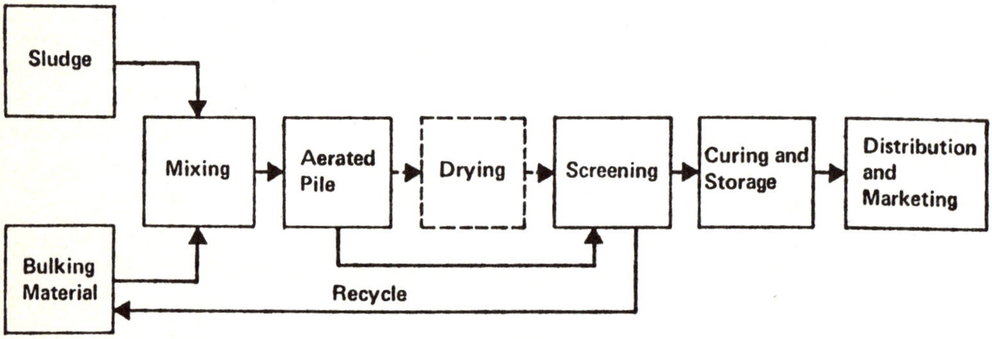

FIGURE 3. Composting With Forced Aeration

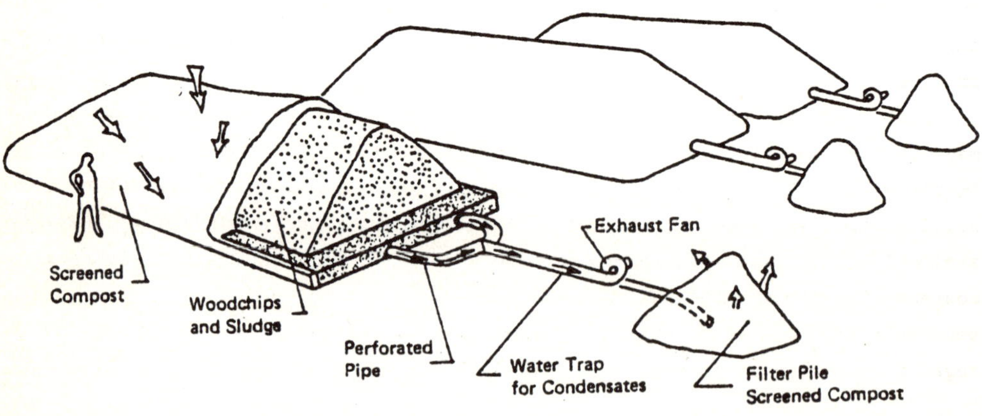

tension of the composting process, although temperatures are generally lower. Drying and screening are often carried out to remove and recycle the bulking material. Screening may also be desired to produce a marketable product.

A major advantage to the Forced Aeration Method is the high temperatures generated and controlled which result in pathogen destruction. Burge, et al. (1978), showed that salmonella, fecal and total coliforms and f2 bacterial virus were destroyed in 10 to 13 days. Temperatures exceeding $55^{0}C$ can be achieved for over 8.4 days at a confidence level of 99.9% and $65^{0}C$ for 2.4 days (Epstein, et al. 1978). U.S. EPA requires $55^{0}C$ or greater for 13 days (40 CFR 257) for the method to be considered a PFRP.

Uses

Sewage sludge compost is a well stabilized organic material whose physical and chemical characteristics depend on sludge, bulking material and processing system. The composition of sewage sludge is greatly affected by the type of community from which it is derived, i.e. industrial or domestic. Industrial discharge into wastewater treatment systems increase the heavy metal and organic chemicals. U.S.EPA is currently preparing regulations for the use of sludge and sludge products (compost) based on the levels of cadmium, lead and PCB's.

Sewage sludge compost is an excellent soil conditioner or soil amendment. The addition of composts to soils improves their physical properties, as evidenced by (a) increased water content, (b) increased water retention, (c) enhanced aggregation, (d) increased soil aeration, (e) increased permeability, (f) increased water infiltration, and (g) decreased surface crusting (Epstein, et al. 1976). The greatest improvement in soil physical properties occurs at the extremes of soil texture. Addition of sludge compost to sandy soils increases their ability to retain water and renders them less droughty. In heavy textured clay soils, the added organic matter increases permeability to water and air, increases water infiltration into the profile, reduces compaction and lowers the bulk density. These soils, therefore, will have a greater water storage capacity to be utilized for plant growth.

Sludge compost contains low levels of the major plant nutrients as well as many of the minor elements. Thus, at application rates often necessary for the improvement of the soil physical conditions sufficient nitrogen, phosphorous, calcium, and magnesium are provided for plant growth. The very low level of potassium in sludge may require the addition of this plant nutrient. The N, P, K, content of compost usually ranges from 1 to 1.5; 1.5 to 2 and 0.1 to 0.2% respectively. Nitrogen mineralization is usually from 6 to 15% of the total nitrogen in the compost (Epstein, et al. 1978).

REFERENCES

1. Anonymous, 1973. Monitoring for polychlorinated biphenyls in the aquatic environment. Mich. Water Resources Comm., Dept. Natural Resources, Lansing.

2. Anonymous, 1980. A guide to regulations and guidance for the utilization and disposal of municipal sludge. U.S. EPA 430/19-80-015, MCD-72, Washington, D.C. 20460.

3. Bitton, G., Adsorption of viruses into surfaces in soil and water. Water Res., 9, 473 (1975).

4. Bouwer, H., J.C. Lance and M.S. Riggs, 1974. High-rate land treatment II: water quality and economic aspects of the Flushing Meadows project. J. Water Poll. Contr. Fed. 46:844-859.

5. Braude, G.L. , C.F. Jelinek and P. Corneliussen, 1975. FDA's overview of the potential health hazards associated with the land application of municipal wastewater sludges. pp. 214-218. In Proc. of the 1975 Nat'l Conf. on Municipal Sludge Management and Disposal. Info. Transfer, Inc., Rockville, Maryland.

6. Burge, W.D., W.N. Cramer, and E. Epstein, 1978. Destruction of pathogens in sewage sludge by composting. Tran. of the ASAE 21(3)510-514.

7. Chaney, R.L. and Giordano, P.M., 1977. Microelements as related to plant deficiencies and toxicities. In Soils for Management and Utilization of Organic Wastes and Wastewaters. L.F. Elliott and J.F. Stevenson (Eds.), Soil Sci. Soc. Amer., Madison, Wisc., 234.

8. Dean, A.B., Hazards from metals and organic pollutants in sludge from municipal treatment plants. Paper presented at conference in connection with the International Water Conservancy Exhibition (Sept.1-5, 1975). U.S. EPA. Cincinnati, Ohio.

9. Duggan, R.E. and P.E. Corneliussen, 1972. Dietary intake of pesticide chemicals in the United States (III). June 1968-April 1970. Pesticide Monitoring Jour. 5:331.

10. Elinder, C.G., T. Kjellstrom, L. Friberg, B. Lind, L. Linnman, 1976. Cadmium in kidney cortex, liver and pancreas from Swedish autopsies. Archives of Environmental Health 31:292-302.

11. EPA, 1978. Sludge treatment and disposal. Vol. 2 Disposal EPA-625/4-78-012. Env. Res. Info. Center. Cincinnati, Ohio.

12. Epstein, E. and G.B. Willson, 1975. Composting raw sludge. In Proc. of the 1975 Nat'l Conf. on Municipal Sludge Mgt. and Disposal. Information Transfer, Inc., Rockville, Maryland 20582.

13. Epstein, E., J.M. Taylor and R.L. Chaney, 1976. Effects of sewage sludge and sludge compost applied to soil on some soil physical and chemical properties. J. Environ. Qual. 5 (4):422-426.

14. Epstein, E., G.B. Willson, W.D. Burge, D.C. Mullen and N.K. Enkiri, 1976. A forced aeration system for composting wastewater sludge. J. Water Poll. Control Fed. 48:688-694.

15. Epstein, E., D.B. Keane, J.J. Meisinger and J.O. Legg, 1978. Mineralization of nitrogen from sewage sludge compost. J. Environ. Qual. 7(2): 217-221.

16. Epstein, E., G.B. Willson and J.F. Parr, 1978. The Beltsville aerated pile method for composting sewage sludge. In New Processes of Waste Water Treatment and Recovery, Ellis Horwood Ltd. Pub. Chichester, England.

17. Foster, D.H. and R.S. Engelbrecht, 1973. Microbial hazards in disposing of wastewater on soil. pp 247-270. In Sopper, W.E. and L.T. Kardos (ed.) Recycling treated municipal wastewater and sludge through forest and crop land. The Penn. State Univ. Press, Univ. Park.

18. Furr, A.K. , A.W. Lawrence, S.S.C. Tong, M.C. Grandolfo, R.A.Hofstader, C.A. Bache, W.H. Gutenmann, and D.J. Lisk, 1976. Multielement and chlorinated hydrocarbon analysis of municipal sewage sludges of American cities. Env. Sci. and Tech. 10 (7):683-687.

19. Hart, R.H. 1974. Crop selection and management. In Factors involved in land application of agricultural and municipal wastes. USDA, ARS, Beltsville, Md.

20. Henry, J.G. and R. Gehr, 1980. Odor control: an operation's guide. J.W. P.C. F. 52(10) 2523-2537.

21. Hori, D.H., 1970. Migration of Polio virus type 2 in percolating water through selected Oahu soils. Tech. Rept. No.36, Water Resources Research Center, Univ. of Hawaii, Honolulu.

22. Jelinek, C.F., and G.L. Braude. 1977. Management of sludge use on land: FDA considerations. p.35-38. In Sludge Management Disposal and Utilization. Information Transfer, Inc., Rockville, Maryland.

23. Jewell, W.J. and B.L. Seabrook, 1979. A history of land application as a treatment alternative. EPA 430/9-79-012, U.S.EPA, Washington, D.C.

24. Kabler, P., 1959. Removal of pathogenic microorganisms by sewage treatment processes. Sewage and Industrial Wastes, 31(6)683-689.

25. Keswick, B.H., and C.P. Gerba, 1980. Viruses in groundwater. Env. Sci. and Tech. 14(11)1290-1297.

26. Olson, G.W., 1976. Significance of soil characteristics to waste management alternatives. In Loehr, R.C. (ed) Land as a Waste Management Alternatives.Cornell Agr. Waste Mgt. Conf. Ann Arbor, MI, Ann Arbor Sci., Inc.

27. Page, A.L. , 1974. Fate and effects of trace elements in sewage sludge when applied to agricultural lands. A literature review study. U.S. Environ. Prot. Agency, Rep. No. EPA-670/2-274-005. 108 p.

28. Pahren, H.R., J.B. Lucas, J.A. Ryan, and G.K. Dotson, 1977. An appraisal of the relative risks associated with land application of municipal sludge. Presented at the 50th Annual Conf. of the Water Poll. Cont. Fed. Phil. Penn.

29. Parsons, D., C. Brownlee, D. Wetter, A. Maurer, E. Haughton, L. Kornder, M. Slezak, 1975. Health aspects of sewage effluent irrigation. Pollution Control Branch, British Columbia Water Resources Services, Victoria.

30. Prince, G., B. Tonetti and C. Metcalf, 1980. Sludge management, a Federal perspective. In Proc. Nat'l. Conf. on Municipal Sludge Composting, pp1-4. Information Transfer, Inc., Silver Springs, Maryland.

31. Romero, T.G., 1970. The movement of bacteria and viruses through porous media. Ground Water, 8,37.

32. Sabey, B.R., and W.E. Hart, 1975. Land application of sewage sludge: Effect on growth and chemical composition of plants. J. Environ. Qual. 4:252-256.

33. Sandstead, H.H., 1974. Cadmium, zinc and lead. Geochem. and the Environment, 1:43-56. Nat'l. Academy of Science.

34. Sepp, E., 1971. The use of sewage for irrigation, a literature review Bureau of Sanitary Engineering.Dept. Public Health, State of California, p. 41.

35. Shuval, H.I., 1977. Night soil composting: state of the art and research-pilot study needs. P.U. Report No. Res. 12, International Bank for Reconstruction and Development, Energy, Water and Telecommunication. Dept. Washington, D.C.

36. Sommers, L.E., D.W. Nelson and K.J. Yost, 1976. Variable nature of chemical composition of sewage sludges. J. Environ. Qual., 5(3):303-306.

37. Taffel, W., J.E. Alpert and E. Epstein, 1980. Technical and economic report for the Ponce composting pilot plant program. Energy Resources Company, Inc., Cambridge, Massachusetts.

38. Walker, J.M., 1979. Overview: costs, benefits and problems of utilization of sludges. In Municipal Sludge Management, Information Transfer, Inc. Silver Springs, Md.

39. Wang, W.L. and S.G. Dunlop, 1959. Animal parasites in sewage and irrigation water. Sewage and Industrial Waste 26(8):1020-1032.

40. Wellings, F.M., A.L. Lewis, C.W. Mountain and L.M. Stark, 1975. Virus consideration in land disposal of sewage effluents and sludge. Florida Sci. 38 (4):202-207.

41. Wellings, F.M., A.L. Lewis and C.W. Mountain, 1976. Demonstration of solids-associated virus in wastewater sludge. Appl. Environ. Microbiol. 31:354-358.

42. Willson, G.B. and J.M. Walker, 1973. Composting sewage sludge: How? Compost Sci. 14:4

43. Wright, et al., 1942. Preliminary observation on the effect of sewage treatment processes on the ova and cysts of intestinal parasites. Sew. Wks. Jour. 14:1274-1280.

44. Yamagata, N., and Shigematsu, I., 1970. Cadmium pollution in perspective. Bull. Inst. Pub. Health, 19, 1.

Developments in Arid Zone Ecology and Environmental Quality
Edited by H. Shuval

27

THE USE OF GEL CHROMATOGRAPHY FOR DETERMINATION OF BOUND Ca AND Mg IN SEWAGE EFFLUENT-SOIL SYSTEMS

A. Acher, Y. Pistol and B. Yaron

Division of Soil Residues Chemistry, Institute of Soils and Water, The Volcani Center, ARO, Bet Dagan, Israel.

ABSTRACT

A method, based on Sephadex G-10 gel chromatography technique, was used for determination of free and bound Ca and Mg ions in model solution, sewage effluents and soil extracts. Since the eluant ionic composition is the predominant factor, influencing the recovery and the resolution efficiency of the gel column, a few eluants containing divalent cations (Ca^{2+}, Ba^{2+}, Mg^{2+}, Sr^{2+}) were tested. The best reproducible results (recovery 96-100% and resolution 68-86%) were obtained using an eluant composed of $CaCl_2$ (0.003 N), $MgCl_2$ (0.001 N) and $MgCl_2$ (0.003 N). The amount of Ca ion bound, found in different sewage effluents, was 10-50% from total Ca present. The amount of Mg ion bound was only 10-20% from total Mg present. In soil-sewage batch experiments the ratio between bound to free ions was significantly increased in the case of Mg.

INTRODUCTION

One of the problems raised in the use of sewage effluents in the crop irrigation is the effect of water composition on the soil physical properties and in particular on the soil hydraulic conductivity. Since the cationic composition of the disposed water affects its movement into the soil, knowledge of the speciation of the Ca and Mg ions in the sewage effluent itslef and in the soil solution, is of major importance. Sposito and Mattigod (1977) already raised the idea of a potential relationship between the sodium adsorption ratio (SAR) and the solution complex formation in the solution phase. Discussing the validity of practical sodium adsorption ratio (SAR_p) - which employs the total concentration of Na^+, Ca^{++} and Mg^{++} instead of their free cationic concentration - they concluded that only a statistical correlation occurs between SAR and SAR_p. Their interpretation was based only on the possible effect of the metal complexing anions and no evidence of organic ligands was discussed. Sposito et al. (1978) proved that calcium ion was complexed by fulvic acid originating from sewage sludge. In this particular case the complexing agent was a well defined one and the complex formation functions were rigorously fixed parameters.

The prediction of changes in the soil physical properties during the process of sewage water disposal or of irrigation with sewage water, requires information on the ratio between free and bound divalent cations in sewage and sewage soil systems. In this paper a method is described which separates Ca and Mg species by avoiding interference of existing components found in the natural systems. This method, based on gel-chromatography techniques was tested for model solution, natural sewage and soil extracts. The results obtained are hereby presented.

EXPERIMENTAL PROCEDURE

Materials

Sephadex G-10 was used for gel chromatography. It had a particle size of 40-120μ, and a bed volume of 2-3 ml/g dry gel.

The chloride salts ($MgCl_2$, $CaCl_2$, $SrCl_2$, $BaCl_2$, NaCl) used for the preparation of eluant solutions and standard samples were of analytical grade. EDTA (disodium salt), used for complexation of Ca and Mg ions, was also of analytical grade.

The municipal sewage effluents and the soils used for column and batch experiments were from different locations; their main characteristics are given in Table 1 and Table 2, respectively.

TABLE 1. Composition of sewage effluents used in the experiments.

Location	Suspended solids, mg/l	pH	EC mmho/cm	COD mg O_2/l	MBAS mg LAS/l	Ca mg/l
Ramat Hasharon 1	180	7.7	1.7	280	4.2	80
Ramat Hasharon 2	110	7.6	1.7	156	2.6	74
Eilat	150	7.2	3.2	136	3.0	140
Rehovot	160	8.35	1.8	140	4.0	101

TABLE 2. Main characteristics of the soil used in the experiments.

Soil type	Locaton	Clay %	OM %	pH	Exch. Cations meq/100 g soil				$CaCO_3$ %
					Ca	Mg	Na	K	
Sandy regosol	Mivtahim	7.5	0.11	8.3	2.8	1.1	0.2	0.2	5.7
Loamy loessial sierozem	Gilat	23.0	0.95	7.5	8.6	3.0	0.7	0.6	12.6
Grumusol	Neve-Yaar	70.0	2.03	7.4	29.4	22.3	5.1	3.2	3.4

The water used for the preparation of solutions was freshly distilled from deionized water containing potassium permanganate.

The columns used for gel chromatography were glass burette of 25 ml (1.0 cm I.D., 42 cm height); the columns used for soil experiments were lucite columns (5.5 cm I.D., 14.0 cm height) mounted vertically and supported on a porous outflow surface of glass wool on a metal screen. The eluant levels over the gel and soil columns were kept constant by using Mariotte flasks as eluant reservoirs for equilibration and elution.

Methods

Standard methods from "Standard Methods for the Examination of Water and Wastewater" (1975) and from "Saline and Alkali Soils" (1954) were used for determination of water and soil characteristics, respectively. The atomic absorption instrument, used for the determination of the Ca and Mg, was a Perkin-Elmer 460 Atomic Absorption Spectrophotometer.

Procedure

1. Separation of complex and free Ca and Mg ions by Sephadex gel filtration. A slurry of Sephadex (12 g) in an eluant (35 ml) was prepared in an erlenmeyer flask (100 ml). The gel was allowed to swell for 5 h at room temperature and, after deaeration by vacuum for 30 min., packed on the column. The column was previously plugged with glass wool and filled with eluant. When all the gel was settled into the column, its upper surface was protected with a piece of filter paper. For equilibration the eluant was run overnight through the gel bed (15 h, 15-20 ml/h). To prevent the microbial growth, 0.2% of sodium acide (NaN_3) was dissolved in the eluant.

The sample (1.0 ml) was applied on the drained bed surface. After the sample was drained into the bed, the column wall and the gel surface in contact with the sample were washed with a small amount of eluant (3 x 0.5 ml). Then the column was filled with eluant, connected again to a Mariotte flask and to a fraction collector through a drop counter (1.0 ml per 4 min per tube). Fifty fractions (tubes) were collected for each sample. The fractions were diluted with distilled water (0.5 ml sample to 9.5 ml water), and the Ca and Mg content was checked on an atomic absorption spectrophotometer.

Standard samples, containing different ratio of free ions (M) to complexed ions (ML) of Ca and Mg (1:1, 1:1.67, 1:2.50), were used in this study. They were prepared by titrating solutions of Ca or Mg

(0.02 N) with the calculated amount of EDTA-disodium solution (0.02 N). The final concentration of total Ca or Mg ions was adjusted with distilled water to 0.01 N. All the experiments were performed in duplicate.

2. Soil effects on the ionic composition of sewage. This effect was checked by leaching soil columns with sewage effluent (a) or by equilibrating the soil with sewage effluent in batch experiments (b).

a) Soil column

Sieved soil (100 g, 250 mesh) and acid wash sand (100 g, fine) were mixed together (to increase the permeability of soil to fluids). After homogenization the mixture was poured at once into the lucite column. The soil column was eluted with sewage effluent or water model solution having the same ionic composition of Ca^{2+}, Mg^{2+} and Na^{+} as the sewage effluent until 500 ml of eluant were collected. Samples of this eluant were analyzed (after dilution, 1:19) for Ca, Mg and Na free and complex ions.

b) Batch experiments

Ten g of the above mixture of soil and sand together with 20 ml of sewage effluent or water model solution were shaken for 60 min at room temperature into plastic bottles (50 ml), followed by centrifugation at 7000 rpm for 10 min. The clear supernatant was analyzed (after dilution, 1:19) for Ca, Mg and Na free and complex ions.

RESULTS AND DISCUSSION

a) Selection of the eluant.

The most important point in practice of the gel chromatography method for separation of the metal complex (ML) from the free metal ion (M) is the selection of a proper eluant. The chosen eluant should avoid the dissociation of ML into its constituents during the passage through the column leading to a complicated elution profile of ML, L (ligand) and M, depending on stability constant of ML and kinetic factors (Yoza, 1977). To minimize the analytical error and for a sensitive detection of ML-M, it is important that the concentration of the electrolyte in eluant be kept higher than the ML concentration. Such a concentration will also avoid the complexity caused by Donnan effect (Jordan et al., 1974; Peterson, 1970).

Using methods described in the literature (Veno et al., 1970), results were obtained which were unsatisfactory because of their poor reproducibility. Thus, various eluants were checked for gel chromatography separation of ML and M. Table 3 shows the effect of the electrolyte in the eluant on the efficiency of ML-M separation. The use of distilled water as eluant is limited, for the same Sephadex bed, to only 2 or 3 samples after which the results became irreproducible. Sodium chloride solution, in spite of literature data (Yoza, 1970) gave poor recovery (the amount of sample ions eluted from column in respect to the initial amount of ions in the sample) and resolution-R-efficiency [$R = 2(V_{ML} - V_M)/W_{ML} - W_M$: V = eluant volume, W = peak width]. The Sr and Ba solutions were chosen as eluants because chemical approach of these elements to Ca, stability constants of Sr(EDTA) and Ba(EDTA) are lower than the stability constants of Ca(EDTA), and they don't interfere in the atomic absorption spectroscopy determination of Ca. However, the results obtained using these electrolytes were far from being satisfactory (Table 3).

TABLE 3. Effect of the electrolyte in the eluant on the efficiency of ML-M separation*.

Electrolyte	Recovery %	Resolution
Distilled water	70-130	65-75
0.08 N NaCl	50-90	separation not good
0.06 N $SrCl_2$	60-125	52-60
0.10 N Sr Cl_2	60-125	53-65
0.02 N $BaCl_2$	65-120	43-65
0.10 N $BaCl_2$	65-120	52-70
0.006 N NaCl 0.003 N $CaCl_2$ 0.001 N $MgCl_2$	87-110	73-96
0.010 N $CaCl_2$	85-115	-

* The sample used contained 0.01 N Ca (ML+M) at a ratio ML:M=1.0.

Eventually, two eluants were found to give good results. One eluant contained NaCl (0.006 N), $CaCl_2$ (0.003 N) and $MgCl_2$ (0.001 N). The recovery of Ca and Mg cations applied with the samples at the top of column was very good (87-110%) and reproducible, while the resolution of the peaks (ML and M) was good enough (73-96%) for practical purposes. The elution profile (Fig. 1) showed two positive peaks (ML and M) which positions and surfaces were reproducible (the elution peaks of Mg ions precede those of Ca ions).

The second eluant used in our experiments was a relatively concentrated solution of $CaCl_2$ (0.010 N). Because the concentration of Ca in the eluant was higher than in the sample, the elution profile (Fig. 2) showed a positive peak for ML followed by a negative peak which corresponded to the amount of M consumed to form ML. The free M could not be determined directly by this method. Because the concentration of Ca in the eluant is higher than the total Ca concentration in the sample, the amount of ML found is also probably usually higher than the actual concentration in sewage effluent (if free ligand is present in the effluent), and less than the maximum binding ability of the ligands present in effluents.

Table 4 shows the efficiency of Ca and Mg complex ion separation, respectively, by Sephadex gel filtration, using model water solutions with different ratios of ML:M. The eluant used in these determinations contained a mixture of Na, Ca and Mg cations. The recovery of Ca and Mg cations applied with the model water samples at the top of column was very good (96-100%) and reproducible, and the resolution of the peaks (ML and M) was good (68-86%).

Using the second eluant, which contains only Ca cations, the ML results obtained were about 20% higher than in the first method. However, if the concentration of the eluant is high enough, the value of ML can show the total binding ability of Ca by ligand from sewage effluents and soils. For such a purpose this method seems to be more suitable than the method using Cobalt complexation for determination of the complexing capacity of natural water (Hanck and Dillard, 1977).

b) Ca and Mg speciation in soil-sewage effluent systems.

When sewage effluent is added to a soil it is assumed that the bound forms of Ca and Mg initially present in the applied sewage water will be affected by the soil composition. Table 5 shows the amount of bound Ca and Mg (ML forms) in the sewage effluent studied, the column effluent and

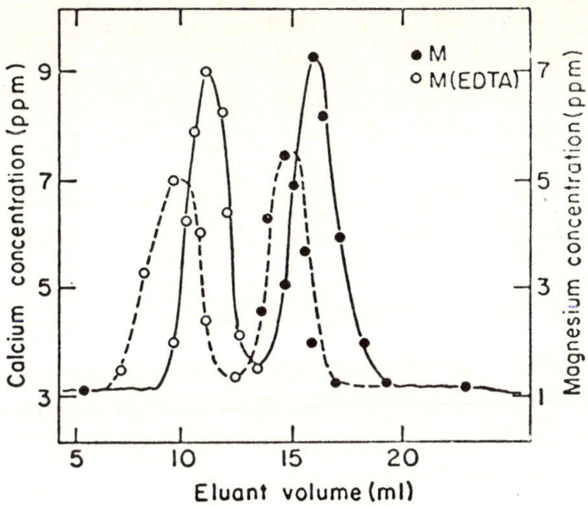

Fig. 1: The elution profile of Ca (solid lines) and Mg (broken lines) species.

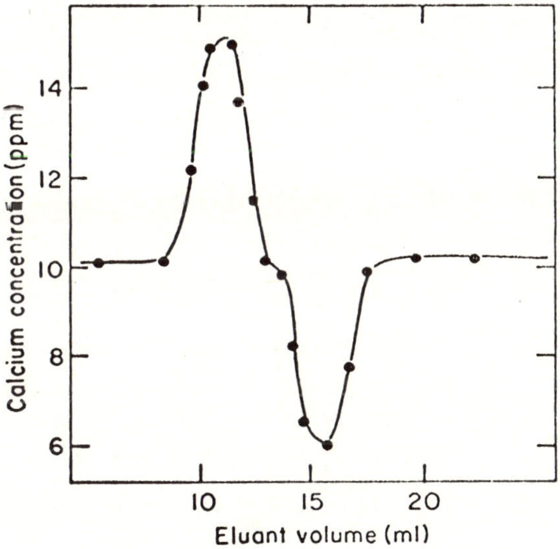

Fig. 2: The elution profile of Ca(EDTA). Eluant: 0.1 N $CaCl_2$.

TABLE 4: Efficiency of Ca and Mg species separation by Sephadex gel filtration

SAMPLE	Sample composition								Efficiency			
	Before separation				After separation				Recovery		Resolution	
	ML + M, 10^{-2}N		ML : M		ML + M, 10^{-2}N		ML : M		%		%	
	Ca	Mg	Ca	Mg	Ca	Mg	Ca	Mg	Ca	Mg	Ca	Mg
Water solution 1	1.37	1.00	1:1	1:1	1.37	0.96	1:1.1	1:1.15	100	96	75	82
Water solution 2	1.66	1.67	1:1.66	1:1.67	1.60	1.58	1:1.7	1:1.45	96	95	94	76
Water solution 3	1.91	1.90	1:2.5	1:2.5	1.85	1.84	1:2.5	1:2.2	97	97	96	73
Sewage RS-1	0.273	0.11	unknown		0.271	0.10	1:9	1:8.5	100	91	84	84
Sewage RS-2	0.384	0.12	unknown		0.384	0.11	1:3	1:5.0	100	92	81	86
Sewage E	0.800	0.82	unknown		0.800	0.78	1:1.9	1:6.5	100	95	71	72
Sewage R	0.295	0.10	unknown		0.285	0.10	1:2.5	1:7.8	96	100	84	68

in the equilibrium solution from the batch studies. It was observed that the bound forms of Ca and Mg were greater in the equilibrium solutions than in the column effluents. This difference may be explained by the binding ability of soil ligands for the metal cations. The least affected soil-sewage system is that equilibrated with Gilat soil. This soil is characterized by a high Ca CO_3 content and soluble Ca and Mg. The dissolution of $CaCO_3$ in the equilibrium solutions affects the ratio ML:ML-M which for the other two soils is determimed mainly by the sewage and soil ligands. As was pointed out, the relative concentration of bound Ca and Mg in the eluant is in general lower than in the initial sewage solution. This effect is strongly observed in the case of the Gilat soil, where the low organic matter content, which can serve as a source of metal cation ligands, is about 50% of the initial ML concentration in sewage effluent. In the column studies, the rate of Ca and Mg dissolution from the soil is greater than the release of potential soil ligands and therefore the M:ML ratio is changed in favour of the free cations.

TABLE 5. Amount of bonded Ca and Mg in sewage-soil solution.

Type of solution	Ca ML:ML+M (%)	Mg ML:ML+M (%)
[1]Sewage effluent	34.5	34.5
[2]Sewage + sandy soil (Mivtahim)	33.7	37.1
[2]Sewage + loamy soil (Gilat)	17.1	17.09
[2]Sewage + clay soil (Neve-Yaar)	33.7	33.3
[3]Sewage + sandy soil (Mivtahim)	43.0	62.9
[3]Sewage + loamy soil (Gilat)	38.8	36.1
[3]Sewage + clay soil (Neve-Yaar)	36.2	53.5

[1]untreated
[2]after leaching, through soil
[3]at equilibrium with soil

ACKNOWLEDGMENT

This study was supported by the research grant WG 129-WT8 of the Israel National Council for Research and Development and the German Bundensministerium für Forschung und Technologie.

REFERENCES

American Public Health Association. (1971) Standard methods for the examination of water and wastewater. 13th ed., APHA, Washington, D.C.

Hanck, K.W., and J.W. Dillard. (1977) Determination of the complexation capacity of natural water by cobalt (III) complexation. Analytical Chem. 49: 404-409.

Jordan, D.O., S. Levell, D. Phillips and D.J. Winzor. (1974) J. Biochemistry 13: 1832-1836.

Peterson, R. (1970) "An Introduction to Ion Exchange." Heyden and Sons, Ltd., London.

Richards, L.A. (Ed.) (1954) Diagnosis and Ioprovement of Saline and Alkali Soils. U.S. Dept. of Agriculture, Washington, D.C.

Sposito, G., K.M. Holtzclaw and C.S. Le Vesque-Madore. (1978) Calcium ion complexation by fulvic acid extracted from sewage sludge-soil mixtures. Soil Sci. Soc. Am. J. 45: 197-198.

Sposito, G. and S.V. Mattigod. (1979) On the chemical foundation of the sodium adsorption ratio. Soil Sci. Soc. Am. J. 41: 323-329.

Veno, Y., N. Yoza and S. Ohashi. (1970) Gel chromatographic behavior of some metal ions. J. Chromatog. 52: 321-327.

Yoza, N. (1977) Determining of stability constant of a metal complex by gel chromatography. J. Chem. Education 54: 284-287.

Yoza, T., T. Ogata and S. Ohashi (1970) The effect of sample concentration on the gel chromatographic behavior of alkaline earth metal ions Sephadex columns. J. Chromatog. 52: 329-338.

Developments in Arid Zone Ecology and Environmental Quality
Edited by H. Shuval

28

ANAEROBIC DIGESTION OF COMBINED AGRICULTURAL WASTES

S. Kimchie and A.M. Wachs

*Department of Environmental and Water Resources Engineering,
Technion, Haifa 32000, Israel.*

ABSTRACT

The reuse through anaerobic digestion of organic waste produced in the agricultural sector is valuable not only for improving environmental quality by eliminating nuisances but also because of the production of energy and livestock feed. Combining various wastes produced on the farm for use as substrate for digestion may improve economic feasibility of the process by making possible increase in plant size, with its inherent economics of scale, while the availability of various waste streams will permit optimizing the chemical composition of the input to the digester.

The ratio of nitrogen to available organic matter (N:AOM) has been used for practical reasons as an indicator of the chemical composition of digester inputs. AOM is defined as the total organic solids less the lignin.

A battery of 7 bench-scale digesters was fed with waste mixtures of different compositions. When the digesters were fed with poultry manure which is characterized by a high N:AOM ratio of 1:16, there was a substantial increase in ammonium as well as free ammonia, both reaching toxic levels. Reducing the N:AOM ratio to 1:27 - 1:40 by mixing poultry manure with paper or cattle manure significantly improved digester performance. Mixing 50% (dry solids basis) cow manure with the poultry manure (1:27 N:AOM) was found to be an efficient way of stabilizing digesters operation even under high loading rates of 8-15 g/l/d. Adding paper to the cattle-poultry manure mixture thus reducing the N:AOM ratio to 1:33 - 1:40 further improved the digestion efficiency. Gas yields in digesters fed with the latter mixture was higher than in highly loaded digesters fed with cow manure only and operated under either mesophilic or thermophilic conditions.

This project was financed by the Kibbutz Industries association as a national R & D project of the Israeli Ministry of Energy and Infrastructure.

Developments in Arid Zone Ecology and Environmental Quality
Edited by H. Shuval

29
DISPOSAL OF URBAN SOLID WASTES

Z. Offer

Municipality of Jerusalem, Israel

ABSTRACT

Israel is a highly urbanized society. By the end of the century 4.8-5.0 million people will be living in 72 towns and urban type settlements. The average solid waste generation in urban areas is 1.8-2.0 kg/day/capit. Of this amount, 40-45% is stable wastes and 60-65% is the so-called "Garbage": a heterogeneous mixture of hundreds of items discarded from household, public, commercial and industrial activities. Of the many pollutants endangering our environment, the garbage is, qualitatively and quantitatively, probably the worst, and by itself can generate all known forms of pollution.

During the next twenty years, our urban population will generate approximately 180 million cubic meters of garbage weighing 35 million tons. The quantity of garbage and the location of our urban concentrations, and the size of the country, preclude the continuation of the present day methods of waste management.

Matter is indestructible and it cannot be disposed of; it can only be converted. That can be done by using advanced methods of materials and energy recovery. In the developed countries, recovery and recycling of waste materials is encouraged and subsidized as the only method of environmental protection from the solid waste-generated pollution. Approximately 45% of our garbage components can be recycled and 85-90% of it can be used for energy recovery. The average Israeli urban garbage has an energy potential of 8.0-8.5 million B.T.U. perton equivalent to 0.154 tons of heavy industrial oil or 0.215 tons of high grade coal. At the present "state of the art", energy and materials recovery is the only viable solution, regardless of economic consideration.

Developments in Arid Zone Ecology and Environmental Quality
Edited by H. Shuval

30

UTILIZATION OF AGRICULTURAL WASTE AS AN ENERGY SOURCE IN ISRAEL

Dan Dvoskin

Heshev - The Inter-Kibbutz Unit for Management Services,
P. O. Box 420021, Tel-Aviv, Israel

ABSTRACT

World energy resources are divided into two major categories: (a) fossil fuels and (b) non-fossil fuels. The fossil fuel energy resources include oil, gas and coal. The non-fossil energy resources are mainly solar, nuclear and geothermal energy. The main characteristic of fossil fuel energy resources is their carbon component. The energy from such resources is released through chemical reaction of carbon and oxygen.

The process of obtaining energy from organic components containing carbon can be applied to any organic matter. Therefore, much attention has been focused on biomass as an energy source.A major advantage of biomass energy is its renewable nature. Thus, if we learn how to utilize biomass, we may be able to alleviate some of the consequences of the energy crisis.

A major source of biomass energy is agricultural waste. These wastes often present a major ecological problem. Thus, the use of these wastes as an energy source might offer a cosiderable amount of energy, as well as a solution to the ecological problems.

The paper describes and analyzes the potential for utilizing agricultural waste in Israel as an energy source. Today, there are three basic ways to utilize these wastes. These are: (a) anaerobic digestion for methane production, (b) direct combustion, and (c) ethanol and methanol production. Each of the above methods is quite different from the others in terms of the waste material that can be used, the investment in the utilization facility, the cost of operation, and the value of the energy prduced.

The paper analyzes the technical and the economic feasibility of utilizing each of the waste materials as a raw material in each of the above three energy methods. Special attention is devoted to the cost of supplying the raw material and the cost of converting the raw waste into useful energy.

The paper emphasizes the importance of the economics of waste utilization as a key factor determining the degree to which energy can be a solution to agricultural waste. Clearly, with increased energy prices, utilization of agricultural waste becomes more and more economical. Thus, the paper demonstrates the close relationships between energy and the environment.

Developments in Arid Zone Ecology and Environmental Quality
Edited by H. Shuval
© 1981 Balaban ISS, Philadelphia, Pa.

31

EFFECT OF SALT ON SURVIVAL AND GROWTH OF *VIBRIO CHOLERAE*

Benzion Cavari* and R. R. Colwell

Department of Microbiology, University of Maryland, College Park, MD 20742 USA

**on sabbatical leave from Israel Oceanographic and Limnological Research, Kinneret Limnological Laboratory, P.O.B. 345, Tiberias, Israel.*

ABSTRACT

Vibrio cholerae are not isolated from Chesapeake Bay water or sediment samples where the salt concentration is higher than 1.7%. *Vibrio cholerae* V-69 serotype 01, isolated from Chesapeake Bay water, was examined for glucose uptake and respiration rates at different salt concentrations. No uptake was found when the salt concentration was ≥ 2%. Sensitivity to high salt concentration decreased with increasing temperature or organic matter concentration.

INTRODUCTION

The recent isolation of 01 serotype of *Vibrio cholerae* from the Chesapeake Bay, the largest estuary on the east coast of the United States, is of great interest for microbial ecology, as well as epidemiology, since strains of *Vibrio cholerae* were isolated from both polluted and unpolluted areas of the Chesapeake Bay (Colwell *et al*., 1977). *Vibrio cholerae* was not recovered from the Chesapeake Bay where salt concentrations were higher than 1.7%. *Vibrio cholerae* V-69 serotype 01, isolated from Chesapeake Bay water, was examined in the laboratory to determine salinity effects on growth in order to achieve a better understanding of the survival of *Vibrio cholerae* under natural conditions.

MATERIALS AND METHODS

Vibrio cholerae cells were incubated in filtered water from the Anacostia River containing different concentrations of NaCl and supplemented with 0.01% yeast extract. Growth of the cells was followed by direct epifluorescent count employing the procedure of Hobbie et al. (1977) and by total viable, aerobic, heterotrophic counts by spread plating on plate count agar (Sayler et al., 1975). Glucose uptake and respiration rates were determined by the method of Vaccaro and Jannasch (1966), with some modification as described by Cavari and Hadas (1979).

RESULTS AND DISCUSSION

As can be seen from Figs. 1 and 2, optimal salt concentration for growth was 0.5%. A less rapid increase in cell number was obtained when no salt was present in the growth medium or when the salt concentration was elevated to 2%. The salt effect was much more drastic when glucose uptake and respiration rates were measured (Table 1). When the assay medium contained 1% NaCl, glucose uptake was only ca. 3% of that measured when no salt was added, and in the presence of 2% NaCl, no activity was detected. It is interesting to note that with increasing salt concentration, the amount of CO_2 released also increased, which suggests that the cells are stressed. The evidence suggests that the cells grew in the presence of 3% NaCl because of the relatively high substrate concentration, i.e., 100 $mg \cdot l^{-1}$ yeast extract, added to the growth medium. Thus, organic matter can act as a protective agent for Vibrio cholerae cells against the effect(s) of salt. Another factor which may provide protection to Vibrio cholerae cells against elevated salt concentration is temperature, as can be seen in Table 2. At 25°, although significant activity was lost when the cells were incubated at increased salt concentration, the remaining activity of the cells in taking up glucose was higher than that obtained when incubation was at 5°. Temperature appears to be a significant factor influencing survival of Vibrio cholerae under natural conditions.

REFERENCES

Cavari B.Z. & Hadas O. (1979) Heterotrophic activity, glucose uptake and primary productivity in Lake Kinneret. *Freshwater Biol.* 9: 329-338.

Colwell, R.R., Kaper J. & Joseph S.W. (1977) Vibrio cholerae, Vibrio parahaemolyticus, and other vibrios: occurrence and distribution in Chesapeake Bay. *Science* 198: 344-346.

Hobbie J.E., Daley R.J. & Jasper S. (1977) Use of nucleopore filters for counting bacteria by fluorescence microscopy. *Appl. Environ. Microbiol.* 33: 1225-1228.

Sayler, G.S., Nelson, Jr. J.D., Justice A. & Colwell R.R. (1975) Distribution and significance of fecal indicator organisms in the upper Chesapeake Bay. *Appl. Microbiol.* 70: 625-638.

Vaccaro R.F. & Jannasch H.W. (1966) Studies on heterotrophic activity in seawater based on glucose metabolism. *Limnol. Oceanogr.* 11: 596-607.

Table 1. Effect of salt concentration on glucose uptake rate and respiration

NaCl concentration %	Glucose uptak μg/l/hr	Respiration %
0	1.73	30.6
0.5	0.48	43.2
1.0	0.048	74.5
2.0	0	

Table 2. Effect of salt concentration on ıcose uptake rates at two temperatures

NaCl concentration %	Incubation temperature °C	Glucose uptake μg/l/hr	Remaining activity (%)
0	5	0.642	100
0.5	5	0.169	26.3
1.0	5	0.0725	11.3
2.0	5	0.0114	1.8
4.0	5	0.0089	1.4
0	25	2.24	100
0.5	25	1.63	72.8
1.0	25	0.49	21.9
2.0	25	0.096	4.3
4.0	25	0.045	2.0

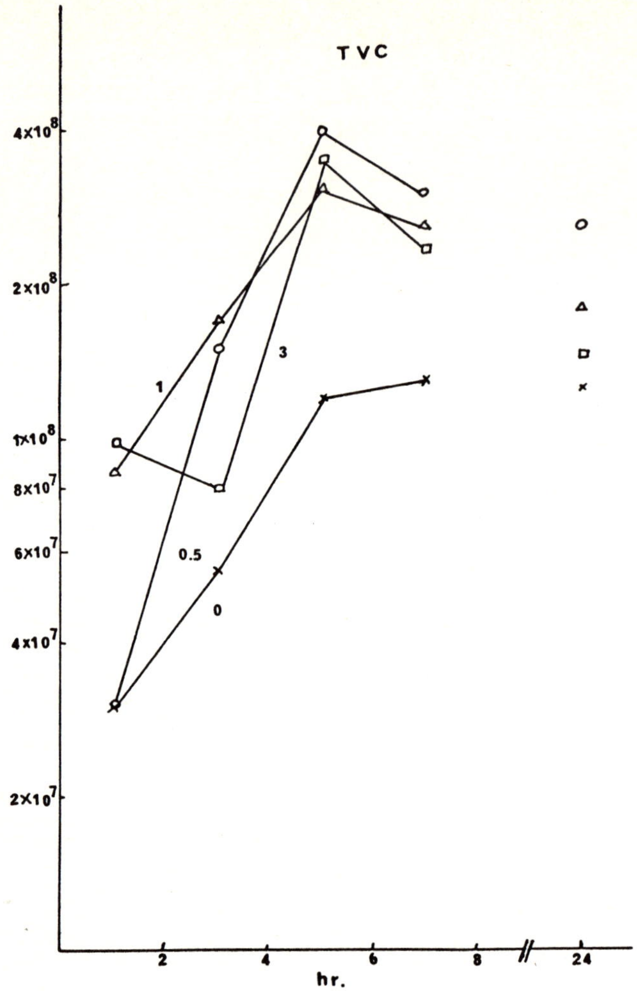

FIG. 1. Effect of salt concentration on total viable count. x——x = no salt added; o——o = 0.5% NaCl; Δ——Δ = 1% NaCl; □——□ = 3% NaCl.

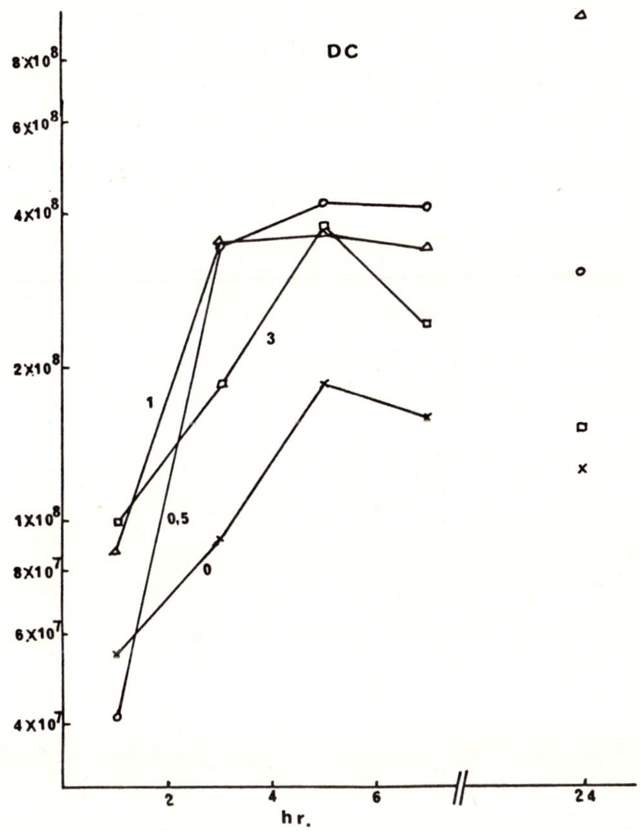

FIG. 2. Effect of salt concentration on cell number. Cells were counted by the epifluorescent method (see text). Symbols as in Fig. 1.

Developments in Arid Zone Ecology and Environmental Quality
Edited by H. Shuval

32

THE FATE OF ENTEROVIRUSES AT HAIFA'S MUNICIPAL WASTEWATER TREATMENT PLANT

Robert Vasl and Yehuda Kott

Environmental and Water Resources Engineering, Technion, Israel Institute of Technology, Haifa, Israel.

ABSTRACT

A simple method for the elution and isolation of enteroviruses from sludge was developed. Results have shown that enteroviruses generally do not survive for more than a month on drying beds. Identification of viruses showed higher numbers of coxsackie B viruses in wastewater effluent than polioviruses whereas polioviruses were the overwhelming majority in all sludges. This may be contributed to different adsorption rates of the various viruses onto the sludge solids. No seasonal variation could be determined from these results. However much more work needs to be done to confirm the results.

INTRODUCTION

It is well known that viruses readily attach themselves to solids in wastewater treatment plants (Gerba et al., 1978). It has also been established that not only different virus types, but also various strains of a particular virus show different adsorption rates onto a wide range of solids (Gerba et al., 1980). Little is known about the fate of the various viral types in wastewater treatment plants. Do all types of viruses leave the plant through the effluent and sludge route in the same ratios that they entered in the influent ? Or perhaps do some types of viruses tend to take the first route and other types the latter ? To demonstrate such a possibility it would be necessary to identify many enteroviruses isolated from sludge and effluent. Seasonal fluctuation in the numbers of viruses in effluent was mentioned in literature (Kott et al, 1978) and it would be interesting to know if this phenomenum exists in the sludge as well.

Any presence of infectious viruses in dried sludge which is eventually applied on land may pose a threat to health either by contamination of surface and ground waters or more directly by contaminating vegetables. In Canada enteroviruses were detected even after 8 months of retention. It is interesting to note the sub-zero temperature during at least 3 of the 8 months in which "the lagoon was inaccessible because of snow" (Sattar & Westwood, 1979).

The climatic conditions in Israel, being different, enable faster die-away of the viruses due to higher temperatures, and sunlight. This should be reflected in the rate of kill in the sludge drying beds, and reduce the time necessary to render the sludge free of viruses.

There are quite a few techniques for the recovery of viruses from sludge but not one has yet to be universally adopted. Many eluents have been tried (Sattar & Westwood, 1979) requiring in most cases many steps (Glass et al, 1978). A simpler method would be most advantageous.

The aims of this study were to isolate and identify enteroviruses from various stages of Haifa's wastewater treatment plant and to look for any trends which may indicate that different viruses take separate routes in the plant, and to see if there are any seasonal fluctuations of viruses in the effluent and sludge. In addition it was decided to investigate the survival of enteroviruses on drying beds.

MATERIALS AND METHODS

The sludge and effluent used in these experiments originated from Greater Haifa's Regional Sewage Treatment Plant which handles 50,000 m^3/d. There are two secondary treatment processes in this plant which run in parallel; an activated sludge unit and trickling filters. There is no further treatment of this effluent. Primary and secondary sludges from both of these processes are mixed and anaerobically digested (mesophilically) with an average detention time of 6-7 weeks in the digesters. After digestion the sludges are pumped out onto drying beds where they are retained for a period ranging from a few weeks to a few months depending on the season and then are composted for a further few months before sale as fertilizer.

The method for isolating enteroviruses from sludge is a simple three step method which was developed for this purpose:

a. 30 ml of wet sludge were spun 10 minutes at 16,000 rpm.

b. 30 ml of Phosphate Buffer Saline (bought up to pH 9.8 before hand by adding NaOH) were added to the sediment after decanting supernatant and then mixed for 30 minutes using a magnetic stirrer.

c. The mixed liquour was spun again for 10 minutes at 16,000 rpm and the supernatant was passed through a 0.45 membrane filter (Millipore). The filtrate was then ready for virological assay. (Final pH is usual about 8.0).

For viral assay of influent and effluent the following method was used. The pH of 50 ml of the sample was brought down to 3.0 - 3.5, was passed through 0.45 membrane filter (Millipore), then eluted with 3% Beef Extract (pH 9.0) and the pH of the filtrate brought down to 7.0 before innoculation.

Virological assay was performed by innoculating filtrate onto a line of BGM tissue culture cells as described previously (Dahling et al., 1974).Where the sludges were dry or almost dry the first stage was omitted. Viruses were identified by the Lim-Benyesh-Melnick pool method.

RESULTS AND DISCUSSION

The viral counts of raw sewage entering the plant, the trickling filter unit effluent and various sludges are listed in Table 1. The numbers of viruses in effluent were constantly lower than those reported previously at the same treatment plant (Kott et al, 1978). This may be explained by the fact that since that work was done at the plant, the activated sludge process has been installed in parallel to the trickling filter process hence reducing the load on the trickling filters.

Table 1 - Numbers of enteroviruses in various samples

Sample	No. of Samples Examined	Enteroviruses/100 ml
Raw Sewage	4	232
Trickling Filter effluent	15	31
Raw mixed sludge	18	2389
Digested sludge	14	104
Sludge after 1 week	5	111
Sludge after 1 month	5	22
Sludge 1 month & over	6	0

The number of viruses in the influent was always larger than in the effluent, but the highest concentration of viruses was found in the undigested sludge which confirms that most of the viruses in the wastewater attach themselves to solids and therefore find their way into the sludge. Digestion of the sludge inactivated a little more than one order of magnitude which corresponds to results published by Berg (Berg & Berman, 1980).

The die-away of viruses on the drying beds is clearly illustrated in figure 1. In all samples that were over one month old no viruses were detected. Up to 750 grams of dried sludge were assayed for virusses and found to be negative.

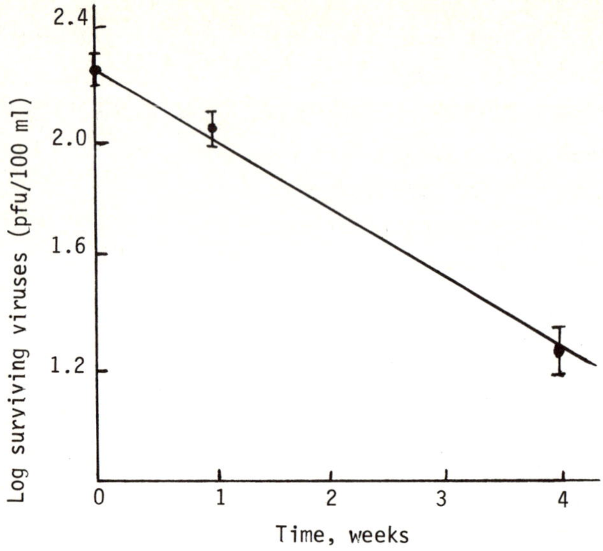

Fig. 1. Effect of retention time in drying beds on the survival of viruses.

Table 2 shows the seasonal dispersion of viruses trickling filter effluent and in undigested sludge. No significant variation between the various seasons could be determined. This does not agree with previous results (Kott et al., 1978) and may be due to the installation of the activated sludge process.

Table 2. Seasonal variation of enteroviruses in trickling filter effluent and in undigested sludge during the years 1979-1980.

		Viruses / 100 ml			
Season	Months	Number of samples examined	Effluent	Number of samples examined	Sludge
Winter	12-2	5	38	6	2127
Spring	3-5	4	42	6	3424
Summer	6-9	5	37	3	3265
Autumn	10-11	1	53	3	1460

Except for one case where one virus was found, no viruses were detected in the effluent from the activated sludge process which certainly shows the preference of the activated sludge process over the trickling filter unit as far as enterovirus removal is concerned. The wastewater treatment plant serves a large community which no doubt should reflect the community's health. It was interesting to find which enteroviruses are excreted and if there is seasonal variation. For this purpose 267 viruses that were isolated were identified and their distribution is shown in table 3. In effluent 64% of the identified viruses were coxsackie B virus as compared to 35% which were polioviruses and only 1% echoviruses. On the other hand in the undigested sludge only 11% were coxsackie B viruses, 8% were echoviruses and a staggering 81% were polioviruses. The digested sludge showed the same proportion of polioviruses (80%) with 18% coxsackie B viruses and 25% echoviruses. The fact that more coxsackie B viruses were found in effluent than polioviruses and in the sludges their numbers were insignificant as compared to polioviruses may shed some light on the tendency of various viruses to take different routes through a wastewater treatment plant.

The results presented in this study indicate that much more work needs to be done in this area.

Table 3. Types of enteroviruses isolated

Virus	Coxsackie virus B					Echovirus						Poliovirus		
type	1	2	4	5	6	6	11	14	19	23	25	1	2	3
Raw sewage				1									5	
Trickling filter effluent		24	30	1	1					1		5	20	7
Undigested sludge	1	11		2		2	3	4			1	20	60	25
Digested sludge		11			1				1			3	29	3

REFERENCES

Berg G. & Berman D. (1980) Destruction by anaerobic mesophillic and thermophillic digestion of viruses and indicator bacteria indigenous to domestic sludges. *Applied and Environmental Microbiology* 39, 361-368.

Dahling D.R., Berg G. & Berman D. (1974) BGM, a continuous cell line more sensitive than primary rhesus and African Green Kidney cells for the recovery of viruses from water. *Health Lab. Sci.* 11, 275-282.

Gerba C.P., Stagg C.H. & Abadie M.G. (1978) Characterization of sewage solid-associated viruses and behavior in natural waters. *Water Research* 12, 805-812.

Gerba C.P., Goyal S.M., Hurst C.J. & LaBelle R.L. (1980) Type and strain dependence of enterovirus adsorption to activated sludge, soils and estuarine sediments. *Water Research* 14, 1197-1198.

Glass J.S., Van Sluis R.J. & Yanko W.A. (1978). Practical method for detecting poliovirus in anaerobic digester sludge. *Applied and Environmental Microbiology* 35, 983-985.

Kott Y., Ben-Ari H. & Vinokur L. (1978) Coliphages survival as viral indicator in various wastewater quality effluents. *Progress Water Technology* 10, 337-346.

Sattar S.A. & Westwood J.C.N. (1979) Recovery of viruses from field samples of raw, digested, and lagoon-dried sludges. *Bulletin of the World Health Organization* 57, 105-108.

Developments in Arid Zone Ecology and Environmental Quality
Edited by H. Shuval

33
METHODOLOGY FOR VIRUS RECOVERY FROM SEDIMENTS

G. Bitton, S.R. Farrah and Y.J. Chou

University of Florida, Gainesville, Florida, U.S.A.

ABSTRACT

Laboratory experiments have been undertaken to study the adsorption - desorption of poliovirus to and from marine and freshwater sediments. It was observed that marine sediments retained 99% of added virus wheras freshwater sediments adsorbed approximately 30% of the added virus.

Ten eluents were investigated for their ability to desorb viruses from Cresent Beach marine sediments. Three to six replicates were run for each eluent. It was found that virus elution from the marine sediment was relatively low and ranged from <1% to 44%. Two eluents, urea-lysine and TCA-glycine at pH 9.0, were found to be the most efficient among the eluents tested. Viruses were more easily released from freshwater than marine sediments.

The urea-lysine method was thus used for the detection of indigenous enteroviruses in sediments from a closed shellfish bed in St. Augustine, Florida. Enteroviruses (Poliovirus, 1, 2 and 3) were recovered with concentrations up to 72 $TCID_{50}$/100 g of sediments.

Developments in Arid Zone Ecology and Environmental Quality
Edited by H. Shuval

34
BACTERIAL QUALITY OF COASTAL WATER IN CERTAIN RECREATIONAL AREAS IN ISRAEL

Yona Yoshpe-Purer

The A. Felix Public Health Laboratory, Ministry of Health, P.O.B. 8255, Tel-Aviv 61082.

ABSTRACT

Beaches of central and southern Israel were monitored for total coliforms, fecal coliforms, fecal streptococci and the presence of salmonellae and vibrio parahaemolyticus, during three summers and two winters, as part of the WHO/UNEF Mediterranean pollution monitoring program. The number of organisms was much higher in the winter. The MPN method gave considerably higher results for total and fecal coliforms and seems to be more suitable for monitoring these organisms in marine water. Two strains of salmonellae were isolated in the winter from fairly clean beaches. V. parahaemolyticus was not detected.

INTRODUCTION

Clean recreational water is part of a healthy environement that promotes the well-being of its inhabitants. Bacteriological monitoring of coastal water is designed to ensure, as far as possible, protection from infection through water contaminated by fecal material. As monitoring for pathogenic organisms is time consuming, the best indicator organism for this purpose should be selected, its permissible level be limited to the number that would most probably exclude the presence of pathogens and the best method for recovering the maximal number of this organism should be selected.

Awareness of the progressive deterioration of the Mediterranean due to man's activities prompted the United Nations Environement Program and the World Health Organization to adopt an intergovernmental Action Plan for protection and development of the Mediterranean. A coordinated Mediterranean Monitoring and Research Program (MED POL) was established in 1975 for assessment of the state of environement. One of its seven pilot projects was the three year Coastal Quality Monitoring of Recreational and Shellfish Areas (MED VII), in which 29 institutions from various countries around the Mediterranean participated. The mandatory bacteriological parameters to be monitored in this project were total coliforms, fecal coliforms and fecal streptococci. Other organisms were optional. The participants were requested to employ the same methods and media, namely membrane filtration (MF) with M-FC medium for the coliforms (incubation temp. 36$\pm$1 C for the total and 44$\pm$0.2 for the fecal) and M-enterococcus agar for the fecal streptococci (incubation at 44$\pm$0.2 C). Analysis of the data was also to be done by the same system (logarithmic median and deviation) so that the results could be compared on the same basis.

Our laboratory has been monitoring the Tel-Aviv beaches for total coliforms since 1963 and since 1975 for fecal coliforms as well, employing the tube fermentation method (MPN) for both (Standard Methods, 1976). In the last 5 years monitoring was extended to other beaches along the shore of central and southern Israel, comprising about 30 beaches. During a study of coastal pollution problems around a sewage outfall north of Tel-Aviv (Yoshpe-Purer & Shuval, 1970) MF with M-Endo broth was also employed. We joined MED VII in 1978. As our previous experience indicated that the MPN procedure gave considerably higher results than the MF in about 50% of samples and the filters on M-FC medium at 35C were often difficult to read due to many interfering organisms, it was felt that a comparative study employing both methods would be beneficial. The results are presented here.

MATERIALS AND METHODS

Area - three beaches in the most populated sections of central and southern Israel were selected for surveillance in the program, namely Nathania, Tel-Aviv and Ashquelon. There were two sampling points in each beach, about 250 m. apart (s.p. 1-6 respectively) and two control points, one in the north, near Nathania (s.p.7) and one in the south near Ashquelon (sp. 8). The land near the beaches is mainly sandy and surface or ground water reaches them only on rainy days in the winter. There is a sewage outfall about 4 km. to the north of the Tel-Aviv beach but none near the others.

Sampling - during the official bathing season (May to October) samples were collected every fortnight (3 seasons) and in the winter - once in six weeks (2 seasons). All samples were collected in the morning and brought to the laboratory within 1-2 hours.

The water temperature was 17-27°C in the summer and around 12°C in the winter.

Total coliforms - MF with M-FC agar and LES Endo agar incubated at 36 ±1 C for 22h. MPN with lectose broth confirmed by plating for isolated colonies on MacConkey agar.

Fecal coliforms - MF with M-FC agar incubated at 44± 0.2 C and MPN by inoculating positive tubes of the total coliforms presumptive test into EC medium incubated at 44± 0.2C .

Fecal streptococci - MF with M-enterococcus agar incubated at 44±0.2C for 48h.

Salmonellae - MF of 500 ml. water, membranes immersed in selenite-cystine broth incubated at 41± 0.5C. Plating was on brilliant green agar after 24 and 48 h., plates incubated at 37°C. Suspicious colonies were picked and identified biochemically and serologically. The final identification of serotypes was done in the National Salmonellae Center in the Central Laboratories in Jerusalem.

Vibrio parahaemolyticus - in the first season enrichement in salt broth (3% NaCl) plated on TCBS agar, later only MF with TCBS agar incubated at 41± 0.5 C.

RESULTS

The prevailing wind direction was south-west and occasionally south. Results of the three indicator organisms were therefore analysed according to season only as presented in tables 1-3.

The data indicate that in most sampling points the number of organisms was much higher in the winter and the efficiency of the MPN procedure was more pronounced in this season (tables 1 & 2).

Table 1. Total coliforms per 100 ml. (logarithmic mean).

Sampling point No.	Summer					Winter				
	No.of samples	MF*		MPN		No. of samples	MF		MPN	
		Mean	Dev.	Mean	Dev.		mean	dev.	mean	dev.
1	32	22	6	40	7	7	243	5	1714	23
2	32	16	6	48	7	7	291	5	1840	9
3	29	53	8	165	20	7	191	3	2594	4
4	29	38	5	122	8	7	230	5	9780	44
5	32	10	13	32	30	7	18	10	141	14
6	32	12	8	23	26	6	8	18	41	11
7	32	13	5	13	12	6	222	7	1270	8
8	32	4	7	8	14	7	22	10	52	8

*) Results with M-FC agar and LES Endo agar were very similar therefore only the first are presented.

Table 2. Fecal coliforms per 100 ml. (logarithmic mean)

Sampling point No.	Summer					Winter				
	No. of samples	MF		MPN		No. of samples	MF		MPN	
		Mean	Dev.	Mean	Dev.		Mean	Dev.	Mean	Dev.
1	32	10	5	19	10	7	58	7	785	20
2	32	7	6	18	8	7	52	10	1115	9
3	29	21	8	82	13	7	51	4	1107	7
4	29	15	5	58	8	7	91	9	1705	108
5	32	5	9	20	19	7	5	13	30	42
6	32	5	8	10	20	6	4	11	16	13
7	32	4	6	9	9	6	56	8	904	11
8	32	3	6	4	7	7	7	11	13	14

Table 3. Fecal streptococci per 100 ml. (logarithmic mean).

Sampling point No.	Summer			Winter		
	No.of samp.	Mean	Dev.	No.of samp.	Mean	Dev.
1	32	6	5	7	19	4
2	32	8	5	7	33	2
3	29	13	6	7	57	3
4	29	13	6	7	100	4
5	32	5	7	7	14	8
6	32	5	6	6	13	11
7	32	8	6	6	18	4
8	32	5	7	7	11	7

When MF counts were related to the 95% confidence limits of the MPN, 64.7% of the 300 samples examined were within the limits, 3.6% were above and 31.7% were below them. However, when compared on a numerical basis, as the number stated for the MPN is employed for evaluating beach quality rather than the confidence limits, 32% of the results were equal, 9.3% were higher by MF and 58.7% were higher by the MPN.

The correlation between the three indicator organisms is given in table 4.

Table 4. Correlation between the indicator organisms.

Sampling point No.	Summer			Winter		
	TC:FC	TC:FS	FC:FS	TC:FC	TC:FS	FC:FS
1	0.922	0.075	0.054	0.867	0.656	0.918
2	0.669	0.019	0.039	0.879	0.613	0.911
3	0.898	0.684	0.804	0.993	0.594	0.581
4	0.975	0.399	0.672	0.971	0.860	0.901
5	0.890	0.827	0.968	0.995	0.983	0.964
6	0.976	0.616	0.628	0.999	0.938	0.935
7	0.986	0.287	0.237	0.944	0.612	0.768
8	0.983	0.575	0.692	0.964	0.986	0.912

There is a fairly good correlation between total and fecal coliforms in both seasons, no correlation between any of them and fecal streptococci in the summer, but a fairly good one between fecal coliforms and fecal streptococci in the winter.

V. parahaemolyticus was not detected but two strains of Salmonella were isolated in 1980, one in February (S. anatum) from the southern point in Nathania, the other in March (S. dublin) from the southern point in Ashquelon.

DISCUSSION

Environemental quality criteria for recreational coastal water, first defined as "100 E. coli per 100 ml." (interim standard), were later modified to "no more than 10% of at least 10 consecutive samples collected during the bathing season should exceed 1000 E. coli per 100 ml." According to the logarithmic median (table 2) all our sampling points complied with these criteria during the bathing season by both methods employed and by MF in the winter as well. However, when the data were tabulated by the range they fell into, all points complied with criteria in the summer by MF but by the MPN results two points would have been disqualified in the summer of 1979, since 3 out of 10 samples in one point and 2 out of 12 in the other exceeded 1000 fecal coliforms. Hence the quality criteria should also define the method of monitoring and set the standard accordingly.

Two of our previous observations were confirmed in this investigation a) the number of organisms was much higher in the winter than in the summer, b) results obtained by the one-step MF procedure were below the 95% confidence limits in a considerable part of the samples - 31.7% in the present study and 46-50% in the previous one (Yoshpe-Purer & Shuval, 1970).

The elevated number of organisms in the winter could be accounted for by two main factors, a) stormwater run-off contaminated by fecal material from animals, as suggested by Geldreich (1965), which occurs in Israel only in the winter as there is no rain in the summer, b) stirring up of organisms that have settled to the bottom by the stormy sea. The latter is borne out by the fact that in clean beaches, like Ashquelon, the number of organisms remained low in the winter.

The efficiency of the MPN method was more pronounced in the winter probably due to the presence of more stressed organisms in this season. In chlorinated wastewater effluents, where stressed bacteria are present, the MPN procedure was also more efficient than MF in recovering fecal coliforms (Lin, 1973). It seems that the MPN method, which in the first step affords favourable conditions for recovery of stressed organisms, may be more suitable for monitoring marine water and when the MF technique is employed, preenrichment to improve recovery efficiency may be desirable.

The number of fecal streptococci was usually quite low and even in the winter, in points with high fecal coliform counts, they did not exceed a few hundred per 100 ml. In our previous work (Yoshpe-Purer & Shuval, 1970) the ratio between TC:FC and FC:FS was 95:1 and 31:1 respectively in heavily contaminated water, while in lightly contaminated water it was 19:1 and 3:1 respectively. The density of fecal streptococci in our coastal water is apparently not high enough to make then a good parameter for assessing the water quality unless monitored together with fecal coliforms as recommended by Geldreich and Kenner (1969). It should be mentioned that in an epidemiological study conducted by Cabelli (1979) in two New York beaches their density was found to be sufficient to index the ratio of gastrointestinal symptoms observed. A similar epidemiological study in our country, where sea-bathing is very popular and the bathing season extends over 7-9 months of the year, would certainly contribute to establishing quality criteria for this region.

Many strains of Salmonella were previously isolated from coastal water up to 2 km. from a sewage outfall (Yoshpe-Purer et al, 1971) but in the present work they were found in water of fairly clean beaches. The presence of Salmonellae in moderately polluted or clean water was reported by several investigators (Dutka & Bell 1973; Seligmann & Reitler,1965). According to the quarterly report of the National Salmonellae Center for January-March 1980, the two Salmonella serotypes we isolated belonged to the five most common serotypes from non-human sources in that period. S. dublin was isolated only from animals, mainly cattle, and S. anatum came from cattle and other animals including one bird, from meat and also from man. As there are no sewage outfalls in the vicinity of the Nathania and Ashquelon beaches it is assumed that the source of those salmonellae was contamination by animals or birds directly or by stormwater run off, over which we have no control.

ACKNOWLEDGEMENTS

The cooperation of Mr. Naphthali Cohen, Assistant District Sanitary Engineer of Tel-Aviv, Mrs. Ayala Gavish and Mrs. Zipora Guttenberg of our laboratory is gratefully acknowledged.

This investigation was done in the framework of MED VII and was partially supported by the UNEP/WHO and by the Environmental Protection Service, Ministry of the Interior, Israel.

REFERENCES

Cabelli, V.J. (1977) Indicators of recreational water quality, Bacterial Indicators/Health Hazzards Associated with Water ASTM STP 635, A.W. Hoadly and B.J. Dutka, Eds., American Society for Testing and materials pp. 222-238.

Dutka, B.J. & J.B. Bell (1973) Isolation of salmonellae from moderately polluted water. J. Water Poll. Cont. Fed. 45, 316-324.

Geldreich, E.E. (1965) Detection and significance of fecal coliform bacteria in stream pollution studies. J. Water Poll. Cont. Fed. 37, 1722-1726.

Geldreich, E.E. & B.A. Kenner (1969) Concept of fecal streptococci in stream pollution. J. Water Poll. Cont. Fed. 41, R337-352.

Lin, S. (1973) Evaluation of coliform tests for chlorinated secondary effluent. J. Water Poll. Cont. Fed. 45, 498-506.

Seligmann, R. & Reitler, R. (1965) Enteropathogens in water with low E. coli titer. J. Amer. Water Works Assoc. 57, 1572-1575.

Standard Methods for the Examination of Water and Wastewater (1976), APHA, 14th ed.

Yoshpe-Purer, Y. & H.I Shuval (1970) Salmonellae and bacterial indicator organisms in polluted coastal water and their hygienic significance. Marine Pollution and Sea Life, M. Ruivo Ed. FAO Fishing News (Books) Ltd. England.

Yoshpe-Purer, Y., Sh. Ricklis & M. Paist (1971) A convenient method for isolation of salmonellae from sewage and contaminated sea water. Water Research 4, 113-120.

Developments in Arid Zone Ecology and Environmental Quality
Edited by H. Shuval

35
BACTERIAL INHIBITION OF ALGAL GROWTH IN PHOTOSYNTHETIC SEWAGE TREATMENT SYSTEMS

Svi Benzion* and Inka Dor+

*Department of Life Sciences, Bar-Ilan University, Ramat-Gan, Israel, and +Human Environmental Health Laboratory, The Hebrew University of Jerusalem, Israel

ABSTRACT

In light of previous results showing growth inhibition of Chlorella vulgaris and Scenedesmus obliquus by bacteria isolated from incidental sewage feeding an algal pond, the possibility of selective repression in the pond in favor of a stimulatory bacterial flora was investigated. Among ten bacterial strains tested none was stimulatory; all of them interfered with the growth of both algae to varying degrees. The algae were axenically grown in organic media previously used for bacterial growth (using four identified bacterial strains) and subsequently reenriched. In all cases but one, algal growth was severely restricted, and in one instance only slight inhibition of algal growth was noted. It is concluded that no selection exists in the oxidation pond for stimulatory bacteria and that the nature of the algal growth inhibiting factor is chemical or biochemical.

INTRODUCTION

The environmental factors wielding the most important influence upon algal growth in aquatic ecosystems are generally accepted as being: light, availability of macro- and micro-nutrients, pH and temperature (Silva et al., 1953). In polluted waters the quality and load of organic substrate (Nemerow & Rand, 1968), presence of toxic substances and availability of inorganic carbon (Goldman et al., 1972) become important parameters affecting algal growth and photosynthesis.

The synergistic effect of bacterial presence on algal growth in polluted waters is widely accepted due to both the unquestionable requirements of the algae for the products of bacterial mineralization processes and to the beauty and simplicity of Schematic Representations showing the biochemical interrelationship between them (Oswald, 1973; Kathryn et al., 1975; Hendricks & Pote, 1974).

The widely accepted dogma of synergism between bacteria and algae has only just been tested experimentally (Dor & Benzion, 1980) and found to be incorrect insofar as organically polluted ecosystems are concerned. All bacteria tested were found to interfere with the growth of the test algae - none was found to be stimulatory.

The objective of the present study was: (1) to ascertain whether the algal-inhibiting strains of bacteria present in the incidental sewage sample, and reported in our previous work (Dor & Benzion, 1980), have undergone selective repression in the oxidation pond in favor of a stimulatory bacterial flora; (2) to elucidate the gross mechanism of bacterial interference with the growth of the green algae in the presence of organic substrate.

MATERIALS AND METHODS

Strains of Chlorella vulgaris Beijerinck and Scenedesmus obliquus (Turp.) Kutz were isolated from a local oxidation pond and kept in axenic cultures on Chu medium (Chu, 1942) enriched with nutrient broth (8 gl^{-1}).

Bacterial strains were isolated from an oxidation pond fed by municipal winter sewage. All were isolated from dilutions to 10^{-7} and maintained on nutrient agar. Axenic separate cultures of Chlorella and Scenedesmus were incubated alone, as controls, and with individual strains of Pond bacterial flora. All cultures were incubated in 100 ml of Chu medium enriched with nutrient broth in 250 ml Erlenmeyers on a shaker at 28°C under white fluorescent continuous illumination of 1500-2000 Lux. Algae were counted on the fifth day. Initial bacterial concentrations were $10^{-7}ml^{-1}$. Initial concentrations of Chlorella (200-400 mm^{-3}) and Scenedesmus (20-50 mm^{-3}) were approximately 10% of axenic 5-day growth.

Four bacterial strains isolated on nutrient agar plates from the incidental winter sewage (diluted to 10^{-7}): Citrobacter freundii, Aeromonas hydrophila, Klebsella sp. and Pseudomonas putida were used for this study. (Bacterial identification by the U.S. Dept. of Health, Education and Welfare, Atlanta, Georgia, courtesy of Dr. Schechter, Israel Ministry of Health).

Batch cultures of bacteria in Chu media enriched with nutrient broth were incubated for 5 days in 1 l. Erlenmeyers vigorously stirred at 28°C under illumination (as above). All visible traces of bacterial biomass were subsequently removed from the culture media either by centrifuging at 9000 RPM for 20 min or by filtration through 0.2 μm membrane filters (Tamar Laboratory Equipment). The pH of the supernatant or filtrate was corrected to its original value with concentrated HCl, reaerated by shaking and divided into 100 ml aliquots.

The aliquots were treated to partial and total reenrichment with the components of Chu media and nutrient broth, using concentrations identical to the original media. After sterilization by autoclaving (1.5 atm, 20 min) they were inoculated with axenic cultures of Chlorella and Scenedesmus (initial concentrations as above). Cultures were returned to previous incubation conditions in 250 ml Erlenmeyers. Algae were counted on the fifth day.

RESULTS

The effect of numerous unidentified bacterial strains isolated from an oxidation pond on the growth of Chlorella and Scenedesmus in the Chu media enriched with nutrient broth is given in Figs. 1 and 2.

Among ten bacterial strains tested, none stimulated algal growth. In the case of Chlorella, algal growth reached at best 20% (P8) of the yield obtained in axenic cultures.

Scenedesmus fared slightly better: Among ten bacterial strains tested, eight allowed algal yields higher than 30% of the axenic culture. The highest yields obtained (P14, P8) were slightly above 50%.

Five-day yields of axenic separate cultures of Chlorella and Scenedesmus in media previously used for bacterial growth is shown in Tables 1 and 2. In all cases Chlorella failed to grow well in media previously used for bacterial growth. The best results were in media used by Klebsiella sp. (19% of axenic growth). Growth of Scenedesmus was equally restricted in media previously used for bacterial growth with one exception.

The media used by P. putida allowed fairly good algal growth (86% of axenic growth).

Except in P. putida, no instance of partial or total reenrichment of the media succeeded in improving algal growth to over 20% of the yield in

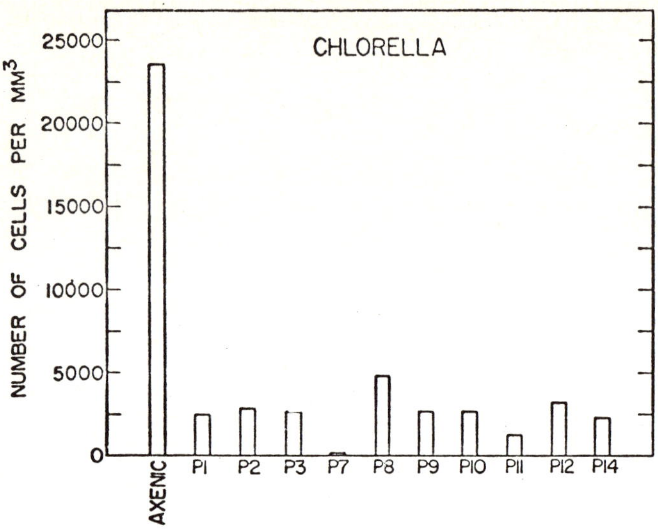

FIG. 1. Growth of Chlorella in Chu-media enriched with nutrient broth with various bacterial strains isolated from an oxidation pond.

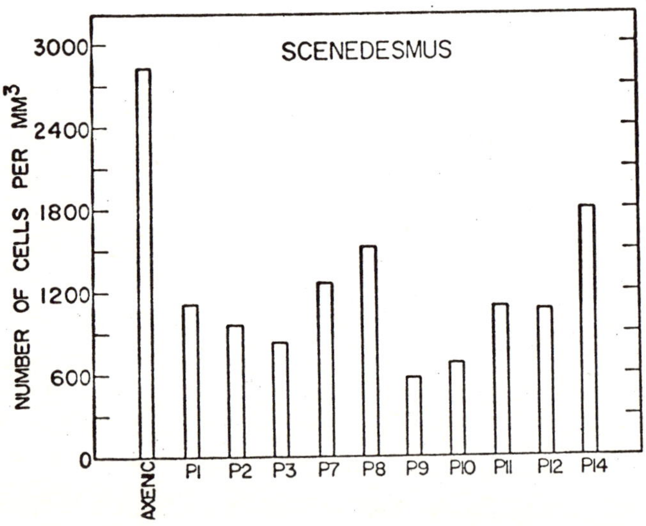

FIG. 2. Growth of Scenedesmus in Chu-media enriched with nutrient broth with various bacterial strains isolated from an oxidation pond.

TABLE 1. Growth of axenic *Chlorella* in media previously used fcr bacterial growth (average of 5 experiments)

Form of Enrichment	Fresh Media	Previous Use of Media: Growth of *Cit. freundii*	Growth of *A. hydrophila*	Growth of *Klebsiella* sp.	Growth of *P. putida*	
Unenriched	26700±4500	2140±1280	2630±1280	5000±1060	2310±300	Yield
	100	8.0	9.0	19	8.7	% Axenic
NO_3	-	2800± 280	2600± 500	2450±1500	2000±280	Yield
	-	10	9.7	9.2	7.5	% Axenic
PO_4	-	2150± 350	1975± 175	1200= 250	1650±280	Yield
	-	8.1	7.4	4.5	6.2	% Axenic
NO_3+PO_4	-	3325± 450	2050± 600	2250= 700	1950±280	Yield
	-	12	7.7	8.2	7.3	% Axenic
Micro Elements	-	1500± 350	1435± 600	1775± 175	1125±175	Yield
	-	5.6	5.3	6.6	4.2	% Axenic
Full Chu Media	-	2425± 100	2250± 750	2500±1000	1900± 90	Yield
	-	9.0	9.6	9.4	7.1	% Axenic
Chu + Nutrient Broth	25900±3000	3490± 625	3780± 265	5350±2000	4890±619	Yield
	97	13	14	20	18	% Axenic

TABLE 2. Growth of axenic Scenedesmus in media previously used for bacterial growth(average of 3 experiments)

		Previous Use of Media				
Form of Enrichment	Fresh Media	Growth of Cit. freundii	Growth of A. hydrophila	Growth of Klebsiella sp.	Growth of P. putida	
Unenriched	2400±400	132± 46	58±22	40±16	2066±300	Yield
	100	5.5	2.4	1.7	86	% Axenic
NO_3	-	190± 41	54±14	160±75	1955±600	Yield
	-	7.9	2.3	6.7	82	% Axenic
PO_4	-	225± 34	49± 6	85±13	1640±155	Yield
	-	9.4	2.0	3.5	68	% Axenic
NO_3+PO_4	-	280± 17	51±11	115±21	2010±245	Yield
	-	12	2.1	4.8	84	% Axenic
Micro Elements	-	140± 21	59±21	110±16	1115± 85	Yield
	-	5.8	2.5	4.6	46	% Axenic
Full Chu Media	-	350± 35	76± 9	112± 3	1775±160	Yield
	-	15	3.2	4.7	74	% Axenic
Chu + Nutrient Broth	2700±350	354±167	48±31	246±76	1442±175	Yield
	112	15	2.0	10	60	% Axenic

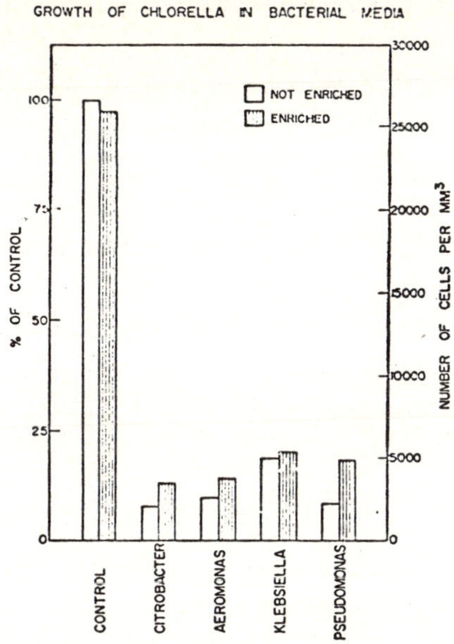

FIG. 3. Growth of Chlorella in organic media previously used for bacterial growth.

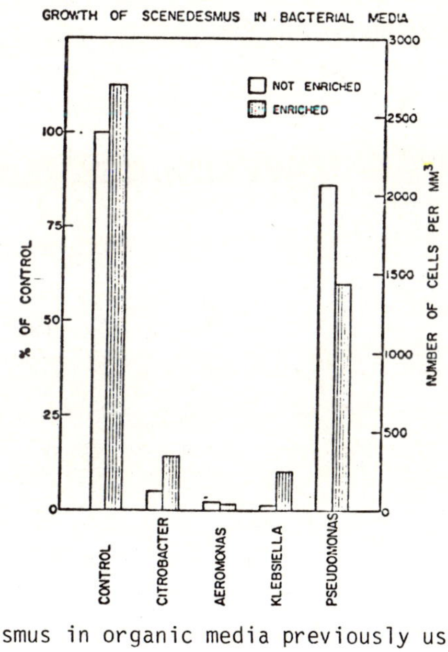

FIG. 4. Growth of Scenedesmus in organic media previously used for bacterial growth.

axenic cultures. Even with P. putida media reenrichment had no beneficial effect.

A comparison of algal growth in previously used bacterial media with and without total reenrichment is summarized in Figs. 3 and 4.

DISCUSSION

Previous work on the effect of heterotrophic bacteria on the green algae growing in wastewater (Dor & Benzion, 1980) had demonstrated inhibitory effects of sewage-isolated bacteria on both Chlorella and Scenedesmus. Among 20 strains tested none was found to be stimulatory.

The first experiment demonstrates that no species selection occurs in the oxidation pond in favor of a bacterial flora stimulating algal growth. However, while all bacterial strains tested severely limited the growth of Chlorella, their effect on Scenedesmus was much less pronounced. This would imply that microbial flora isolated from the oxidation pond by the domestic sewage of Jerusalem in winter was more compatible with Scenedesmus than with Chlorella. That this indeed is the case is borne out by the fact that Scenedesmus obliquus is the only dominant alga in the pond during the winter months. The possibility that changes in the quality of the microbial flora of the pond are responsible for algal species succession requires further study.

The second experiment, algal growth in previously used bacterial media, demonstrates several points regarding bacterial inhibition of algal growth.

(1) the inhibitory effect of bacteria is neither due to turbidity of the growth media, decreasing availability of light for photosynthesis, nor is it due to changes in the oxygen content or the pH of the media.

(2) the fact that the inhibitory effect was consistently present after reenrichment of the media proves that competitive removal of nutrients was not the factor responsible for poor algal growth.

(3) bacterial growth in the media causes formation of some substance(s) toxic to algae. This substance(s) is both thermostabile and filterable.

Our present results prove that the theory of synergism between bacteria and algae is not applicable to the oxidation pond strains of Chlorella vulgaris, Scenedesmus obliquus and the concomitant bacterial flora. The effect of the bacteria on the algae is clearly inhibitory and mediated by some chemical or biochemical substance.

The ecological significance of this relationship is as yet unresolved; however, a similarly antagonistic relationship has been noted in natural aquatic ecosystems; both marine (Berland et al., 1972) and freshwater (Chrost, 1975a,b). In contrast to our results, however, Berland found the algal inhibiting factor to be thermolabile.

Further studies are required to identify the algal growth inhibiting factor and to discover its mechanism of action. It is hoped that with this information we can improve the performance of oxidation ponds and algal sewage treatment systems.

ACKNOWLEDGMENTS

Dr. Schechter is thanked for bacterial identification. This study was supported in part by Grant No. 1756 of the Israel National Council for Research and Development, financed by and in cooperation with GSF, Federal Republic of Germany and in part by the A.D. Bergman Foundation for Applied Science and Technology of the Hebrew University, Jerusalem.

REFERENCES

Berland B.R., Bonin D.J. & Maestrini S.Y. (1972) Etude des relations Algues-Bacteries du milieu marin: Possibilite d'inhibition des algues par les bacteries. *Tethys* 4, 339-343.

Chrost R.J. (1975a) Inhibitors produced by algae as an ecological factor affecting bacteria in water ecosystems. I. Dependence between phytoplankton and bacteria development. *Acta Micro. Polonica Ser. B.* 7/24, 125-133.

Chrost R.J. (1975b) Inhibitors produced by algae as an ecological factor affecting bacteria in water ecosystems. II. Antibacterial activity of algae during blooms. *Acta Micro. Polonica Ser. B.* 7/24, 167-176.

Chu S.P. (1942) The influence of the mineral composition of the medium on the growth of planktonic algae. I. Methods and culture media. *J. Ecol.* 30, 284-325.

Dor I. & Benzion S. Effect of heterotrophic bacteria on the green algae growing in wastewater. *Algal Biomass* (Edited by Shelev G., Soeder C.J. & Balaban M.). Elsevier-North Holland Biomedical Press, Amsterdam.(in press)

Goldman J.C., Porcella D.B., Middlebrooks E.J. & Toerien D.F. (1972) The effect of carbon on algae growth - its relationship to entrophication. *Water Res.* 6, 637-652.

Hendrick D.W. & Pote W.D. (1974) Thermodynamic analysis of a primary oxidation pond. *J. Water Poll. Control Fed.* 46, 333-339.

Kathryn B. (1975) Ecological aspects of used water treatment. In: *Algae and Bryophytes* (Edited by Curds C.R. and Hawkes H.A.) Academic Press.

Nemerow N.L. Rand, M.C. (1968) Algal growth affected by degree and type of wastewater treatment. In: *Algae, Man and Environment* (Edited by Jackson F.D.). Syracuse Univ. Press.

Oswald W. (1973) Productivity of algae in sewage disposal. *Solar Energy* 15, 107-117.

Silva P. (1953) Report on a systematic study of the algae of sewage oxidation ponds. *State Water Poll. Control Board, Sacramento, Calif.* 7, 34 pp.

Developments in Arid Zone Ecology and Environmental Quality
Edited by H. Shuval

36

LEVELS OF FECAL POLLUTION INDICATORS IN FISH PONDS ENRICHED WITH SEWAGE EFFLUENT AND THEIR HEALTH IMPLICATIONS*

Naomi Guttman-Bass, Norma Noe Moccetti and Badri Fattal

Environmental Health Laboratory -School of Public Health, Hebrew University-Hadassah Medical School, Jerusalem, Israel

ABSTRACT

Edible fish have been bred in sewage-enriched fish ponds in Israel for several decades, yet the levels of human pathogens in the pond water or in the fish themselves have not been thorougly investigated. In order to determine whether or not there is a health rish associated with growing fish on human waste, we have conducted a preliminary survey of the concentration of fecal pollution indicators in effluent enriched and non-enriched fish ponds.

Initially, pond water was obtained from seven kibbutzim and the total coliform levels found were somewhat higher in the enriched ponds. Subsequently, more intensive sampling over a period of four months was performed on two kibbutzim, one of which used no effluent for fish pond enrichment while the second had both effluent-enriched and non-effluent enriched ponds.

Samples were tested for fecal and total coliform levels and for the presence of salmonella. The fish (carp) were also examined and intestinal tract contents and muscle tissue were assayed.

Overall, the levels of coliforms and fecal coliforms in the water and fish samples were not strickingly higher in the effluent-enriched ponds in comparison with the non-enriched ponds. Furthermore, the five salmonella isolates were obtained from water and fish intestinal tract samples from non-enriched ponds, and none from enriched ponds. Thus, from these preliminary results, it was not found that the supplementation of effluent in fish ponds greatly increases the level of fecal pollution indicators in the water or the fish which are subsequently marketed for human consumption. Communicable disease rates from kibbutzim with sewage enriched ponds will be analyzed in relation to disease rates in other kibbutzim.

*Financed in part by EPA Cooperative Agreement #CR806-416. W. Jakubowski, project officer.

Developments in Arid Zone Ecology and Environmental Quality
Edited by H. Shuval

37
QUALITY ASSURANCE IN WATER LABORATORIES

Arnold E. Greenberg

Assistant Chief, Laboratory Services Branch, California Department of Health Services, Berkeley, California

ABSTRACT

To provide meaningful data in ecological research, in the assessment of drinking water quality or water pollution control, and in a variety of water-related regulatory activities, requires that suitable water laboratory quality assurance programs be established and operated. While laboratory scientists recognize the need for producing high quality data, this need is seldome formulated. Often, the procedures by which good data are obtained are not known or, at least, not formally described. Neither good laboratory practices nor common sense, in the analytical laboratory, is an adequate substitute for a definitive statement of the principles and practices of quality assurance.

A laboratory quality assurance program begins with the collection of samples which must truly represent the universe being sampled, continues with analysis by proven methods of known accuracy and precision, and concludes with reporting results showing the level of error. Utilization of laboratory results often is beyond the scope of the laboratory, but the analyst should be prepared to participate.

Frequently the laboratory quality assurance program is confused with water quality assurance. Good laboratory work serves to describe water quality but does not produce good quality water.

Developments in Arid Zone Ecology and Environmental Quality
Edited by H. Shuval

38

DETECTION OF HALOFORMS IN THE ISRAEL NATIONAL WATER CARRIER: 1978-79

Aharon Eisenberg,* Emil Weissenberg,** Yehudit Shoenberg** and Zoya Sahm**

*Ministry of Health, Jerusalem, **Institute of Control and Standardization of Drugs, Ministry of Health, Jerusalem.

ABSTRACT

A survey was carried out during 1978-79 to detect the presence of haloforms in the Israel National Water Carrier including the Yarkon Negev lines. Haloform levels, especially bromoform increased after chlorination treatments. The presence of inorganic bromine in raw Lake Kinneret water is sufficiently high to account for the formation of bromoform and the other bromine containing haloforms. Total haloform levels approach and occasionally exceed the official maximum USEPA level.

INTRODUCTION

Since the appearance of the papers by Rook (1) and Bellar et al (2) in which they reported the presence of haloforms in drinking water many studies have been published which confirm these findings. The most important single work on this subject has been the 80 City study carried out by the EPA (3). It is now quite clear that these substances are formed as a result of chemical reactions following the chlorination of raw water.

The public health significance of the presence of haloforms in drinking water (primarily chloroform, bromoform, dibromochloromethane dichlorobromomethane) has been the subject of a heated debate especially in the U.S.A. However this matter has been resolved by the EPA which has set a maximum level for chlorinated community drinking water of 100 parts per billion. These regulations were made official in March 1980. Many papers have subsequently appeared describing various aspects of haloform formation in drinking water systems and under controlled laboratory conditions. Methods for the prevention and removal of haloforms have been developed and of special interest to this work, the influence of inorganic bromines has been studied, (4) (5).

MAP: SAMPLING POINTS ON THE NATIONAL WATER CARRIER.

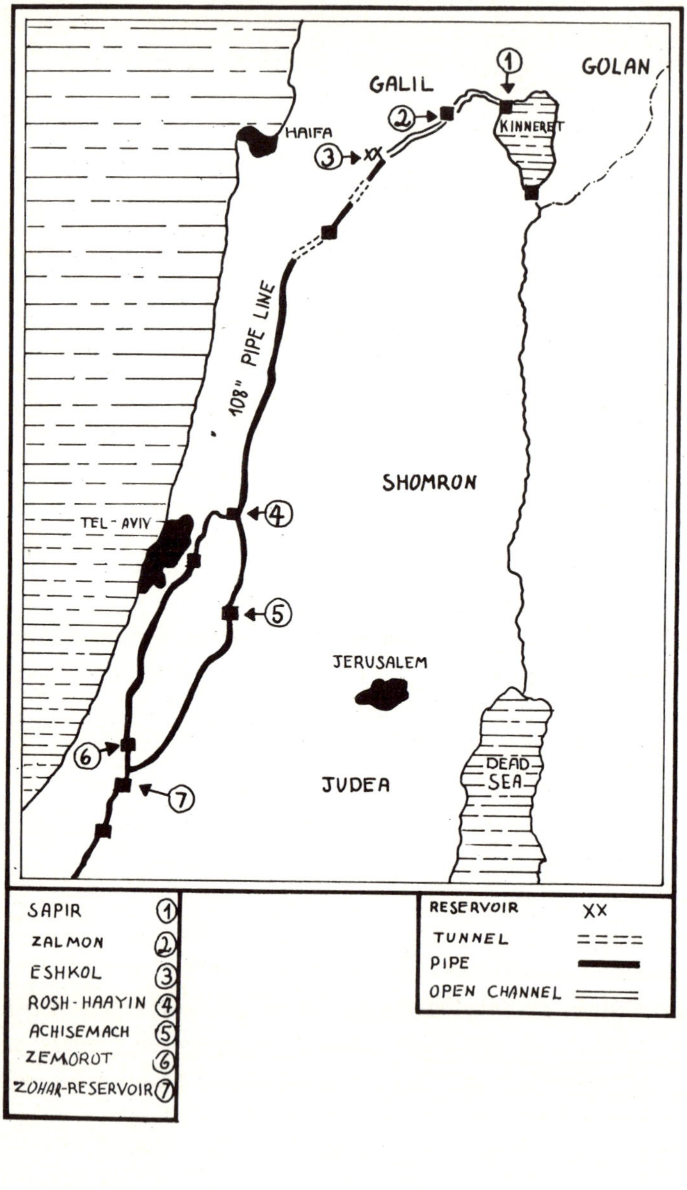

This problem was the subject of extensive discussion and consultation in Israel and the Ministry of Health concluded that the National Water Carrier System should be examined first, for the following reasons:

1. These are primarily surface waters having significantly higher concentrations of organic substances (such as humic and fulvic acids) than water originating in ground water.

2. The waters of the National Water Carrier are subjected to a number of chlorination treatments before reaching the consumer.

In a series of meetings caried out during 1978 between representatives of the Ministry of Health and the Mekoroth Water Company it was concluded that this problem should be jointly studied and the basis for co-operation was laid that enabled the carrying out of this project.

SAMPLING PROGRAM

Permanent sampling points were established on the National Water Carrier including the Yarkon Negev lines (see Map). During the survey special care was taken to sample only from these points. The stations were selected in order to enable sampling before and after chlorination. The following data were recorded at the time of sampling: residual chlorine, water temperature, and operation conditions. Special efforts were made to follow the sampling procedures described in the EPA study (3). The same sampling bottle with special teflon coated closure was used in order to assure hermetic sealing of the bottle. In addition the samples were kept cool until examined. These measures minimized the danger of escape of volatile substances.

METHODS OF ANALYSIS AND RESULTS

All the chemical analyses were carried out at the Institute of Control and Standardization of Drugs, Ministry of Health, Jerusalem. The methods used were those described by Rickard and Funk (6). The results of this study have been published in a joint report prepared by the Ministry of Health and the Mekoroth Water Caompany September 1980, (in Hebrew) (8).

DISCUSSION

In the initial stages of the study only chloroform was detected by the analytical methods used. Chloroform levels for the period April - September 1978 are presented in Fig. I. The chloroform levels detected

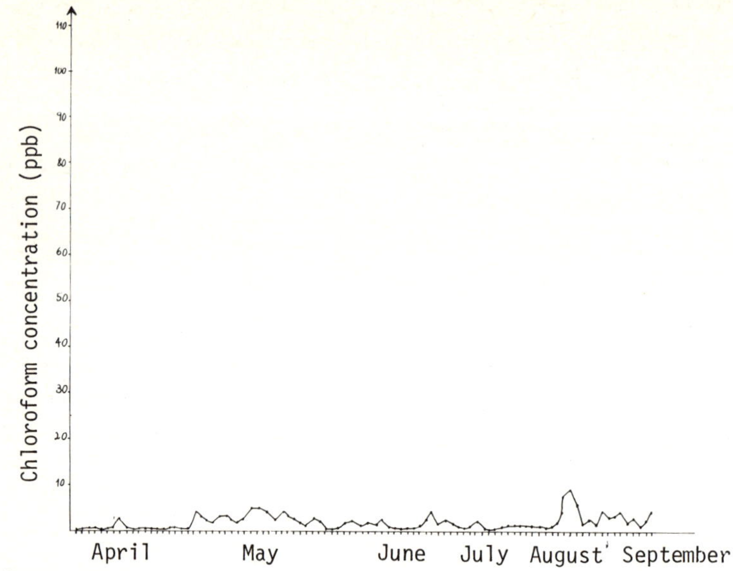

fig 1:CHLOROFORM LEVELS IN THE NATIONAL WATER CARRIER: APRIL- SEPTEMBER 1978.

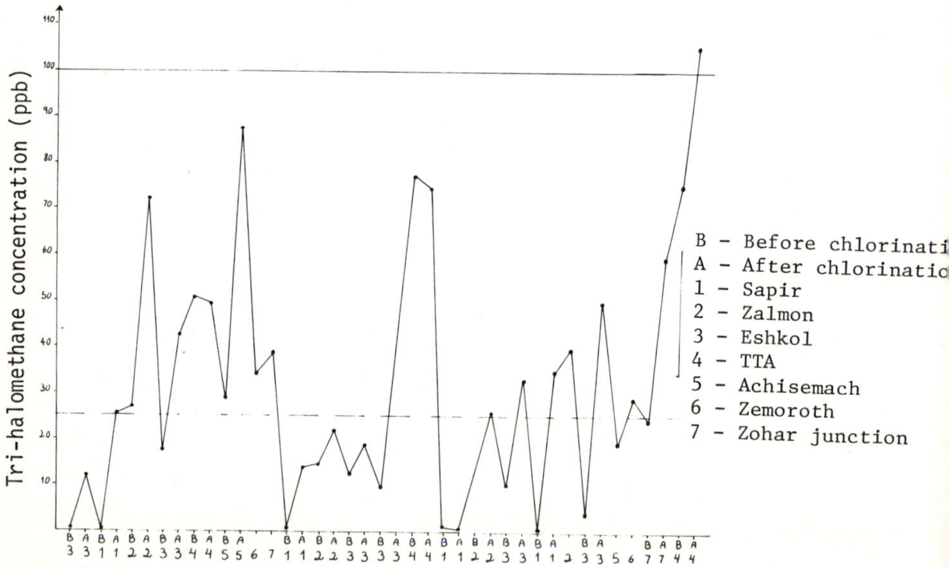

fig 2: TRI-HALOMETHANE LEVELS IN THE NATIONAL WATER CARRIER.

throughout the survey were low and the highest level detected was 13.3 ppb. Subsequently the analytical methods were expanded to include the other haloform compounds. The general picture changed drastically and total haloform concentrations were significantly higher as compared to chloroform. The total haloforms detected in some samples reach and even exceed the official US EPA tolerance of 100 ppb. Fig. 2. Even when only the low chloroform levels were detected, it was apparent that it was formed as a result of chlorination. In Fig. 3 data are presented which demonstrate this effect for chloroform and more dramatically for the individual and total haloforms. The fluctuation of haloform levels in the National Water Carrier System is of great significance especially since there is a gradual buildup of haloform levels due to serial chlorination treatments. An attempt to follow haloform levels was carried out several times during periods of increased chlorination of the system. These results are presented in Figure 4 and should be interpreted with caution. However it is clear that rises in haloform levels are as a result of chlorination. Reductions of haloform levels observed can be associated with flow in open channels and storage in reservoirs and dilution due to the introduction of ground water into the system.

The special significance of the presence of inorganic bromides in raw water is now recognized. Bromine is oxidized in the presence of chlorine to form intermediate complexes which are involved in the haloform formation reactions and also increase the yield (5). Bromine levels in Lake Kinneret were determined at our request by the Nessin Laboratory of the Mekoroth Water Company and found to be in the vicinity of 2 ppm. This level is sufficient to account for the results obtained in this study.

During this period other investigators in Israel detected haloforms in water. Miss Y. Yungster at the Hebrew University in Jerusalem (8) and a group at the National Physical Laboratory. Their published results are in accordance with those presented in this paper.

ACKNOWLEDGEMENTS

A large number of people were involved in the planning and carrying out of this study. It would not have been possible without the active assistance of the following:

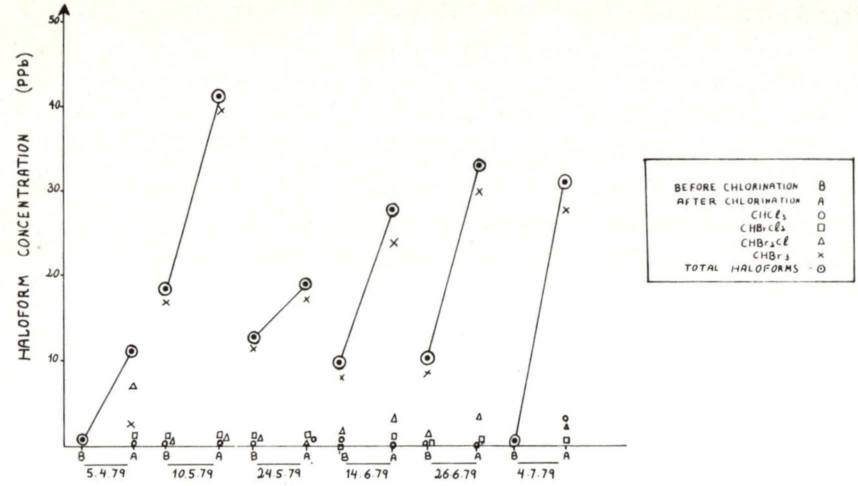

fig 3: INFLUENCE OF CHLORINATION ON HALOFORM FORMATION IN WATER.

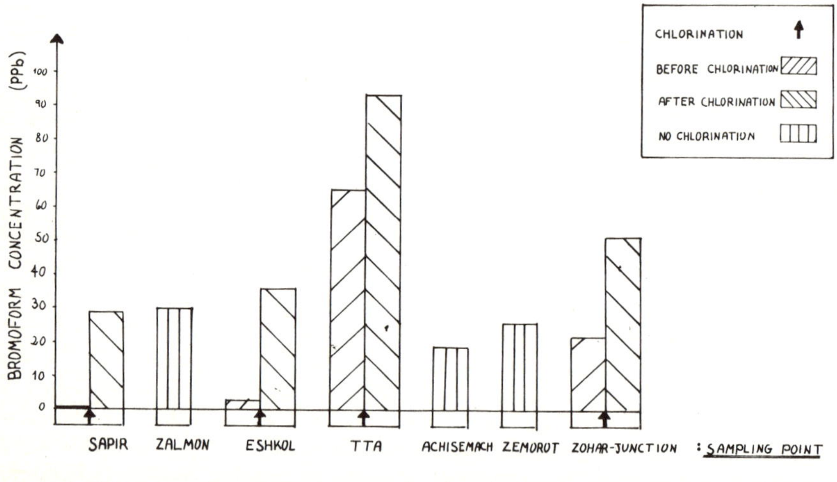

fig4:BROMOFORM LEVELS IN THE NATIONAL WATER CARRIER -4 TO 11 JULY 1979.

Eng. R. Halperin, Eng. Y. Zohar, Miss Naomi Burger, Miss Sara Bernstein, Mrs. Z. Balzer of the Ministry of Health.

Eng. Y. Ben Zur, Eng. L. Halperin, Y. Peleg, P. Berkovits, Dipl. Chem., Eng. Z. Stuhl, Eng. I. Mantel, San. Eng. Ran Zur and D. Ziskind of the Mekoroth Water Company.

REFERENCES

1. Rook, J.J., (1974), Formation of Haloforms during chlorination of natural waters, Wtr. Trtmt & Exam. 23, 234.

2. Bellart, T.A., Lichtenberg, J.J., & Kroner, R., (1974) The occurrence of Organohalides in finished drinking waters. Jour. A.W.W.A. 66, 703.

3. Symons, J.M., et al (1975), National Organics Reconnaissance Survey for Halogenated Organics, Jour. A.W.W.A. 67, 11, 634-647, ibid 67, 12, 708, 209.

4. Trussel, R.R., Umphres, M.D., (1978), The formation of Trihalomethanes. Journal A.W.W.A. 70, 589-664.

5. Arguello, M.D., et al (1979), Trihalomethanes in Water: A Report on the occurrence, seasonal variation in concentrations and precursors of Trihalomethanes. Jour. A. W. W. A. 71, 9, 504-508.

6. Rickard, J.J., Funk, G.A., (1977), Liquid Extraction for the rapid determination of Halomethanes in water. Jour. A.W.W.A. 5, 69, 62-67.

7. Yungster, H., (1980), The formation of organo chlorine compounds as a result of chlorination of water from various sources in Israel. M.Sc. Thesis, Hebrew University in Jerusalem.

8. Report of a joint project carried out by the Ministry of Health and the Mekoroth Water Company on the subject. "The Detection of Haloforms in the National Water Carrier - 1978-79 (in Hebrew).

Developments in Arid Zone Ecology and Environmental Quality
Edited by H. Shuval

39

THE FORMATION OF ORGANO CHLORINE COMPOUNDS AS A RESULT OF CHLORINATION OF WATER FROM THE NATIONAL CARRIER IN ISRAEL

M. Peleg, H. Yungster and H. Shuval

Human Environmental Sciences Division, School of Applied Science and Technology, The Hebrew University of Jerusalem.

ABSTRACT

Measurements have been carried out on the formation of organo chlorine compounds in the National Water Carrier System. A pyrohydrolysis technique was implemented for the study of the total organo chlorine content (TOCl) of the waters and this parameter together with values of the volatile haloganated organo compounds (THM) and dissolved organic carbon content (DOC) were measured. The results indicate that TOCl and THM values can reach values of up to more than 350 μg Cl/liter and 150 ppb respectively.

INTRODUCTION

Concern has recently been expressed over the possibility of the existence of trace amounts of organo-chlorine compounds, such as chloroform, in the nation's drinking water supplies. Certain of the organo-Chlorine compounds are known or suspected as carcinogem, Searle (1976) and Cantor (1977).

The first reports of the formation of chlorinated organics during treatment of drinking water came simultaneously from Rook (1974) in the Netherlands and from Bellar (1974) in the U.S.A. Both found that chlorine used for disinfection reacted with organic substances, so called precursors, to form chloroform and other halogenated organics.

The U.S. Environmental Protection Agency (1978) has recently proposed standards for the maximum allowable concentrations of trihalomethanes (THM) in drinking water. The suggested maximum contámination level for total THM has been set at 100 μg/liter.

The water flowing in the National Carrier System was suspected as being a possible source of chlorinated organics since it contains known precursers such as humic and fulvic acids due to the nature of the vegetation rich Lake Kinneret water. In addition, the water is chlorinated a number of times during its flow in the pipeline from the Kinneret to the south of Israel. Further, the relatively long contact time of the precursers with the chlorine assists in the organo-chlorine formation.

In the present study the total organic chlorine (TOCl) content of the water was measured together with the total trihalomethane value (THM) and the dissolved organic carbon content (DOC). The TOCl value is a group parameter intended to be a measure of the total amount of chlorinated organics present in water. The above parameter includes the non-volatile organic chlorine compounds not measured by the THM value. Because chlorinated organics are not naturally occurring, and because many chlorinated organics are known to be detrimental to the health of the water consumer, a TOCl measurement has been suggested as a useful indicator of water quality.

MATERIALS AND METHODS

a) Total Organic Chlorine - TOCl

The technique used in the present study was based on the pyrohydrolysis method developed by Kuhn (1973, 1974) together with the adaption suggested by Dressman (1977).

In order to measure the TOCl value of an aqueous sample it is necessary to preconcentrate the organic content of the water. This is carried out by absorption of the organics onto powdered activated charcoal in the presence of inorganic nitrates which prevents the absorption of inorganic chloride onto the charcoal. The powdered activated charcoal is then flocculated and the washed and dried sample is placed in the pyrohydrolysis system as shown in Figure 1. The sample is slowly heated to 900°C (1 hour) in a stream of oxygen and water under which condition the organic compounds are decomposed and any chlorine present is converted into HCl which is then trapped in aqueous solution. The chloride content of the water, after appropriate treatment, can be measured using an ion sensitive electrode or any other suitable technique.

The sensitivity of the technique is as low as 20 ppb chloride ion and depends on the volume of the original water sample used in the concentration stage. The efficiency of the method is at least 90%.

Fig. 1 Pyrohydrolysis Apparatus

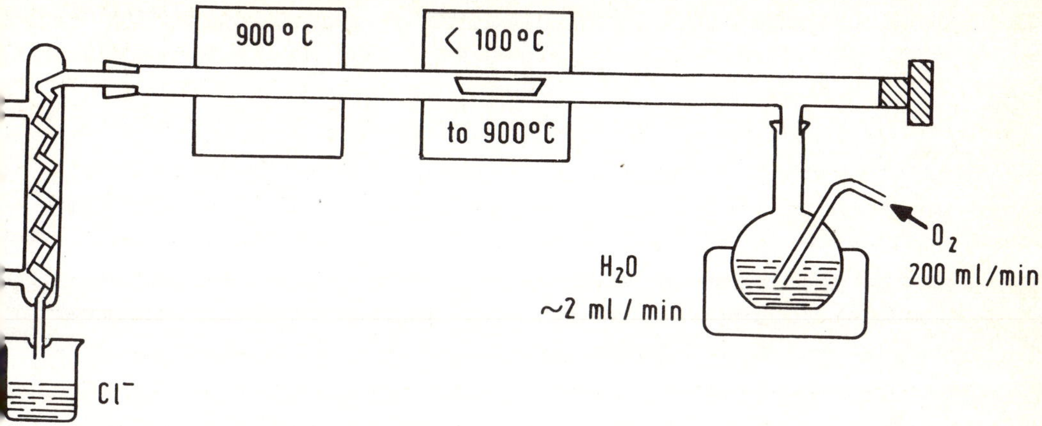

b) Trihalomethanes (THM)

The above parameter was originally used to designate the measurement of trihalomethanes such as chloroform, etc. However, with the discovery of other volatile halogenated organics such as carbon tetrachloride, these have now been included in the so-called THM value

The analysis of the volatile halo-organics was performed using a gas chromatograph with an electron capture detector. A 5% DC 500 Chromosorb G on AW DMCS (80/100 mesh) was the column employed in the present study.

Although a number of methods have been used to preconcentrate the sample, it was found that a simple liquid-liquid extraction technique, such as that employed by Henderson (1977), using pentane as the extraction agent, was very efficient.

A series of standard solutions were prepared containing the following compounds which are most commonly found in chlorinated water systems. $CHCl_3$ - chloroform, CCl_4 - carbon tetrachloride, C_2HCl_3 - trichloroethylene, C_2Cl_4 - tetrachloroethylene, $C_2H_2Cl_4$ - 1,1,1,2 tetrachloroethane, $CH\ Br_3$ - bromoform, CBr_4 - carbon tetrabromide.

Since it was observed that no carbon tetrabromide was formed in the natural waters it was possible to use the above compound as an internal standard.

c) Dissolved Organic Carbon (DOC)

The DOC value of the water is an important parameter since it gives an indication of the concentration of the precursers. The techniques used in the present study employed UV oxidation of the organics to CO_2 and measurement of the CO_2 formed by Infra Red absorption.

d) Sampling

Immediately on sampling any residual chlorine was neutralized in order to terminate the chlorination reaction with the organics. The water samples for the THM measurements were kept in completely filled BOD bottles to prevent loss of the volatile organics. Since the TOCl measurement required large volumes of up to 10 liters of water it was necessary to use plastic containers. Preliminary experiments showed that waters stored in containers made of PVC gave erroneously high results. It was necessary, therefore, to use containers made of polyurethane which gave a blank value for water stored in that type of container of less than 20 ppb chlorine.

RESULTS AND DISCUSSION

Initial experiments were carried out on waters available from the Jerusalem Municipality water supply. The results obtained showed values close to the minimum detection limit of 20 ppb chlorine and DOC values of less than 0.5 ppm carbon. This indicates that water supplied to Jerusalem during the period under examination (12/77 to 2/78) was of high quality as to be expected of ground water sources.

A series of experiments were performed on waters from the National Carrier System. A typical set of results is shown in Table 1 for samples taken during the fifth and sixth of April 1978. The duration of the water in the pipe line is an approximation based on the flow rates. No account is taken into consideration of the possible time lag spent by the water in the various reservoirs on the pipe line.

The results obtained in the present study are in agreement with a later more extensive study carried out by the Ministry of Health together with Mekoroth (1980) for the period between September 1978 to August 1979. Figures2 and 3 show the influence of time on the formation of the chlorinated compound. Both the results for the TOCl and THM indicate that there is an increase in the organochloro compounds in the water as both the time of contact between the chlorine and precursers increases and the amount of chlorine added to the water increases.

Table 1

TOCl and THM Formation in the National Water Carrier System

Sampling site	Chlorination	Duration in pipe line hrs.	DOC ppm	TOCl ppb	$CHCl_3$ ppb	C_2HCl_3 ppb	C_2Cl_4 ppb	CH Br_3 ppb	THM ppb
Tabha	before	0	1.8	10	n.d.	n.d.	n.d.	n.d.	0
Tabha	after	0.2	1.2	75	n.d.	0.8	7	4	11.8
Zalmon	before	3.5	1.9	130	1.3	1.5	8	55	65.8
Eshkol	before	7.5	2.6	85	0.6	0.1	1.2	7	8.9
Eshkol	after	8.0	2.3	190	1.4	3.7	10.5	117	132.6
Rosh HaAyin	before	20.0	2.0	250	2.5	1.5	12.0	102	118
Rosh HaAyin	after	20.5	2.3	185	3.6	1.4	7.0	38	50

Sampling carried out during 5-6 April 1978

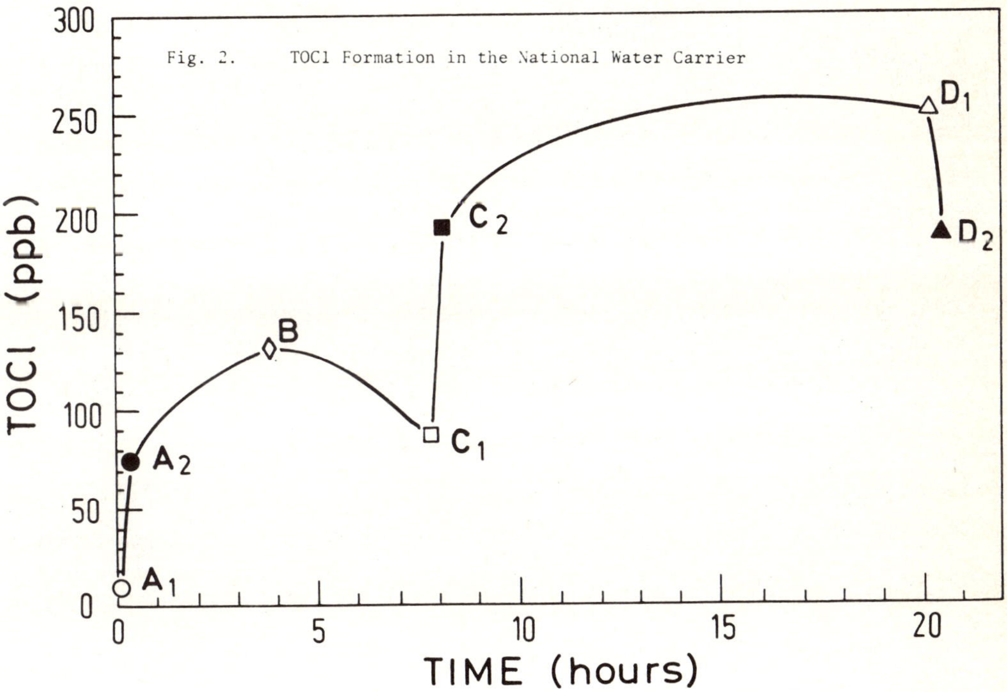

Fig. 2. TOCl Formation in the National Water Carrier

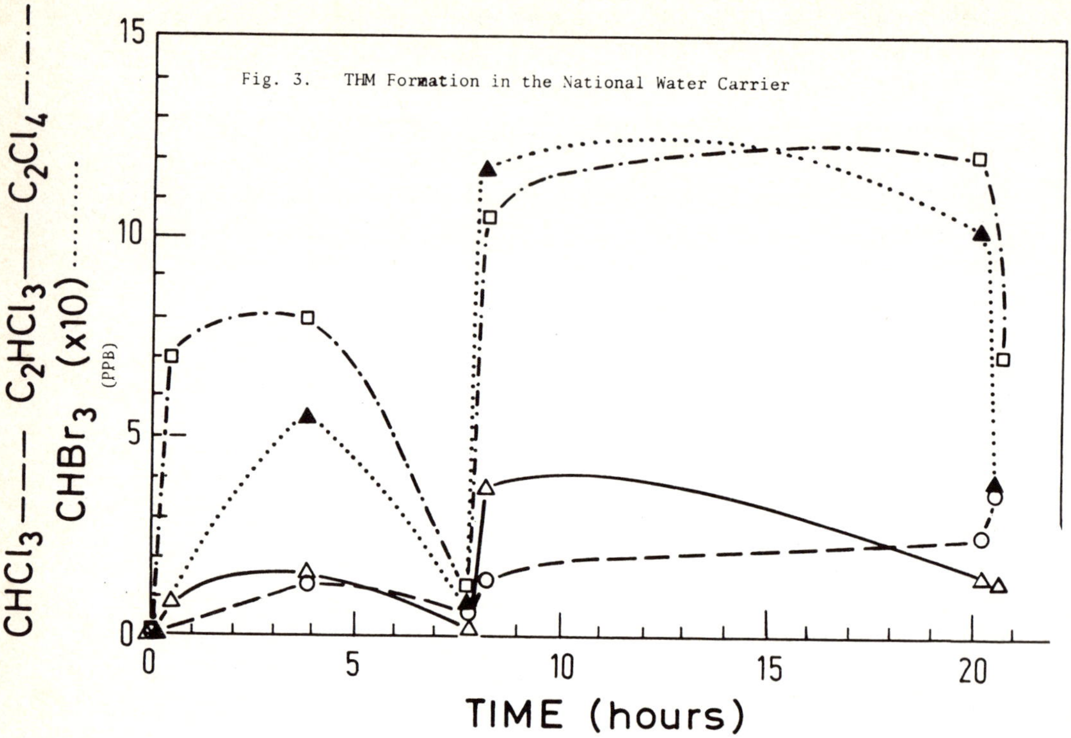

Fig. 3. THM Formation in the National Water Carrier

In order to determine the effect of time on the organochloro formation a study was performed on the potential of the water to form organochloro compounds by not neutralizing the residual chlorine immediately on sampling. The TOCl value was measured when no residual chlorine remained in the water. The results are shown in Table 2.

Table 2

Organo Chlorine Potential

Sampling Site	Residual Chlorine at time of Sampling	TOCl (ppb)	Contact Time (days)
Rosh HaAyin	0.78	377	7
Achizemach	0.30	155	7
Zemorot	0.35	208	8
Zohar Reservoir	0.90	322	9

The above results indicate that the water in the National Carrier contains the necessary precursors which on contact with chlorine give rise to substantial amounts of chloro-organic compounds of which up to around fifty percent will be volatile THM compounds.

The amount of bromoform present in the National Carrier waters was at first thought to be unusually high. However, other studies both in Israel, Europe and the U.S.A. have recently reported similar results.

The bromide ion is present in most natural waters in concentration generally less than one ppm. When chlorine is added to water containing bromide, hypobromide is formed: $Br + HOCl \rightarrow HOBr + Cl$.

HOBr like HOCl is an electrophilic agent but reacts kinetically two hundred times quicker than HOCl. Thus even minute amounts of bromide will on chlorination of water give rise to organobromo compounds.

CONCLUSIONS

The present study indicates that chlorination of the National Carrier System will give rise to relatively substantial amounts of organo chlorine compounds, especially under conditionsof repeated chlorine dosage and long retention times of the water before usage. Continuous monitoring, such as undertaken by the Ministry of Health, is necessary to ascertain the quality of the water. If the proposed U.S. standard of 100 μg/liter is accepted in Israel it may be necessary to undertake measures to remove the precursers or the halogenated organics from the water before use.

Caution will also be required if renovated affluents such as from the Dan Region Project are to be reused. These waters may have a relatively high dissolved organic carbon content and chlorination, if employed, will certainly give rise to high THM and TOCl values. It is therefore suggested that a TOCl potential be employed in order to ascertain the organo chlorine formation limits of the water.

REFERENCES

Bellar T.A. & Lichtenberg J.J. (1974) Determination Volatile Organics at Microgram-per Litre Levels by Gas Chromatograph, J.A.W.W.A. 66, 739-744 (1974).

Cantor, R.H., Mason T.J. & McCabe L.J. (1977), Association of Cancer Mortality Rates and Trihalomethane Levels in Municipal Drinking Water Supplies, Proc. 10th Ann. Mtg. Soc. Epidemol. Res.

Dressman R.C., McFarren E.F. & Symons J.M. (1977). An Evaluation of the Determination of TOCl in Water by Absorption onto Ground Granular Activated Carbon, Pyrohydrolysis and Chloride - Ion Measurement. Water Quality Tech. Conf. Kansas City, Dec. 5-6.

Henderson J.E., Peyton G.R. & Glape W.H. (1977). Identification and Analysis of Organic Pollutants in Water, Ed. L.M. Keith, Ann Arbor Science.

Kuhn W. & Sontheimer H. (1973), Several Investigations on Activated Charcoal for the Determination of Organic Chloro compounds. Vom Wasser, 41, 65-77.

Kuhn W. & Sontheimer H. (1974). Analytic Determination of Chlorinated Organic Compounds with Temperature Programmed Pyrohydrolysis. Vom Wasser, 43, 314-327.

Ministry of Health and Mekoroth (1980). Detection of Haloforms in the National Water Carrier for 1978-1979, Internal Report, Sept.

Rook J.J. (1974). Formation of Haloforms during Chlorination of Natural Waters, Water Treat. and Exam. 23, 234.

Searle, C.E. (1976). Chemical Carinogens. Monograph, 173, ACS, Washington.

U.S. Environmental Protection Agency Federal Register (1978). Interim Primary Drinking Water Regulation. Control of Organic Chemical Contaminants in Drinking Water, Feb. 9, Part II.

ACKNOWLEDGEMENT

We would like to thank Dr. Wiessenberg and Mr. A. Eisenberg of the Ministry of Health and Mekoroth for their assistance in obtaining the necessary water samples.

Developments in Arid Zone Ecology and Environmental Quality
Edited by H. Shuval

40
PROBLEMS RELATED TO THE MONITORING OF NITRATES IN GROUNDWATER

Yacob Kanfi and Daniel Ronen

Israel Water Commission, Tel Aviv, Israel

ABSTRACT

Surveys of nitrate in groundwater have been conducted over a number of years. The cummulative data still evade rigorous interpretation due to the variation in the number and the identity of the wells sampled and the utilization of different analytical techniques.
Sampling a large number of wells in a region once each year can reflect the average nitrate content of groundwater at the time of sampling, but this is not sufficient to disclose the nitrate variation in individual wells or the geographical distribution of the variation.

INTRODUCTION

The changes in the nitrate level of a groundwater basin can be determined through the comparison of statistical parameters, such as the mean regional concentration. However, variations in the number and the identity of the wells sampled in different years, the problem of selection of random samples and the utilization of different analytical methods may complicate the interpretation of results.

Due to budgetary constraints, the sampling of a large number of wells is some times limited to one sample per year per well. According to Ronen and Kanfi (1979) this sampling pattern does not reflect either the nitrate content in individual wells or the geographical distribution of nitrate concentration.

In the present work the nitrate variation in the Rishon Le Zion - Rehovot region is estimated by means of stable and nonstable samples of wells analyzed by different analytical methods in 1970 and 1973. The implication of these factors on groundwater quality monitoring is discussed.

METHODOLOGY

The Rishon Le Zion - Rehovot region is a part of the shallow sandy and phreatic coastal plain aquifer of Israel. In this region approximately four hundred wells were sampled once in 1970 and once in 1973 by Ronen (1972) and Kanfi and Ronen (1976), respectively. All the samples were obtained during the summer months. This program covered over 80% of the pumping wells of the region. The 1970 sample consisted of 376 wells and the nitrate was determined by the Devarda method, Black et al. (1965). In 1973, 396 wells were sampled and the nitrate was determined by the Cadmium method, Henriksen and Selmer - Olson (1970). The precision of the Devarda and Cadmium methods as expresed by the coeficient of variation is 5% and 2%, respectively, Ronen (1972) and Kanfi and Ronen (1976). The two methods were compared by analyzing 17 samples ranging in nitrate content from 10 mg/l to 100 mg/l. The following linear relationship (r = 0.98) was found:

$$NO_3^- \text{ (mg/l)}_{\text{Devarda meth.}} = 1.05 + 0.95 \times NO_3^- \text{ (mg/l)}_{\text{Cadmium meth.}}$$

The variation in the nitrate content of groundwater was determined by five different procedures:

a) A comparison of the mean nitrate concentration calculated from the data obtained from 320 identical wells sampled in both periods ("stable sample" of wells). Nitrate concentrations corrected according to the relationship found between the two analytical methods ("analytical correction").

b) Same as above without analytical correction.

c) A comparison of the mean nitrate concentration calculated from all the wells sampled in each period ("nonstable sample" of wells) after analytical correction.

d) Comparison of the statistical distribution of nitrate concentration in the stable sample of wells after analytical correction.

e) Comparison of the nitrate variation in the individual wells of the stable sample and their geographical distribution, after analytical correction.

RESULTS

1) Change in the mean nitrate concentration

The change of the regional mean nitrate concentration and the statistical significance of this change analyzed by the t test is presented in Table 1.

Table 1 - Change in the mean nitrate concentration of groundwater in the Rishon Le Zion - Rehovot area, 1970 - 1973

Type of Sample	Year	Number of Samples	Mean NO_3^- Concent. (mg/l)	Standard Deviation (mg/l)	NO_3^- Change (mg/l)	Significance Level %
Stable with analyt. correc.	1973 1970	320 320	58.8 54.5	29.6 28.8	4.3	7
Stable without analyt. correc.	1973 1970	320 320	60.8 54.5	31.0 28.8	6.3	1
Nonstable with analyt. correc.	1973 1970	395 376	55.1 55.3	30.2 30.5	0.2	not significant

The analytical correction reduces the magnitude and the significance level of the mean nitrate variation computed from the stable sample. No significant variation is observed from the comparison of the nonstable sample. It should be stressed that in this case although the numerical difference between the samples is only 19 wells, the comparison is based on 131 non identical wells.

2) Statistical distribution of nitrate concentration in the stable sample

The distribution of nitrate concentration in 1970 and 1973 is presented in Fig. 1. A decrease in the number of wells with a nitrate concentration ranging from 0 mg/l to 60 mg/l and an increase in the number of wells with a nitrate concentration above 60 mg/l is observed in 1973. The general trend of nitrate increase is also observed by the shifting of the acummulative curve with an increase of 4.3 mg/l in the median value.

3) Variation of the nitrate concentration in each well of the stable sample

The distribution of the 320 wells according to the average yearly variation in their nitrate content (total variation in each well divided by the lapse of

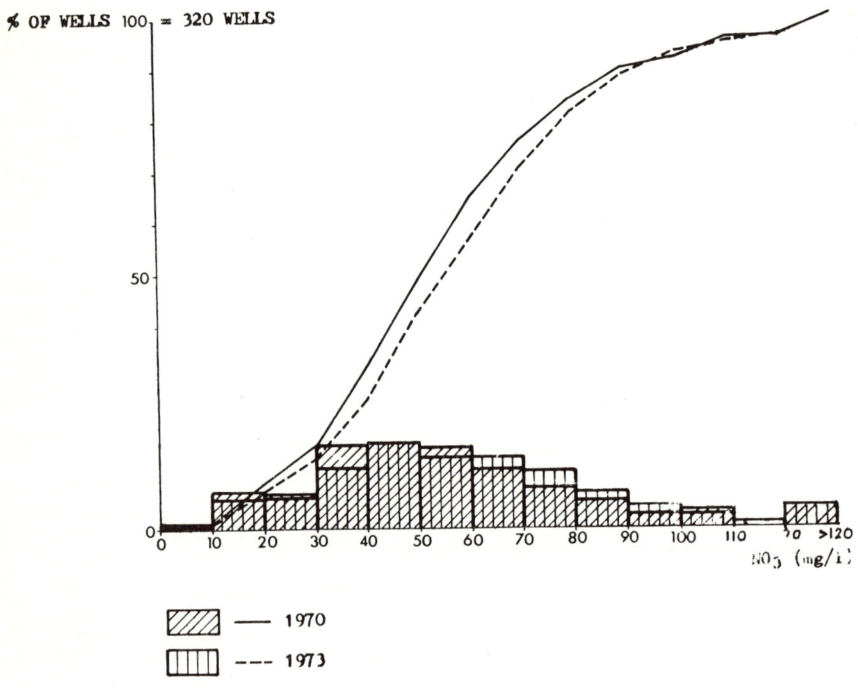

Fig. 1. Change in nitrate concentration in Rishon leZion - Rehovot area.

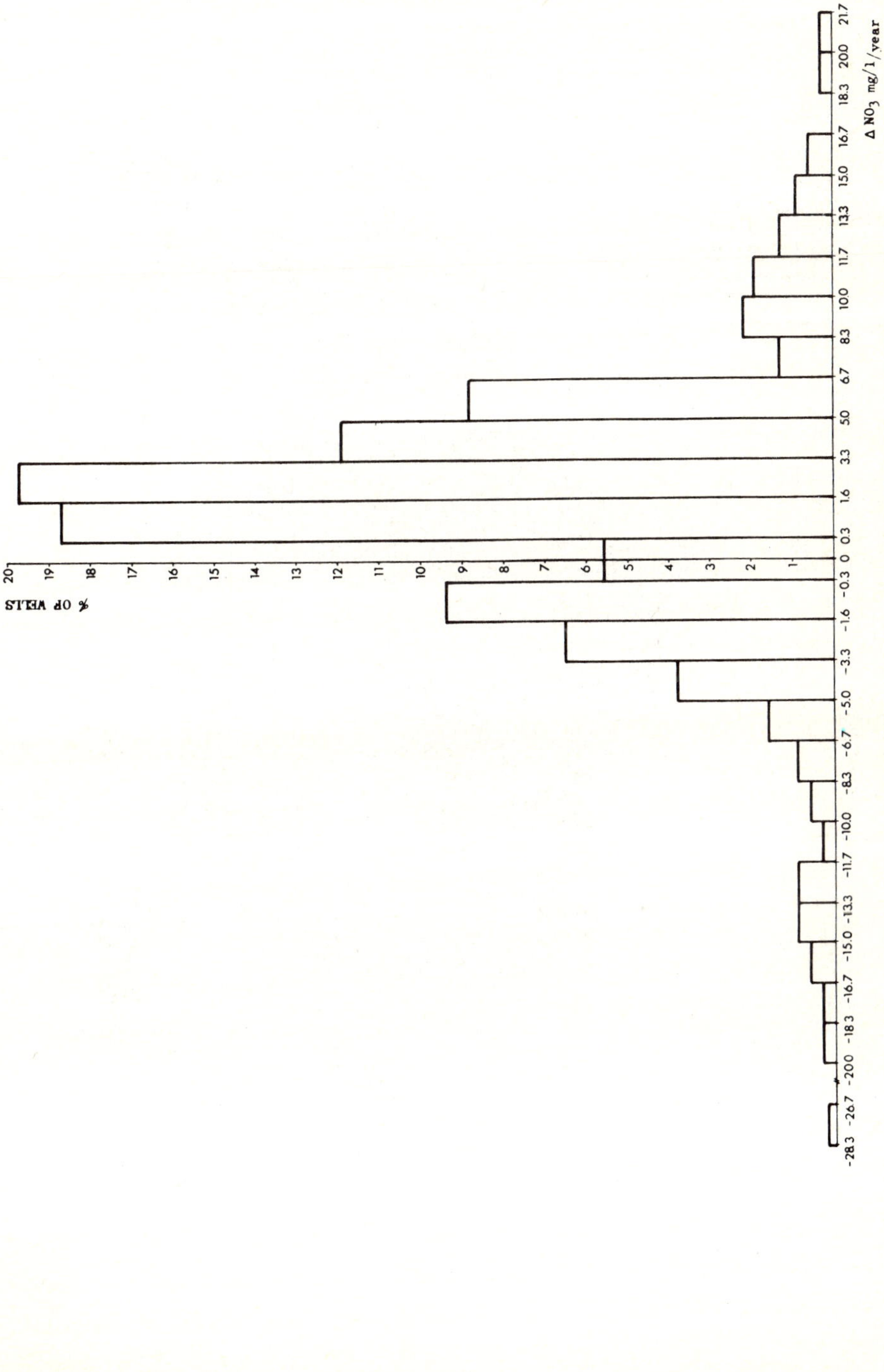

Fig. 2. Distribution of NO_3^- variation in individual wells.

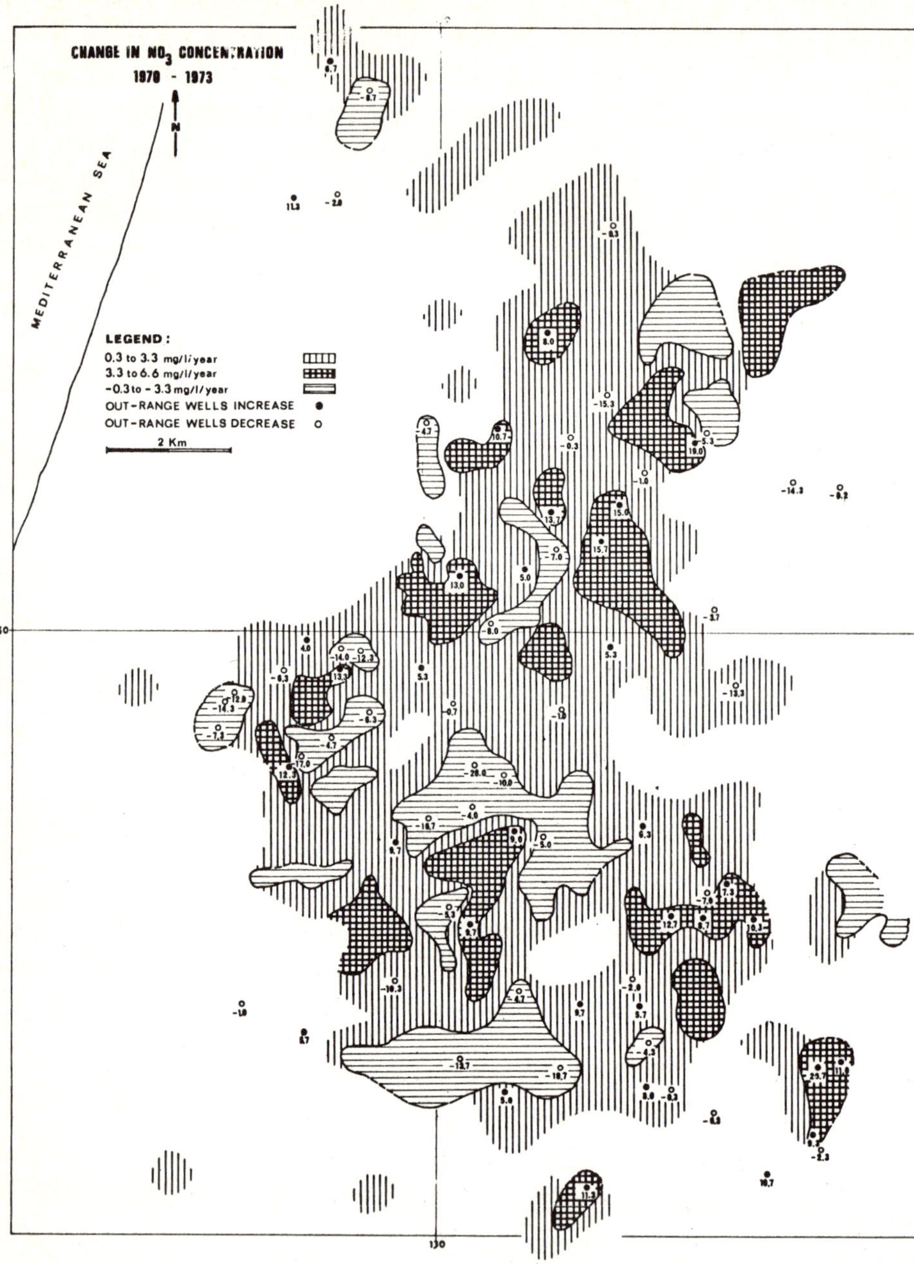

Fig. 3. Change in NO_3 concentration 1970-1973.

time between each sampling) is presented in Fig. 2. Three types of variations were evident: â) An increase in the nitrate content in 68% of the wells. The average increase was 3.7 mg/l/year. b) A decrease in nitrate content in 26% of the wells averaging 4.4 mg/l/year. c) No nitrate change in 6% of the wells. The geographical distribution of the nitrate variation is presented in Fig. 3. A zonal distribution of wells having a similar trend of nitrate variation is observed. Zones with a nitrate decrease are scattered all over the region.

DISCUSSION

The function of monitoring groundwater quality has been decentralized for many years. Sampling has been performed by private water companies, governamental offices and research institutes according to their specific goals and needs. Coincident with increasing pollution of the aquifers, mainly by nitrates, studies were conducted in order to define the state of pollution, the pollution sources and to elucidate the past and future trends of water quality, Saliternik et al. (1972), Ronen (1972), Mercado (1973), Kaanovitch and Blank (1974). In each of these studies the analysis of trends was generally based on inconsistent chemical data, nonstable samples being analyzed by different analytical methods of varying accuracy.

The importance of analytical correction to permit the comparison of data obtained by different analytical methods is obvious. Kaanovitch and Blank (1974) reported on the relatioship between four analytical techniques for nitrate analysis. They studied the variation in the average nitrate content of the coastal plain aquifer of Israel. They estimated that the high nitrate value calculated in 1972 is the result of the use of the ultra-violet method for the analysis of most samples.

The analysis of the average nitrate variation in a region, by means of stable and nonstable samples, raises several questions. In a stable sample the average nitrate content of groundwater is estimated from a smaller number of wells than in a nonstable sample. Therefore the stable sample would be expected to be less representative. Moreover, if a stable sample is utilized over a long period to monitor a groundwater basin, some variations in the quality of groundwater may be disregarded. On the other hand, the variation in the nitrate content calculated from a stable sample can not result from the inclusion of non-identical wells in the analysis.

In the present study, a rise in nitrate level of 4.3 mg/l was calculated by means of a stable sample. The analysis of the nonstable sample shows no nitrate change. A closer look at this phenomenon discloses that the 75 non-identical wells sampled in 1973 are located in regions of low nitrate content. Therefore, their addition to the sample reduces the average nitrate value by 3.7 mg/l.

In the 1970 and 1973 studies no preselection was made of the number or identity of the wells sampled. Still, the randomness of the samples may be questioned. It is possible to select random samples of existing wells, but it is questionable whether the wells are truly randomly scattered in relationship to the nitrate distribution in groundwater. Kaanovitch and Blank (1974) faced a similar problem while analyzing the average nitrate variation in the coastal plain aquifer between 1961 and 1972. During this period a nitrate rise of 0.85 mg/l/year and 2.70 mg/l/year was calculated from stable and nonstable samples, respectively. The stable sample included 223 wells, whereas in the nonstable sample 891 non-identical wells were compared. These authors attributed the high nitrate rise in the nonstable sample to the inclusion of wells with high nitrate content during the sampling period 1966 to 1972.

The variation of the mean nitrate concentration and the acummulative distribution curve(Fig. 1) of the stable sample reflect the regional change in the nitrate content of groundwater. From this data, it could be mistakenly concluded that the nitrate concentration increases in each one of the wells of the area. However, Fig, 2 shows that a nitrate decrease was measured in 83 wells. The maximum nitrate decrease was 85 mg/l in the course of the three year period. The frequent variations in the concentration of chemical components in wells reported by Ronen and Kanfi (1979), can explain the increase and decrease in the nitrate content of individual wells; it should be stressed that the wells were sampled only once in 1970 and once in 1973. The zonal distribution of wells having a similar variation trend (Fig. 3) may have both a statistical explanation, based on the probability of aggregation of wells with the same trend, or a hydrochemical explanation such as a decrease in the nitrogen input to the aquifer and/or changes in the pattern of groundwater flow. However, no support was found for this second hypothesis. Areas with nitrate decrease are located under nitrogen fertilized land as well as under populated areas with septic tanks. No correlation was found between the areas exhibiting nitrate decrease and the places where groundwater is artifitially recharged with surface water of low nitrate content. A decrease in the nitrate content of groundwater was reported by Ronen (1972) in the western region of the area studied. This decrease was attributed to the easterly flow of non-polluted natural recharge water from the sand dunes. This explanation can not account for the decrease in nitrate content in the central part of the area.

CONCLUSIONS

1. The problems faced when analyzing the nitrate variation of a groundwater basin by means of stable and nonstable samples of wells may not be disregarded.
2. The same analytical methods should be used to enable comparisons of the concentration of chemical components. As old analytical methods are improved and new ones developed, a relationship between the methods must be established.
3. Due to the frequent changes in the concentration of chemical components in groundwater, any analysis of their variation in individual wells or the geographical distribution of this variation, must be based on more than one sample per year.
4. The analysis of the variation of chemical parameters other than nitrates in groundwater may pose difficulties similar to those exemplified in this work.

REFERENCES

Black C.A. et al. (1965) Methods of soil analysis, chemical and microbiological properties. Am. Soc. of Agronomy, No.9, Part 2, pp.1191 - 1201.

Henriksen A. and Selmer - Olson A.R. (1970) Automatic methods for determining nitrate and nitrite in water and soil extracts. Analyst, Vol.95, pp. 514 - 518.

Kaanovitch Y. and Blank D. (1974) Groundwater pollution by nitrates in Israel. Tahal, Report No. 01/74/47. (in Hebrew)

Kanfi Y. and Ronen D. (1976) The chemical quality of groundwater in the coastal plain aquifers of Israel. Water Commissioner's Office, Report No. 76/1. (in Hebrew)

Mercado A. (1973) Pollution and salinization of groundwater in Rishon Le Zion-Rehovot. Tahal, Report No. 01/73/46. (in Hebrew)

Ronen D. (1972) Nitrate contamination of groundwater in the Rishon Le Zion - Rehovot district. Tahal, Report No. HR/72/046. (in Hebrew)

Ronen D. and Kanfi Y. (1979) Frequent variations in the chemical quality of groundwater. Tenth Scientific Conference, Israel Ecological Society, Sede Boker, Israel.

Saliternik C., Kaanovitch Y., Rave I., Shevach I., Ronen D., Shvartz I. and Blank D. (1972) Groundwater contamination in Israel, collection and processing of data. Tahal, Report No. HR/72/043. (in Hebrew)

Developments in Arid Zone Ecology and Environmental Quality
Edited by H. Shuval

41

"Granular Activated Carbon and Preozonated Granular Activated Carbon Treatment for Biologically and Physically-Chemically Treated Wastewater Effluents"

A. Netzer and J. McNutt
Environmental Sciences
The University of Texas at Dallas
Richardson, Texas 75080

ABSTRACT

The Duck Creek Wastewater Treatment Plant is a combined municipal-industrial wastewater treatment facility. The municipal effluents are from a population of over 200,000 and the industrial discharges are from over 600 various light industries, ranging from paint manufacturing and metal plating to food and dairy processing. The Duck Creek Wastewater Treatment Plant is a 30 MGD split-stream treatment facility. 7.5 MGD is treated with biological processes and 22.5 MGD is treated with physical-chemical processes. Final treatment for both systems consists of granular activated carbon adsorption and chlorine disinfection.

The University of Texas at Dallas has operated two pilot plant systems to evaluate granular activated carbon (GAC) and preozonated granular activated carbon (BAC) treatment for biologically and physically-chemically treated wastewater effluents at the Duck Creek Wastewater Treatment Plant since November, 1979.

INTRODUCTION

The City of Garland, Texas, is an industrial, suburban community of over 200,000 located on the northeast perimeter of Dallas, Texas. The industrial community located in Garland, Texas, consists of over 600 various light industries, ranging from paint manufacturing to metal planting and dairy processing. These industries, some with very little, but most with no pretreatment, discharge their effluents into the municipal system. The influent wastewater treatment at the Duck Creek Wastewater Treatment Plant is composed of approximately 70-75% municipal and 30-25% industrial wastes.

The City of Garland elected to expand the 10 MGD biological facility with the addition of 22.5 MGD physical-chemical treatment processes(1).

The pretreatment processes include trash removal with trash screens, a 60 MG equalization basin with brush aerators, bar screening, and grit removal.

The pretreatment process then splits, with up to 7.5 MGD directed to biological treatment processes and up to 22.5 MGD directed to physical-chemical treatment processes.

The biological treatment processes consist of primary clarification, primary trickling filters, intermediate clarification, secondary trickling filters, and final clarification.

The physical-chemical treatment processes consist of coagulation-flocculation, clarification, recarbonation, and ultra-high rate sand filtration.

The final treatment for the combined effluents is accomplished with the use of granular activated carbon adsorption and chlorination[2].

Studies by the City of Garland Laboratory at the Duck Creek Wastewater Treatment Plant found the following influent characteristics for the year 1979 (Table 1).

Table 1: Average Influent Wastewater Characteristics, 1979

Parameter	Averaged Result
Total BOD_5 (mg/l)	239
Total COD (mg/l)	540
Suspended Solids (mg/l)	233
pH (mean)	7.10

The State of Texas discharge permit for the Duck Creek Wastewater Treatment Plant allows for discharge of a maximum of 10 mg/l BOD_5, 10 mg/l of Suspended Solids, with a 1 mg/l minimum chlorine residual in the effluents.

Previous operational data for the City of Garland Duck Creek Wastewater Treatment Plant has indicated that the granular activated carbon facility located on-site would not meet the EPA discharge requirement for the plant. (Tables 2,3).

The University of Texas at Dallas was asked to construct, operate, optimize, study, evaluate, and compare the performance of preozonated and non-preozonated granular activated carbon adsorption on the wastewater effluents from both physically-chemically and biologically treated wastewaters. The project has allowed for study and comparison of the existing, on-site upflow-

Table 2: Duck Creek Wastewater Treatment Plant Operation July, 1977*

Parameter	Raw Wastewater from the equilization basin	Effluent after biological treatment	Effluent after physical-chemical treatment	Combined Effluent stream	Effluent after GAC treatment
Total COD mg/1	375-855	134-261	151-275	141-268	54-137
Total BOD_5 mg/1	202-369	22-63	69-168	79-165	35-103
Total SS mg/1	188-744	38-84	21-156	41-125	7-24

Table 3: Duck Creek Wastewater Treatment Plant GAC Operation - September, 1980**

Parameter	Effluent after GAC treatment
Total COD mg/1	41-186
Total BOD_5 mg/1	12-78
Total SS mg/1	4-20

*July, 1977 was the only full month that the carbon systems operated.
**September, 1980 was a period of 10 days in which the carbon systems operated.

downflow treatment design granular activated carbon adsorption (GAC) and the implementation of preozonation prior to upflow-downflow granular activated carbon adsorption treatment (BAC)*.

BAC has been practiced with great success in some European facilities since the early 1960's. The method has proven to be more efficient and cost effective than activated carbon adsorption without preozonation ([3]).

PILOT PLANT FACILITIES

The effluents for the UTD**-Duck Creek Pilot Plant operation are pumped from two locations at the Duck Creek Wastewater Treatment Plant. The biologically treated effluents are pumped from the biological lift station, immediately following the final clarification in the biological processes. The physical-chemical effluents are pumped from the final stand-pipe manifold immediately after the ultra-high rate sand filtration.

There are two original pilot plant systems. One is for biologically treated effluents and one for physically-chemically treated effluents. Each system consists of one 4.0 GPM GAC pilot plant and one 4.0 GPM BAC pilot plant. Each BAC pilot plant consists of an ozone generator, one two-stage ozone contacting unit, one retention tank, two sand filtration units, five pumps, and two granular activated carbon columns (Figure 1).

The ozone generators in use are Union Carbide model number SG-4060, capable of delivering ozone dosages from 1-50 mg/l under current test conditions. The ozone contacting units consist of two PVC columns, both 8" in diameter. The first is 20' in height, the second 18' in height with a specially designed and fabricated foam removal system at its top. Ozone is introduced through French flat porcelain diffusers, 8" in diameter at the bottom of each of the columns. The first column is operated in a counter-current flow, while the second is in a co-current flow. The flow through the ozone contacting columns is 5.0 GPM, giving an average residence time of 20 minutes. Ozone dosages were determined in order to provide 20 mg/l of dissolved oxygen to the BAC systems. The current ozone dosages are 7.5 mg/l for the biological system and 15.0 mg/l for the physical-chemical system.

The post ozonation retention tanks are of treated fiberglass with a capacity of 320 gallons. A special overflow drain has been provided to insure residence times of 65-70 minutes.

The sand filtration units are of downflow type, stainless steel designed for 85-90% suspended solids removal efficiency. The units have 1.8 ft^2 of surface area, and a maximum designed flow rate of 36 GPM.

Four granular activated carbon columns for each pilot plant system are of PVC, 8" in diameter, and 14' in height. The granular activated carbon used is currently ICI Hydrodarco 3000. The carbon bed is 5' high (43.6 lbs),

*BAC stands for "Biological Activated Carbon
**UTD stands for The University of Texas at Dallas

Figure - UTD-Duck Creek Pilot Plant Facility. Foreground shows four sets of upflow-downflow carbon columns. Upper right shows two ozone contacting systems and post ozonation retention tanks. Upper left shows modified Calgon Filtersorb unit adapted to upflow-downflow treatment and blending system.

on top of 1' of sized gravel. The maximum residence time in the carbon columns is approximately 20 minutes. The carbon contacting time is approximately 8 minutes. The carbon columns are paired, operating in the upflow-downflow mode. One pair of carbon columns is used for contacting with the effluent without preozonation (GAC), the other for contacting with preozonated effluents (BAC).

A fifth pilot plant system has been operated to study blended effluents (biological and physical-chemical). This system is a modified Calgon Filtersorb manufactured unit, converted to two column upflow-downflow. The unit is operating at 0.40 GPM with a 42" granular activated carbon adsorption bed. The current blended effluent consists of 66% physical-chemical and 34% biological effluents (duplication of the current Duck Creek Wastewater Treatment Plant operation.)

EXPERIMENTAL

Daily samplings are collected at 14 sampling points. These points are located before and after each unit process. Immediately after collection, the pH, dissolved oxygen, and temperature parameters are determined according to "Standard Methods" in the UTD Field Research Laboratory on-site. The samples are then taken to the City of Garland Duck Creek Analytical Laboratory for the daily determination of TSS, COD, TOC, and BOD_5. Weekly analysis are performed for TKN, NH_3, NO_2, NO_3, Alkalinity as $CaCO_3$, chlorides, and fecal coliform levels. Three times weekly, conductivity and turbidity are determined. Every two weeks, total and soluble ortho-phosphate are measured. Once a month, a heavy metal analysis (Cd, Pb, Zn, Cr, Cu, Ag, Mn, Ni, As, and Hg) is performed.

RESULTS

As of December 31, 1980, the UTD-Duck Creek Pilot Plant Facility has been operating continuously for over 13 months (Table 4). Overall, the pilot plant has functioned according to design specifications, with the only downtime of length due to interruption in flow to the pilot plant facility from the Duck Creek Wastewater Treatment Plant operations.

The data described below represents the last quarter of 1980 and reflects the operations throughout the calendar year.

pH: The pH of the physical-chemical influent was dependent on the Duck Creek Wastewater Treatment Plant chemical operations. The values ranged daily, but remained near the 7.5-8.0 range. After GAC treatment, the ranges decreased to 7.3-7.8. After BAC treatment, the pH ranges decreased even further, averaging 7.0-7.5. The average pH reduction for GAC was 0.15-0.25 pH units, and the average pH reduction for BAC was 0.30-0.50 pH units.

The pH of the biological influent was stable, averaging 7.35-7.55. After GAC treatment, the effluent range was 7.05-7.25 pH units. After BAC treatment, the effluent pH range was 6.55-7.00. The average pH reduction for the GAC systems was 0.25-0.45. The average pH reduction for the BAC systems was 0.60-0.70 pH units.

Table 4: GAC and BAC Carbon Replacements

Date	GAC	BAC
November 15, 1979	Initial operation	Initial operation
February 10, 1980	Carbon replaced after 70 days	Carbon replaced after 70 days
May 5, 1980	Carbon replaced after 70 days	
July 3, 1980	Carbon replaced after 50 days	
November 7, 1980	Carbon replaced after 64 days	
December 31, 1980	49 days of operation on current carbon	234 days of operation on current carbon

The blended system influent pH values were the calculated average of the biological and the physical-chemical effluents, 7.3-7.8 pH units. The pH reduction after GAC treatment averaged 0.15-0.45 pH units.

Total Suspended Solids: The TSS of the physical-chemical influents averaged 10-20 mg/l. When the Duck Creek Wastewater Treatment Plant ultra-high rate sand filtration units were taken off-line, the TSS increased to 30 mg/l. The GAC effluents TSS averaged 4-10 mg/l, with a 50% average reduction. The BAC effluents TSS averaged 8-18 mg/l, many times exceeding the influent levels. This phenomenon has been attributed to chemical precipitation after ozonation and biological growth in the BAC beds.

The biological influents TSS averaged 8-18 mg/l. After adsorption, the effluents from GAC averaged 0-6 mg/l. The BAC effluents TSS averaged 2-6 mg/l, usually 1-2 mg/l higher than the GAC effluents. This, again, can be explained by the increased biological growth in the BAC beds.

The blended system had an average influent TSS similar to that of the calculated blend of the biological and physical-chemical system. The influent TSS levels averaged 10-20 mg/l, with GAC effluent TSS levels of 4-9 mg/l.

Chemical Oxygen Demand: The physical-chemical influent COD averaged 120-220 mg/l, with occasional peaks as high as 295 mg/l. The effluent COD levels from the GAC systems averaged 60-200 mg/l, an average reduction near 20%. The initial COD removal in the GAC system was 58% on day 1, and the removal on day 48* was 11%. The BAC effluents COD averaged 60-100 mg/l, an average

*December 30, 1980.

reduction of 55%. The initial removal was 60% on day 1 and 55% on day 233*.

The biological influent COD averaged 70-130 mg/l, with occasional spikes of 150 mg/l. The GAC effluent COD averaged 50-90 mg/l, an average reduction of 30%. Initial removal was 37% on day 1 and 25% on day 48. BAC effluent COD averaged 30-70 mg/l, an average removal of 45%. The removal was 36% on day 1 and 38% on day 233.

The blended system influent was similar to the calculated average of biological and physical-chemical influent COD values. The average influent COD was 80-180 mg/l, with occasional spikes of up to 260 mg/l. The average effluent COD from the GAC system was 50-130 mg/l, an average removal efficiency of 35%. The removal was 50% on day 1 and 25% on day 48.

There has been a gradual decrease in the COD removal efficiency in the GAC systems since day 1. The system will reach an expected point of exhaustion near day 60-80. The removal efficiency, then, should be near 10%. The BAC systems are continuing to operate well on day 233.

Total Organic Carbon: The physical-chemical influent TOC averaged 45-65 mg/l, with occasional spikes of over 90 mg/l. The GAC effluent TOC averaged 40-60 mg/l, an average reduction near 28%. The initial removal was 57% on day 1 and 15% on day 48. The BAC effluent TOC averaged 20-45 mg/l, an average removal efficiency near 45%, a 55% removal on day 1 and 38% removal on day 233.

The biological influent TOC averaged 20-30 mg/l, with occasional spikes to 45 mg/l. The GAC effluent TOC values were 15-25 mg/l, an average reduction near 26%. The initial removal was 48% on day 1 and 8% removal on day 48. The BAC effluent TOC averaged 10-20 mg/l, an average removal efficiency of 28%. The removal efficiency on day 1 was 55% and 19% on day 233.

The blended TOC averaged 25-50 mg/l, with spikes to 75 mg/l. The GAC effluent TOC averaged 15-35 mg/l, with spikes to 45 mg/l. The average removal was 36%, 63% removal on day 1 and 23% removal on day 48.

Biological Oxygen Demand: The physical-chemical influent BOD_5 averaged 45-90 mg/l, with spikes to 128 mg/l. The GAC effluents BOD_5 averaged 35-65 mg/l, with spikes to 101 mg/l. The removal efficiency averaged 28%, 57% removal on day 1 and 12% removal on day 48. The BAC effluent BOD_5 averaged 15-40 mg/l, with spikes to 55 mg/l. Removal efficiency averaged 58%, 76% removal on day 1 and 55% removal on day 233.

The biological influent BOD_5 averaged 12-35 mg/l, with spikes to 45 mg/l. The GAC effluent BOD_5 averaged 4-15 mg/l, with spikes to 27 mg/l. The removal efficiency averaged 48%, 60% on day 1 and 50% on day 48. The BAC system effluents averaged 6-26 mg/l, with spikes to 40 mg/l. The removal efficiency averaged 49%, 66% on day 1 and 20% on day 233.

The blended system influent BOD_5 averaged 12-40 mg/l, with spikes to 60 mg/l. The removal efficiency averaged 37%, with 66% removal on day 1 and 23% removal on day 48.

---

*December 30, 1980.

DISCUSSION

The BAC and GAC systems have been operating at the University of Texas at Dallas-Duck Creek Pilot Plant Facility for over 13 months in continuous operation. The current data available from operations indicates the BAC systems are superior to the GAC systems in all parameters except BOD_5 removal. The BAC systems have been operating continuously for over 234 days. During this period the GAC systems have been replaced 3 times (Table 4).

Operation of the UTD-Duck Creek Pilot Plant Facilities will continue in order to gain more data and better understanding of the various operation parameters.

ACKNOWLEDGEMENTS

This research has been supported by the Municipal Environmental Research Laboratory, U.S. Environmental Protection Agency under Grant No. CR 806466-01.

The authors would like to acknowledge the great assistance and support of the City of Garland, Texas, in the construction, operation, and performance of these pilot plants.

REFERENCES

(1) McDuff, D.P. and Chiang, W.W.J. "Physical-Chemical Treatment Design for Garland, Texas. Applications of New Concepts of Physical-Chemical Wastewater Treatment." (1972)

(2) Strudgeon, G.E. and Earens, B. "Upflow-Downflow Carbon Adsorption." AIChE Conference, Los Angeles, Calif. (1975)

(3) Miller, G.W. and Rice, R.G. "European Water Treatment Practices - The Promise of Biological Activated Carbon." Civil Engineering - ASCE, 81-83. (1978)

Developments in Arid Zone Ecology and Environmental Quality
Edited by H. Shuval

42
CLOGGING OF STRAINERS BY ALGAE IN DRIP IRRIGATION SYSTEMS

Giora Alon* and Avner Adin+

‡ Human Environmental Sciences, School of Applied Science and Technology, Hebrew University of Jerusalem. + Environmental Protection Service, Ministry of the Interior, Jerusalem.

ABSTRACT

The clogging phenomena in strainers was investigated using different screens, filtration velocities and algal concentrations. Clogging rate was found to be dependent on the above parameters, but not always as expected by classical filtration formulae. A phenomenon of larger particles leaving the screen points out a possible detachment mechanism which may bring about further clogging of drippers.

INTRODUCTION

The design and manufacture of filtration equipment for low rate application, and particularly for drip irrigation systems, has lately become an integral part of the water systems industry. The reason for this development is that the clogging of orifices, holes or lateral lines of the systems has turned to be the main obstacle for further use of low rate applicators in irrigation. The problem is becoming even worse now, mainly in arid and semi-arid countries, as a result of increasing exploitation of marginal waters of secondary quality such as wastewater effluents following various treatment methods.

Different filters are being offered nowadays to cope with the clogging problem. Most of them are based on straining mechanisms, where the filter (strainer) pores are smaller than most particles to be removed. This is to differentiate them from granular filters, which operate on the principle of in-depth filtration. The most sophisticated filters are designed to relieve the farmer from undesirable cleaning work by incorporating automatic cleaning mechanisms.

Nevertheless, there is still much to be done in the development of basic criteria and parameters for filter design and classification.

In its first part, this paper describes the present knowledge of the main causes for clogging, and the principles of the straining process. In its second part, the paper presents and discusses results of a study of the filtration screens used in the production of filters for drip-irrigation lines.

CAUSES OF CLOGGING

Recent work and field observations concerning the performance of drip irrigation systems, utilizing either fresh water or wastewater effluents, point out that the causes of clogging of low rate applicators may be divided into three main categories: a) suspended matter, b) chemical precipitation; and c) bacterial growth, out of which the first seems to be the main cause in cases. (Adin 1978)

a) Suspended matter - these are particles of size range varying from colloidal (1 μm>dp) up to coarse (1mm.). Three different types of particles may be observed:

1) Silt and clay - released from surface water and sometimes from ground water.
2) Algae - originated in surface water and in aerobical treated wastewater, mainly from stabilization pond effluents.
3) Organic suspended solids - mainly in effluents. The elasticity of these particles gives them the ability to penetrate through openings smaller than their original sizes. They also tend to encourage bacterial growth in the systems.

b) Dissolved matter - these are salts precipitating in the pipes and fittings. The different factors causing precipitation are:

1) Sealing due to the application of hard water with low saturation index (particularly at high temperatures).
2) Corrosion products from steel pipes.
3) Precipitated salts due to the application of dissolved fertilizers through systems utilizing hard water.

c) Biological growth - that is the development of bacterial or algal colonies in the systems. This phenomenon may primarily be observed when dealing with systems utilizing effluents from activated sludge plants, trickling filters, or stabilization ponds.

It is often found that the cause of clogging is actually a combination of, at least, some of the factors mentioned above; e.g. clay and corrosion products entrapped within a biological mass cemented with $CaCO_3$ precipitate. At any rate, it seems now beyond doubt that the clogging problem is mostly associated with the presence of suspended particles in the irrigation water.

PRINCIPLES OF IN-LINE FILTERS

In order to reduce frequent recurrence of dripper-clogging, several producers developed filters which are characterized by self-rinse devices. (Amiad Catalogs 1977)

Filters of up to 600 m^3/hr can currently by obtained, the sophistication of which depends largely on filter dimensions and flowrates.

Filters designated for high flowrates consist of two screens. The external screen is coarse, usually made of stainless steel, and serves as a shield for the more delicate screen which is the main filter medium. Such a screen is usually made of synthetic materials as polyester or nylon. Aperture diameters of shielding screens range from 500 to 3500 microns, whereas the corresponding figures in the main media are 80-300 microns only.

The supervising mechanism for the flushing process is based on headloss, and is actuated when headloss has reached a pre-set value. Flush mechanisms based on time elapsed, do exist as well.

The filter cake is removed by the unfiltered water, which brings about a 5% loss of the total volume.

Most problems associated with such filters, as seen until now, stem from incomplete flushing. This causes clogging phenomena that can be corrected only by individual filter servicing.

Since irrigation waters contain considerable amounts of suspended matter (silt, algae, etc.), problems associated with the incomplete functioning of flush mechanisms tend to be frequent.

As a result of above-mentioned causes, and marginal manometric sensors' performance, head losses on the filters can not be considered as the sole measure for flushing time and frequency.

It is obvious that regional and temporal data should be collected to monitor changes in water quality. Thus flushing intervals and type of filters can be selected for specific use and location.

FILTRATION PRINCIPLES

A practical definition of filtration for this subject can be the removal of suspended matter from water by letting it flow through a porous medium. Two main categories of filters may be defined, based upon filtration mechanisms:

a) Strainers - where the main filtration mechanism is straining (also named mechanical or surface filtration) e.g. a screen.

b) Granular filters - also named deep bed, rapid or high rate filters, e.g. sand filters. Here the particle size dp is smaller than the grain size d_0 and the removal is dominated by physical-chemical mechanisms.

Since this paper deals primarily with filter strainers used in drip irrigation lines, the subject of straining theory will be elaborated in greater detail.

By current definition, the straining process is based on the principle that the medium's pores are smaller than the particles' diameters. A mathematical expression of exponential growth in pressure losses as a function of volume filtered through a steel screen (d_0 = 35µm) is known as Boucher's law which assumes the form

$$H = Hoe^{IV}$$

where Ho is the initial pressure loss, and V is the filtered, volume per surface area of screen.

This kind of filtration depends on two simultaneous processes:
1) Deposition of material within the pores of the screen, and filter cake, thus, contracting the passage of the suspension.
2) Deposition of material at a constant rate on the surface of the screen and later of the filter cakes. Thus an "extension" of pores is caused, with the formation of capillaries, the diameter of which is constant and does not change with length. Thus a perforated spongelike filter cake is formed, the pores of which correspond to the pores on the filter screen.

Ives (1960) regards Boucher's equation as a resultant representing two processes:
a) Deposition on the pores, and b) Deposition on the surface. Both, when combined and weigh equally, yield the abovementioned mathematical form of Boucher's equation.

The results obtained in this work indicated clearly, that in the initial stages of the filtration process, both mechanisms (a) and (b) were equal in their influence on the total process. Yet, upon its propagation, the deposition in the pores significantly gained in influence. This fact as has been stated by Ives (1960), yielded the curved shapes as shown in Figures 5-7, rather than the straight line as suggested by Boucher's equation.

The fouling rate of a filter, besides being affected by mechanisms such as the deposition within the capillaries or deposition on them in the form of a filter cake, depends to a considerable extent on the distribution of capillaries diameters, and particle's dimensions. Both items being subject for statistical definition.

Boucher's formula provides solely the connection between the pressure drop and the flow rate, (or volume filtered). Therefore for the definition of the whole filtering process, a connection has to be found between the factors expressed in Boucher's law and the filter area in any specific filter, establishing constants to characterize each filter in question.

Usually, a filter can serve a wide range of flowrates (6). As the flow rates passing through a given filter increase, the work cycle is shorter.

Experimental and practical evidence indicate that increasing the flowrate in a granular medium when particles are deposited already in the pores, causes a detachment process as a result of which the particle concentration in the effluent increases (Adin et al 1979). This is caused by extended shear forces on the particles in a granular medium. The same forces act on the filter cake in straining, as described by Rushton (1979). It has been postulated by Chi-Tien et al, (1979), that the detachment phenomenon in strainers is common, and contributes largely to the relative inefficiency of this process, particularly in its initial stage. Besides, Colloidal particles which do not penetrate into the effluent in granular filters due to attachment processes, succeed in doing so when straining is being applied. Thus it can be concluded that it is impossible to establish a clear optimum for the operation of screen filters.

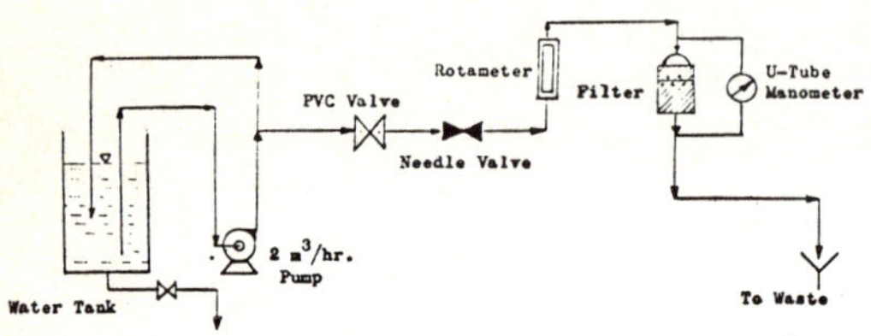

Fig. 1: The system for the evaluation of filter screens

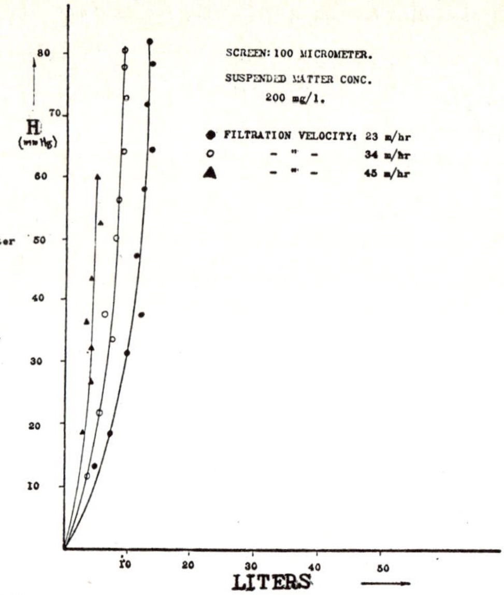

Fig. 2: Developing headloss in the screens with rising filtration velocities

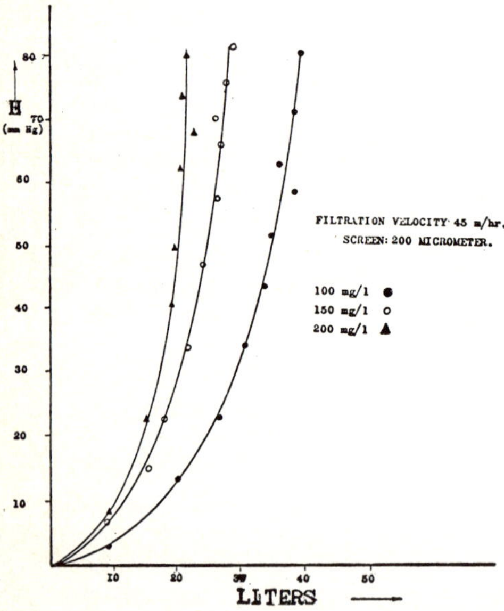

Fig. 3: Developing headloss in the screens with rising concentrations of suspended matter

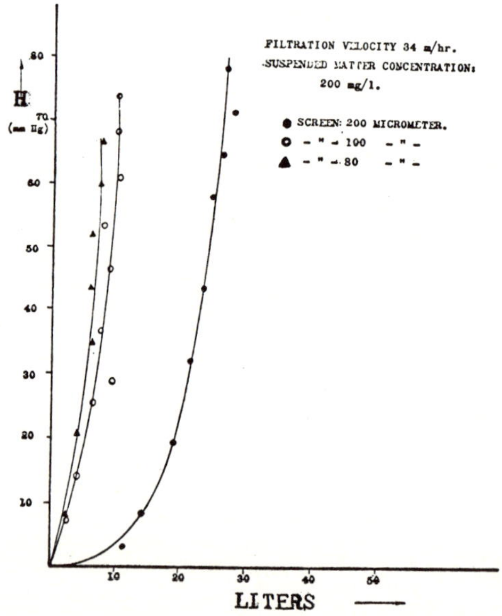

Fig. 4: Developing headloss in various

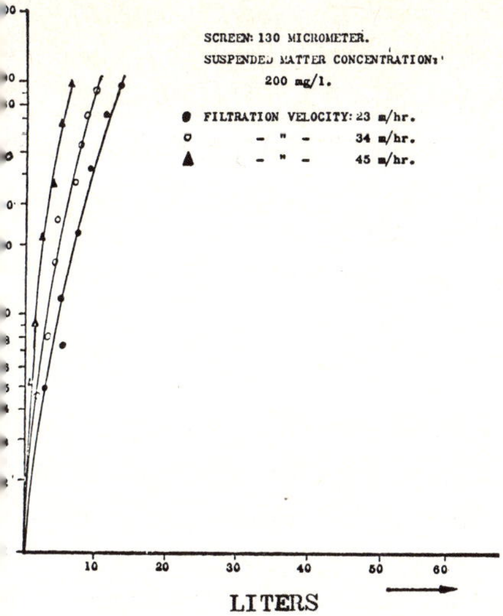

g. 5: Developing headloss in the screen with rising filtration velocities

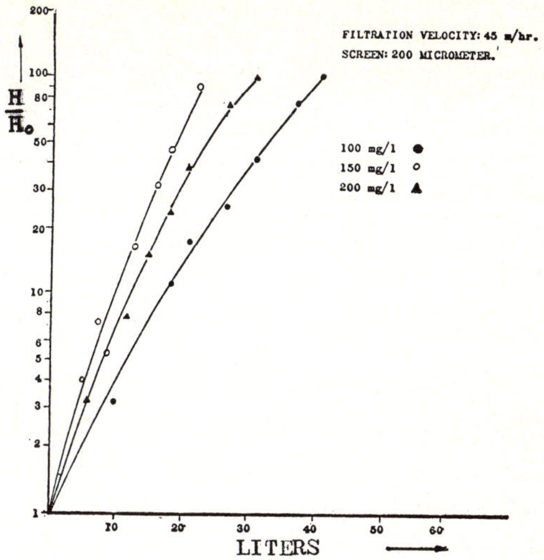

Fig. 6: Developing headloss in the screens with rising concentrations of suspended materials.

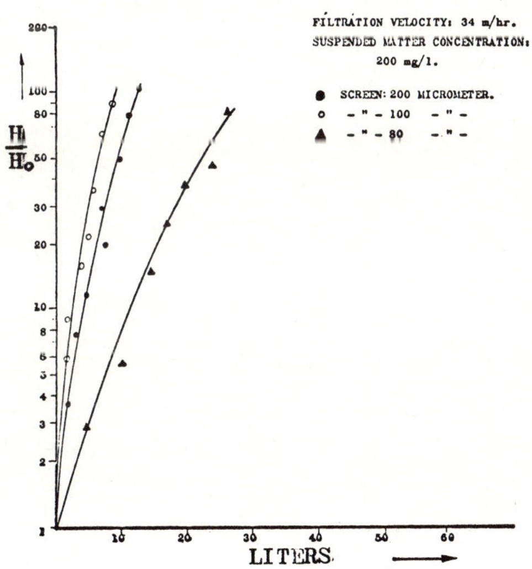

Fig. 7: Developing headloss in various screens

MATERIALS AND METHODS

The experimental system was based on the assumption that field conditions such as suspended material concentration, filtering velocity and identical filter media should be closely simulated.

The requirement to achieve this goal included:

1) Homogenious suspension
2) Constant and controlled flowrate throughout the experiment.
3) Uninterrupted sampling.
4) Precise measurements of headloss on both sides of the filter.

The experiments were carried out using polyester media received from "Amiad" Co. The media chosen were of mesh numbers 75, 130, 160, and 200. Those figures correspond to 200, 130, 100 and 80 micrometers respectively.

The filter runs were carried out at a constant rate, while headloss served as the measure for the clogging rate.

The suspended matter:

The representative suspension to be used in the investigation was chosen according to the following considerations: suspended solids concentration of 50-70 mg/l as VSS, Total suspended solids conc. of 100-500 mg/l, and particle distribution in the range of 1-100 micrometers. These values are typical to some waters used in irrigation in Israel.

The suspended matter consisted of sediment drawn from a freshwater pond. Microscopic observation revealed detritus, algae (mostly chlorella and scenedesmus), and various mineral and organic particles, up to 100 micrometers in size, using the Feret diameter determination (Yamate & Stockham 1975). This sediment served as a stock solution with an initial concentration of 33,000 mg/l expressed as VSS. It was preserved as a 3% Formaline solution and stored in $4^{o}C$.

The suspension was prepared in a plastic tank by dilution with tap water. The screen to be tested (out of the four sizes listed in the previous section) was installed in the filter septum which was then filled with tap water to release air bubbles entrapped underneath the screen.

The needle valve (Fig. 1) was preset at the desired flow rate, and when headloss had reached 85 mm Hg, the experiment was terminated. The constant rate was controlled, as abovementioned, by manual manipulation of the needle valve.

The microscopic analysis was carried out as follows: 10 ml of suspension (before and after filtration) were centrifuged in a Clay-Adams centrifuge for 20 minutes at 3500 r.p.m. A sample of the 1 ml centrate was observed under a microscope in a hemocytometer using a 10 x 40 magnifiation.

RESULTS

Runs were performed using the above screens, applying suspensions of 100, 150, 200 mg/l and superficial velocities of 23, 34, 45 m/hr. The runs were based on the principle of letting two parameters constant while the third served as a variable. Figures 2, 3, 4 present some typical results. Figure 2 depicts developing headloss with water filtered in three ascending filtration velocities 23, 34 and 45 m/hr; when the screen used in the experiment was of #160 (100 μm pore diameter) and suspended material concentration of 200 mg/l.

In Figure 2 one can observe a steep rise in headloss when the most moderate rise out of the three is the one resulting from the lowest filtration velocity, and the steepest rise is the one resulting from the highest velocity. The amount of water passed until a 60 mm Hg headloss was reached was 6, 8 & 10 liters with filtration velocities of 23, 34 & 45 m/hr respectively.

Figure 3 depicts developing headloss in the filter with water filtered in three ascending values of suspended solids 100, 150, & 200 mg/l. When filter velocity was kept constant at 45 m/hr, and the screen used was of #75 (200 μm pore diameter). The rise in headloss here is milder than in the former case, yet a trend can be observed in which the rise in headloss becomes steeper with concentration. Thus it can be seen that at 100 mg/l suspension concentration the rise is the smallest while at the 200 mg/l it is the largest.

It can be observed from the graph, that while filtering the 100 mg/l suspension, at 60 mmHg headloss, 32 liters were filtered whereas only half that quantity was filtered when the suspension was of 200 mg/l.

Figure 4 depicts developing headloss in three screens: 200 μm, 100 μm and 80 μm pore diameter at construt filtering velocity of 34 m/hr, and a constant suspension concentration of 200 mg/l. From the graph, a considerable rise in headloss rate between the 80 μm and 100 μm screen on one hand, and the 200 μm screen on the other can be observed. At 60 mmHg headloss, 24 liters were filtered using 200 μm screen as compared with 7 liters with the 80 μm screen.

Particles of 100 mg/l suspensions were counted in two separate analyses and the results were compared with counts of particles that remained in filtrates of two different screens.

The range of particle dimensions was from under 1 μm up to 60 μm. The results are summerized in Table 1.

It can be seen, that the 80 μm screen yielded a 9% removal efficiency based on number of particles before and after filtration. The 130 μm screen yielded a "negative efficiency". 1245 particles per mm^3 were counted before filtration, but 1710 particles per mm^3 were counted after filtration.

TABLE 1.

130 Micrometers		80 Micrometers		
Before filtration	After filtration	Before filtration	After filtration	Range (μm)
800	970	800	550	5-10
330	270	330	400	10-20
120	270	150	200	20-40
25	40	20	35	40-60
1245	1710	1300	1185	

DISCUSSION

In order to verify the compliance of the results to Boucher's law (Boucher 1967), they were plotted so that the abscissa expressed the volume filtered (V), and the ordinate represented the natural logarithm of H/H_0 at any given point of measurement.

The logarithmization of Boucher's formula yields the form:

$$\ln \frac{H}{H_0} = IV$$

and plotting it on a semi-log paper shows a straight line, the slope of which is "I" - known also as Boucher's constant.

Representative results are shown in Figures 5-7, ouf of which it is obvious that Boucher's formula is valid only in the beginning of the filter run. Only then, a straight line was formed which later became convex as the process propagated.

According to Ives (1960), two deposition mechanisms prevail in straining. They are a) Deposition within the pores and b) Deposition on the screen. It was assumed there that the combined effect of both mechanisms yield the straight line. When one of the above mechanisms does prevail, the line assumes the form of convex line for deposition on the surface, and concave line for the case when particles are deposited within the pores.

The experiments in this work yielded all, when plotted on a semilog paper, a line which initially was straight and assumed later a concave shape. The point where it started was different in each run, and could be easily defined.

Two explanations are offered for this phenomenon:
a) Ives' theory is based on a cake formation in which the capillaries are of a constant diametrical cross-section, and the cake's porosity does not change until the process has been terminated. Thus the porosity change during the process is not considered, as well as capillaries tortuosity. By the same token, the change in pore diameter along the cake is not considered either in Ives' theory. Those "unconsidered" factors do cause serious deviations from Ives' mathematical analysis, and change completely the flow regime. Which, in turn, alters the balance between the mechanism of deposition on the pores, and the mechanism of deposition on the surface.

All considered, the convex line towards the end of the experiment, raises doubts about the literal applicability of Ives' model to the case in consideration. Along with the filter cake thickening, it would be logical to assume that the tortuous non-regular in diameter capillaries, might cause a higher headloss. Yet, it proves to be the opposite. It can be assumed that this phenomenon can be explained by recognizing the domination of the detachment mechanism. As a result of this domination aggregates leave the cake, penetrating the screen into the filtrate, thus forming pores in the filter cake. These pores enable free water flow which, in turn, bring about the gradual decrease in final headloss.
b) Microscopic analysis confirms the formation of larger particles in the filtrate. As mentioned in (a) a detachment phenomenon does exist; which prevails towards the end of the experiment runs. In this case, the build-up of particles on the cake competes with detachment of particles from the cake, (sometimes labeled "chanelling phenomenon"). Thus causing the decrease in headloss rate.

Additional phenomenon shown in Figures 2-4 and 5-7 is the increase in headloss rate caused by:

1) Increase in suspended material.
2) Decrease in pore diameter in various screens.
3) Increase in filtration velocity.

It is worth mentioning that the detachment phenomenon causes clogging of drip irrigation system components which are placed behind the filters, and in the drippers themselves.

Evidence for this was received from farmers and other users who evaluated the reasons for clogging.

The impact of particle size and number on clogging mechanism

The pore size in the narrowest screen was 80 micrometers. From Table 1 it was obviously seen that most particles in the filter suspension did not exceed 20 micrometers in diameter. Even though, the dominating mechanism was straining, although the particle diameter was smaller than pore diameter. This did not correspond to the proper definition of straining (where the particle diameter ought to exceed the pore diameter).

Although many researchers claim the advantage of granular filtration over straining in filtering colloidal and submicronic particles (Adin 1978, Rothwell 1978), the possibility to retain such particles is not to be excluded in the cake formed in straining.

The particles' profile shows a decrease in the number of particles in the range of 5 µm, within which the particles smaller than 1 µm are contained. From here it can be concluded that the bridging phenomenon along with the formation of 20 micrometer aggregates, and their detachment later on through the screen to the filtrate, thus increasing the numbers of such particles in the filtrate as compared to their number in the influent suspension. The existence of bridging in strainers is mentioned by Ives (1975) and Dickey (1961), though no mention had been made concerning the formation of aggregates in the filtrate.

This is evidenced by microscopic examination of the aggregates in the effluent. They are compacted amorphous particles typically formed on a filter cake compared to the "original" particles in the suspension. All considered, the factor "I" can be looked upon as an index for the filter resistance to the filtration process.

With the increasing "I" values, one can expect more frequent clogging.

CONCLUSIONS

1. Although the major part of particles in the suspension was smaller in diameter than the smallest pore within an order of magnitude - clogging of screens did occur. The time elapsed for a complete clogging was a function of suspended material concentration, filtration velocity and pore diameter.
2. Clogging was a result of the bridging mechanism in the screen, and attachment mechanisms in the filter cake. The clogging process was more rapid in higher suspended material concentration, higher filtration velocities and small pore screens.
3. As a result of particles released from the screen while filter cake was being formed, a suspension resulted in which particle size distribution and number, differed from these respective parameters before filtration. Larger number of larger particles thus generated, might increase the clogging phenomenon down the irrigation line, and in the systems' various component that were to be protected by those filters.
4. These forces on the filter cake brought about the formation of aggregates in the filtrate, which were released from the cake through the screen. Thus a decrease in the smaller than 10 micrometers was observed, in parallel to an increase in the 10-60 micrometer range.
5. It can be concluded that due to chanelling and detachment of particles, the filter cake serves as a source for the formation of particle aggregates within 10-60 μm. Thus a new source for clogging problems was generated, added to those of the original suspension.

REFERENCES

Adin A. (1978) Progress in Filtering Methods for Low Rate Applicators. *Israel Center of Waterworks Appliances, Proceedings, Israqua.*

Adin A. et al (1979).The Application of Filtration Theory to Pilot Plant Design, *JAWWA (12), 17.*

Amiad - Irrigation Systems, M.P. Korazim, Catalogs No. 39-0-39-15 (1977).

Boucher, P.L. (1947) A New Measure of the Filtrability of Fluids with Applications to Water Engineering *J. Inst. Civ. Engrs. (27).*

Chi-Tien et al.(1979) Simulation of the Dynamic Behavior of Deep Bed Filters *A.I. Ch.E. Journal (5), 385.*

Dickey G.D. (1961)Filtration *Reinhold Publishing Corp., New York.*

Ives K.J. (1975) The Scientific Basis of Filtration *Noordhof Leiden, 278-283.*

Ives K.J. (1960) Filtration Through a Porous Septum: A Theoretical Consideration of Boucher's Law *J. Inst. Wat. Engrs. (14).*

Rothwell E. (1978) Fabric Filter Failures *Filtration & Separation (6), 15.*

Rushton A., et al (1979) Shear Effects in Cake Formation Mechanisms. *Filtration & Separation (5), 16.*

Yamate G., Stockham J.D. (1975) Sizing Particles Using the Microscope *Filtration Day. The Filtration Society Proceedings at IIT, Chicago, Ill.*

This paper is based on M.Sc. work of Giora Alon.

Developments in Arid Zone Ecology and Environmental Quality
Edited by H. Shuval

43
THE USE OF SOLAR ENERGY FOR WASTEWATER TREATMENT

A.J. Acher

Institute of Soils and Water, The Volcani Center, Bet Dagan, Israel

ABSTRACT

The material presented represents three years of ongoing research on a new method for using solar energy for municipal and industrial wastewater treatment. This research seeks to develop economical methods of disinfection and detoxification of wastewaters which would results in the reuse of effluents for crop irrigation.

Aerated municipal or industrial wastewaters containing dye-sensitizers were exposed to solar irradiation for various periods of time. The influence of the solar energy, absorbed by these sensitizers in the visible range, on the organic matter and anionic surfactants in secondary effluents was checked by determining the COD and MBAS values of treated wastewater.

The disinfection potential of this method, of running water or secondary effluents, previously contaminated or enriched, respectively, with laboratory cultures of E. coli, bacteriophages, polio virus and salmonella, will be presented in a separate paper.

The influence of this photooxidative method was also studied on eutrophic algae present in the Lake of Galilee. It was found that the conditions under which disinfection proceeds also support algicidal processes, causing lethal damage to algal cultures.

Further investigations on stable pesticides (uracil derivatives) in surface water or industrial wastewater showed that such a method can also be used for detoxifying pesticides in these waters. Frozen aqueous solutions, directly exposed to sunlight or covered by different layers of ice, could be also detoxified.

Developments in Arid Zone Ecology and Environmental Quality
Edited by H. Shuval

44
THE USE OF ROTOR STRAINING FOR PRETREATMENT OF SEWAGE

J. Sheinberg, A. Lavy and A. Reisser

Dan Region Association of Towns, Tel-Aviv

ABSTRACT

An industrial full-scale Rotor strainer was tested for possible use of several units of this kind for preliminary treatment of Dan Region wastewater. The work was based on several previous successful pilot scale operations. The design, construction and mechanical operation of the test unit were accomplished without major difficulties.

Different parameters were evaluated during the investigation: mechanical parameters i.e. effluent appearance, BOD, COD, suspended matter, settlable material, heavy metals, effect on primary treatment; and bacteriological parameters (fecal coliforms). The sludge was analyzed for chemical constituents emphasizing heavy metal concentrations.

All results show that the effect of Rotor Strainer filtration of sewage is similar to the effect of primary treatment. It was found that using this process brings about land and energy savings as well as improvement of the biological process. It removes non-degradable particles and enables smooth and constant operation of the treatment plant.

Developments in Arid Zone Ecology and Environmental Quality
Edited by H. Shuval

45

EFFECT OF ORGANIC LOAD, ALGAE AND PRIMARY CONSUMERS ON THE COMPENSATION DEPTH IN A WASTEWATER RESERVOIR

I. Dor*, H. Schechter* and T. Bromly-Shur**

*Human Environmental Sciences Laboratory and **Department of Zoology, Hebrew University, Jerusalem, Israel

Deep wastewater reservoirs are being operated in Israel, serving as treatment plants and resources of inexpensive water for agriculture. An interdisciplinary team investigated one of the best planned and carefully operated reservoirs in order to estimate performance efficiency. The above type of water body is characterized by its considerable depth (10m) connected with a high content of organic matter (BOD=50-90mg/l). After a short period of rearation by mixing in winter, the reservoir becomes thermally stratified and supply of oxygen depends chiefly on photosynthesis. Accordingly, the thickness of the oxygenated layer is closely related to the depth of compensation. During the rainy season (Dec. - Feb.) characterized by relatively high BOD and high turbidity, depth of compensation was 20-30 cm only. After establishment of the spring stratification and development of a heavy algal bloom of chlorella (10^6cells/ml) compensation depth increased slightly to 30-40cm. Appearance of rotifers (10^4 animals/l) grazing on the algae biomass decreased turbidity and the resulting depth of compensation reached 1 - 1.2 m. These biological transformations had a decisive effect on the oxygen regiment within the system and on the water quality used in the summer for irrigation.

Developments in Arid Zone Ecology and Environmental Quality
Edited by H. Shuval

46

THE EFFECT OF SMOKE IN THE BEDOUIN COOKING AREA ON THYROID METABOLISM

S. Walfish, U. Abu-Rabiya, Z. Etzion and R. Yagil

Soroka Teaching Hospital and Medical School, Faculty of Health Sciences, Ben-Gurion University of the Negev, Beer-Sheva

ABSTRACT

The cooking area in Bedouin encampments is filled with smoke. Laboratory rats placed in this area for two months were smaller than their control siblings. They had pneumoconiosis and great changes in thyroid morphology and metabolism. The glands showed both atrophic areas and hypertrophic areas. Cysts were found in all glands. When compared with the body weight of the animals the thyroids were significantly enlarged. There were significant elevated blood levels of thyroxine, triiodothyronine and Thyroid Stimulating Hormone. These data were consistent with the findings of thyroids and lungs in Bedouin (Hirsh et al, 1974; Walfish, 1978) and with anoxia (Canhase, 1980). It is postulated that the smoke caused a relative anoxia which initiated the changes in thyroid metabolism via the physiological feed-back mechanism.

Developments in Arid Zone Ecology and Environmental Quality
Edited by H. Shuval

47

CANCER AND THE ENVIRONMENT

Jerome B. Westin, M.D., Ph.D.

Department of Preventive and Social Medicine, Sackler School of Medicine, Tel-Aviv University, Ramat Aviv, Israel

ABSTRACT

Some 90% of human cancer is related causally to environmental agents, primarily chemicals, and is thus theoretically preventable. The vast majority of chemicals tested have shown no evidence of carcinogenetic potential. However, for proven carcinogens there is no way to set a dose, however small it may be, that is assuredly safe for man. Animal bioassays are universally considered predictive for man, and the use of large doses in these assays is both necessary and scientifically valid. Epidemiological evidence of carcinogenicity is extremely difficult to obtain, and the absence of such evidence should never be construed as proof of safety.

INTRODUCTION

In the Western world, cancer is the only major lethal disease whose age-specific incidence and mortality rates continually increase from year to year. In Israel and elsewhere the *overall*, age-adjusted increases of approximately 1% per year in mortality rates during the last decade are in themselves cause enough for alarm (Devesa and Silverman 1978; U. Golburt, personal communication). Overall rates, however, markedly understate the severity of the problem: The rates of many of the most common types of cancer are increasing in Israel at some 5% per year or more, with at least one, bladder cancer in males, increasing by 150%(!) in a recent 5-year period (Israel Cancer Registry 1977a, 1977b). When internationally respected authorities (Doll 1977) believe that some 90% of the disease is theoretically preventable and that in practice we are equipped today with sufficient knowledge to eliminate some 50% of human cancer (Epstein 1979; Lancet 1977), the seemingly inexorable yearly increases in incidence and mortality seem all the more anomalous.

Perhaps the major contributory factor to the seemingly paradoxical continued increase of an essentially preventable disease has been the failure of large segments of the medical and scientific communities to understand fully the environmental origins of cancer and to relate to the disease as they would to any other potentially preventable environmental scourge. This failure has engendered two phenomena in scientific circles: an inordinate emphasis on *cure* (both in research and in practice) to the detriment of preventive action, and a tendency to minimize and misinterpret environmental carcinogenic threats, thereby delaying public health decisions involved in their control and eradication. It is primarily the latter point that will be addressed here.

CANCER FROM MAN'S ENVIRONMENT

It is widely accepted that some 90% (Higginson 1979; Fraumeni 1979) or more (Hoover 1979) of all human cancer is related causally to environmental agents (whence the belief that virtually all human cancer is theoretically preventable). Recent epidemiological investigations in Israel demonstrate this principle: It has been found, for example, that the rate of lung cancer among residents of Haifa is *four times* that of Jerusalem residents (A. Tamir, personal communication). Only environmental factors (*viz.*, air pollution) seem to be able to account for this enormous difference. Worldwide comparisons of this type (where rate differences may exceed 100-fold! [Doll 1979; Muir 1975]), along with studies of immigrant populations and of changes in cancer rates following environmental alteration, make up the basis for the conviction that human cancer is essentially an environmental disease.

The "environment," however, is a big place. In the present context it means, almost exclusively, chemical entities in man's surroundings. (Only a very small percentage of human cancer seems to be caused by environmental factors other than chemicals: Ionizing radiation, for example, appears to be responsible for no more than 2-3% [Fraumeni 1979].) In this regard, it is vital to stress from the outset that the overwhelming majority of chemicals tested have *not* been shown to be carcinogenic, regardless of the concentration to which test animals have been exposed: Of the 30-60 thousand chemicals considered to be in common use, some 7,000 of the most suspicious of them have undergone animal bioassays for carcinogenicity and, of these, only about 700 are currently considered to be carcinogenic (IRLG 1979). On the other hand, and in clear contradiction to what various industry-connected scientists would have us believe, there is no scientific evidence that small doses of a proven carcinogen are "safe." During the past 20 years literally dozens of expert committees have repeatedly emphasized that there is no way to establish a "safe" level of exposure to a carcinogen for man, that is, a "threshold" dose (TSSC 1980, p.133; IRLG 1979; Epstein 1979 pp.62, 518ff). A growing body of scientific data indicates further that such thresholds simply do not exist (Schneiderman et al. 1979; Hooper et al. 1979; Scherer & Emmelot 1979; Preussman 1978).

The reasons for the inability to establish threshold values for carcinogens are several. For example, it is now well known that many substances, not necessarily carcinogenic initiators themselves, enhance the action of carcinogens. Thus n-dodecane, itself devoid of carcinogenic activity, increases the carcinogenic activity of benzo(*a*)pyrene and benzo(*a*)anthracene by a factor of 1,000 (Bingham & Falk 1969). Additionally, man may be much more sensitive to a given substance than the test animal---again the difference in response may approach 3 orders of magnitude. Also, the human lifetime is 35 times that of the mouse, and it is recognized that the length of exposure to a carcinogen increases the cancer risk. These factors, along with several others, result in our inability to extrapolate *quantitatively* from animal tests to human life situations with any degree of accuracy.

The theoretical and experimental considerations that militate against the very existence of a threshold are based on the conclusions that in most, if not all, cases the carcinogenic process begins with a heritable genetic (DNA) or epigenetic aberration *in a single cell* that, after one or more additional steps, results in a tumor of *clonal* origin (Weinstein et al. 1979; Scherer & Emmelot 1979). Clearly, a single molecule of the initiating substance may be sufficient to begin the inexorable process of tumor formation. Those advocating the continued "moderate" use of known carcinogens may wish to recall, for example, that *a single tablet* of saccharin-based artificial sweetner contains some 6×10^{19} molecules of saccharin---enough to surround every cell in the human body with more than half a million molecules of the carcinogen (Westin 1977).

THE ANIMAL MODEL

Industry-connected scientists have repeatedly challenged the explicitly stated, reiterated, and strongly supported conclusion of the independent scientific community that "if a substance has produced positive results in a single adequately designed and conducted animal bioassay...the conclusion is that the substance is likely to pose a risk of cancer to humans" (IRLG 1979; cf. also TSSC 1980, p.125ff; NRC/NAS 1978, pp.TS-3, 3-61). Industry's "heads-I-win-tails-you-lose" contention that the results of animal bioassays are acceptable if they indicate "safety" but unacceptable if they point to danger is perhaps more amenable to treatment by social scientists than by life scientists. The fact that in 20 different countries nearly 100 scientific institutions are currently conducting animal bioassays on some 1000 chemicals (Boyland 1980) underscores the singular importance of animal testing in the detection of potential *human* carcinogens. Moral considerations aside, and even if it were possible to isolate large groups of human beings for 20, 30, or even 50 years to conduct human bioassays, conclusions might be several generations in coming, resulting, possibly, in interim exposure of large populations to chemicals that only decades hence would finally be proven to be carcinogenic. We see, therefore, that animal experimentation is not the best alternative to human testing (which is, as noted, impossible), but is in fact clearly superior to it in many respects.

This lesson has been hard in learning. Only weeks ago, at Bellinson Hospital, the cancerous uterus of a 16-year-old high-school girl was surgically removed. Since her mother, during her pregnancy, had been given the synthetic estrogen diethylstilbestrol (DES), this tragedy underlines the terrible danger involved in ignoring the results of animal experimentation: Bioassays in a variety of animal species over the last *42 years* have repeatedly demonstrated the carcinogenicity of DES (NIOSH 1977, p.872; Epstein 1979, p.220ff; Wellford 1973, p.157ff). Despite that, 2 million daughters of DES-treated women in the United States and an unknown number in Israel live today with a sword of Damocles suspended above their heads. The cancer, a rare type affecting the cervix and vagina, tends to strike young women 15-22 years of age. And DES is not the only compound shown conclusively to cause cancer in humans only *after* positive animal bioassays had been disregarded (or suppressed by industry); other agents in this category include vinyl chloride and bis(chloromethyl)ether.

With regard to animal testing, perhaps the most misunderstood point of all---and the one that has been most abused by uninformed or industry-connected spokesmen---concerns the large doses normally given the test animals. We may recall at the outset that if a chemical is incapable of carcinogenesis, even the heaviest exposure to it has not been shown to produce cancer. As for carcinogens, it is true that as the dose of a carcinogen is increased the *rate* of carcinogenesis in the exposed population increases. But there is not a shred of scientific evidence for the contention that high doses can change a non-carcinogen into a carcinogen. Indeed, available evidence points in the opposite direction: It has been shown both that at excessively high levels of exposure to a carcinogen the cancer rate increases less than would be expected (Hooper et al. 1979), and also that the higher the daily dose, the *greater* the total dose of carcinogen administered until the appearance of cancer (Schmähl 1979; Preussman 1978). Interestingly, radiation has recently been found to behave in a similar manner, such that *low* doses may be 50-75% more biologically effective than higher ones (Dobson 1978).

Far from representing a "flaw" in experimental design, exposures to maximal doses of test substance are an agreed and *essential* part of bioassay procedures and have been explicitly and repeatedly recognized as such by countless committees of internationally respected experts. As recently noted by a blue-ribbon Presidential scientific taskforce in the United States, protocols that include the "administration of high test doses...are appropriate and scientifically valid test methods for identifying carcinogens" (TSSC 1980; also cf. IRLG 1979; NRC/NAS 1977, p.788; NIEHS 1980, p.12f). The scientific bases for this practice are several, the most important being statistical constraint.

Let us suppose for a moment that man and mouse are equally sensitive to a given carcinogen. (We have already seen that this supposition is far from true in many cases.) If we should then determine that a given dose of carcinogen will result in one case of cancer per 10,000 mice, it follows that exposure of the entire Israeli population to the same dose would produce some 400 cases of malignant disease. Clearly however, giving such a dose to a test group comprising only 50 animals (the standard today) would almost certainly fail to produce even a single case of cancer; we should need some 10,000 animals in the group to demonstrate the disease in one of them. For a number of technical and economic reasons, test groups of this size are impracticable (if not impossible). However, since it has repeatedly been shown that the dost-response curve to a carcinogen is linear, it follows that by increasing the dose we can linearly increase the percentage of animals developing cancer.

Let us say that we can show that the equivalent of 1,000 saccharin tablets per day results in one case of cancer in a group of 50 mice. We may infer from this that 100 tablets would give rise to one case in 500, and 10 tablets, to one in 5,000. In Israeli terms, we should expect some 800 excess cases of cancer from exposing the entire population to a daily 10-

tablet dose of saccharin. Again, this mathematical exercise is based on at least two major assumptions: first, that man and mouse are equally susceptible to the carcinogenetic effects of saccharin (which as we have seen may not be the case), and second, that the linear dose-response relationship holds true in the very low dose range. Although there are many indications that this may be true, it has not been firmly established. In the absence of final proof, however, health authorities have clearly chosen to err (if at all) in the direction of greater public safety, rather than greater public danger, and use linear extrapolation in estimating the extent of potential hazards to man (TSSC 1980, p.134f; IRLG 1979; Schneiderman & Brown 1978).

VANITY OF EPIDEMIOLOGY

Epidemiology, the study of populations in order to discover causes of disease, is both a fascinating and extremely useful tool. Nonetheless, in the study of cancer etiology epidemiology has some serious limitations, limitations that, despite their gravity, are often completely ignored by scientists untrained in this discipline. The results are often that these scientists make statements that confuse or, worse, are totally at odds with the facts. It should be remembered that in man cancer is generally a disease with an extremely long latent period. For epidemiological purposes, therefore, it is often necessary to follow tens of thousands of persons for 20 years, or even longer. Furthermore, the rarer the type of cancer, the larger the population base needed to arrive at statistically significant conclusions. For example, only about one person in 100,000 dies from primary liver cancer. In order to document a doubling (*i.e.*, 100% increase) in the death rate from this disease, a prospective epidemiologic study would have to include nearly *six million* persons.

Even with regard to more common forms of cancer the numbers are huge. To discover a 100% rise in mortality from bladder cancer, for instance, a prospective survey would have to encompass some 760,000 persons. Were our aim to detect a rise of less than 100%---and clearly even a 10 or 20% rise in rates may be far from insignificant with regard to the health of the public---we should again be forced to expand our population base into the millions. It must be stressed here that the above figures apply only to flawlessly conducted surveys: Dozens of types of errors or confounding circumstances often creep into the most carefully planned epidemiological studies, and any one of them may blur, or even conceal from the epidemiologist, a strong association that in reality exists.

Much smaller numbers are involved in retrospective surveys. Here, as opposed to the following of large population groups for years as in the prospective studies, groups of persons that already have the disease in question are interviewed or surveyed in the hope of finding a common denominator among them. The results of such surveys, although much easier and faster to carry out than the prospective ones, are usually much less decisive. From here it should be clear that the chances that a given substance will be suspected of carcinogenicity due to the "clinical impression" of a practicing physician are virtually nil. A case in point: Mortality rates from lung cancer, some 90% of which is caused by cigarette smoking, had increased 25-fold(!) between the beginning of the century and the time (1947) that a concerted attempt was made to look into the matter (Doll 1977).

For our purposes here, we may conclude that the establishment of a causal relationship between a chemical and cancer by epidemiologic means is in many cases difficult in the extreme. If too small a population is used, or if the survey is not continued long enough, or if some unsuspected confounding factor is present, a strong, existing causal relationship may not be detected at all. For these reasons epidemiologists continually caution that "a carcinogenic effect can *easily* be missed by epidemiologic methods [emphasis added]" (IRLG 1979; also cf. TSSC 1980, p.125ff). Put differently, *failure* to discover epidemiologically that a given substance is carcinogenic can in no way be construed to represent proof of its benignity; *lack* of epidemiological proof of carcinogenicity should under no circumstances be taken as a proof of safety. In many cases (where adequate studies have not been conducted), it is not even an indication of safety. And certainly, in the face of suitable laboratory evidence of carcinogenicity, the contention that a chemical may be safe because "there is no evidence that the substance is carcinogenic *in man*" is as irrelevant as it is irrational.

We may summarize our findings as follows:

1. The vast majority of human cancer has an environmental component that is causal in nature.
2. Only a very small portion of all tested chemicals has proven to have carcinogenic potential.
3. There is no scientific basis for the contention that carcinogenicity is dose determined; only the *rate* of the carcinogenetic response depends upon dose.
4. There is no way of determining a dose of a carcinogen, however small, that

is completely safe for man.

5. Properly conducted animal studies are relevant to man.
6. Maximally tolerated doses in animal bioassays are both necessary and scientifically valid.
7. The absence of epidemiological proof of carcinogenicity is by no means proof of safety.

These principles have been formulated over the past 20, and especially the last 10, or so years. They represent the work of the finest scientific minds in the field. Not only do they form the basis of cancer-prevention legislation in numerous countries, but they have repeatedly withstood the sophisticated assaults of industry-related scientists in courts of law. It is vital that they be used as guidelines by public-health authorities.

REFERENCES

Bingham, E. and H.L. Falk (1969). Environmental Carcinogens. Modifying Effects of Carcinogens on Threshold Response. Arch. Env. Hlth. 19:779-784.

Boyland, E. (1980). History and Future of Chemical Carcinogenesis. Brit. Med. Bull. 36: 5-10.

Devesa, S.S. and D.T. Silverman (1978). Cancer Incidence and Mortality Trends in the United States: 1935-74. J. Nat. Cancer Inst. 60: 545-571.

Dobson, R.L. (1978). How Toxic is Tritum? Relevance of High-Dose Results and Gamma-Ray Data to Evaluating Low-Level, Chronic Exposure. Env. Hlth. Prospec. 22: 145-147.

Doll, R. (1977). Strategy for Detection of Cancer Hazards to Man. Nature 265: 589-596.

Epstein, S.S. (1979). The Politics of Cancer. Anchor Press/Doubleday, Garden City, N.Y.

Fraumeni, J.F., Jr. (1979). Epidemiological Studies of Cancer. In Carcinogens: Identification and Mechanisms of Action, A.C. Griffin and C.R. Shaw (eds), pp.51-63. Raven Press, New York.

Higginson, J. (1976). A Hazardous Society? Individual Versus Community Community Responsibility in Cancer Prevention. Am. J. Pub. Hlth. 66: 359-366.

Higginson, J. (1979). Perspectives and Future Developments in Research on Environmental Carcinogenesis. In Carcinogens: Identification and Mechanisms of Action, A.C. Griffin and C.R. Shaw (eds), pp.87-208. Raven Press, New York.

Hooper, N.K., R.H. Harris, and B.N. Ames (1979). Chemical Carcinogens (letter). Science 203: 602-603.

Hoover, R. (1979). Environmental Cancer. Ann. N.Y. Acad. Sci. 329: 50-60.

Interagency Regulatory Liaison Group (1979). Scientific Bases for Identification of Potential Carcinogens and Estimation of Risk. J. Nat. Cancer Inst. 63: 241-268.

Israel Cancer Registry (1977a). Cancer in Israel. Facts and Figures 1967-1971. Ministry of Health, Jerusalem.

Israel Cancer Registry (1977b). Cancer in Israel. Facts and Figures 1971-1976. Ministry of Health, Jerusalem.

Lancet (1977). Are 90% of Cancers Preventable? (editorial) i: 685-687.

Muir, C.S. (1975). International Variation in High-Risk Populations. Chapter 18 in Persons at High Risk of Cancer, J.F. Fraumeni, Jr. (ed). Academic Press, New York.

National Institute for Occupational Safety and Health (1977). Registry of Toxic Effects of Chemical Substances, vol.II. U.S. Government Printing Office, Washington, D.C.

National Institute of Environmental Health Sciences (1980). Basic Concepts of Environmental Health. NIH/DHEW, Washington, D.C.

National Research Council/National Academy of Sciences (1978). Saccharin: Technical Assessment of Risks and Benefits, Report No. 1. NRC/NAS, Washington, D.C.

Preussman, R. (1978). Toxicological Aspects of Food Safety---Carcinogenicity and Mutagenicity. Arch. Toxicol. Suppl. 1: 69-84.

Scherer, E. and P. Emmelot (1979). Multihit Kinetics of Tumor Cell Formation and Risk Assessment of Low Doses of Carcinogen. In Carcinogens: Identification and Mechanisms of Action, A.C. Griffin and C.R. Shaw (eds), pp.337-364. Raven Press, New York.

Schneiderman, M.A. and C.C. Brown (1978). Estimating Cancer Risks to a Population. Env. Hlth. Prospec. 22: 115-124.

Schneiderman, M.A., P. Decoufle, and C.C. Brown (1979). Thresholds for Environmental Cancer: Biologic and Statistical Considerations. Ann. N.Y. Acad. Sci. 329: 92-130.

Toxic Substances Strategy Committee (1980). Toxic Chemicals and Public Protection. U.S. Government Printing Office, Washington, D.C.

Wellford, H. (1973). Sowing the Wind. Bantam, New York.

Westin, J.B. (1977). Saccharin Spells Danger. Consumer Shield, Tel-Aviv.

Developments in Arid Zone Ecology and Environmental Quality
Edited by H. Shuval

48

PREGNANCY OUTCOME, SOCIO-ECONOMIC FACTORS AND POLLUTION

Ecological Analysis Strategy Based on Data for a California County

John R. Goldsmith, M.D., M.P.H., Professor* Eva R. Glazer, M.D.+

**Epidemiology and Health Services Evaluation Unit, Faculty of Health Sciences, Ben-Gurion University of the Negev, P.O.B. 653, Beer Sheba*
+California Department of Health Services, 2151 Berkeley 94704 California

ABSTRACT

In industrialized countries, it is commonly observed that poor people live in locations with high levels of pollution, relative to people with better incomes who live in less polluted areas. When we come to examine the possible associations between pollution exposures and health, our analysis is likely to be confounded by the effects of poverty on health. Using a data set based on a partly industrialized California county, the use and interpretations of partial correlation analysis strategy is presented.

The rates for pregnancy outcome used include the maternal age-adjusted infant mortality rate (AAIMR), the age-adjusted fetal death rate (AAFDR), and the age-adjusted low birth rate ratio (AALBWR). Data for the white population for 21 contiguous census tract areas in the county were used and the pregnancy outcome data were for 1968-1972. Data on socio-economic variables were derived from the 1970 census, and include median family income (MFI), percent of the male population who were blue collar workers (PBC), and percent of the population who were Black (PB). Estimated exposures for each area were prepared for Particulate matter from industrial sources (PM) and for total exposures to nitrogen oxides (NOX) and carbon monoxide (CO).

The correlation matrix for these variables is presented, along with the partial correlation of health variables on pollution variables with "control of" socio-economic economic variable effects.

A positive correlation (r=0.57) between industrial particulate matter exposure and fetal death rate was found. Partial correlations with three socio-economic variables reduced this only to between 0.43 and 0.51). The rates for 10 central residential areas, 6.1-12.9, mean 8.0±2.306 is significantly less than for the 11 peripheral, more industrialized areas, 7.8-15.8, mean 11.52±2.67 (t=2.95).

These findings, like other ecological studies must not be interpreted as showing a causal relationship between industrial particulate pollutants and fetal death rate, nor between proximity to industrial emissions and fetal death rates. They do indicate a need for a case-control study and this is recommended, to examine a variety of personal, social, and exposure variables among families with and without a history of fetal death.

Developments in Arid Zone Ecology and Environmental Quality
Edited by H. Shuval

49
HISTORICAL PROSPECTIVE EPIDEMIOLOGICAL STUDY OF WASTEWATER UTILIZATION IN KIBBUTZIM IN ISRAEL: 1974-77

Badri Fattal & Hillel Shuval

Environmental Health Laboratory Hebrew University-Hadassah Medical School of Jerusalem

ABSTRACT

An historical prospective epidemiological study was carried out on the association between enteric disease incidence and wastewater utilization in agriculture in 81 kibbutzim in Israel (cooperative agricultural settlements) having a population of 34,492.

Medical records on disease incidence were collected directly from the patient's files at the clinic of each kibbutz. These records contained information on twenty-four defined diseases, laboratory confirmed or clinically diagnosed, half of which were enteric diseases which might possibly be transmitted by wastewater, while the other half were non-enteric diseases which served as controls.

Environmental data concerning sewage utilization such as the location of the sewage irrigated tracts, wind directions and other possible sources of pathogens such as barns and chicken coops was collected for each kibbutz.

The kibbutzim included in the study were divided into four categories:
1. Thirty kibbutzim irrigating with wastewater effluent in addition to irrigating with freshwater. Population: 13,531.

2. Twenty-eight kibbutzim not utilizing effluent for any purpose, which served as a control group. Population: 11,096.

3. Ten kibbutzim using wastewater effluent as feed-water for fish ponds. Population: 5,005.

4. Thirteen kibbutzim which practiced two consecutive years of effluent utilization and another two consecutive years in which no effluent was utilized for any purpose or vice versa. Population: 4,860.

In the first three groups, we compared the enteric and control disease incidence in kibbutzim using effluent irrigation and/or effluent in fish ponds with the enteric and control disease incidence in kibbutzim not using effluent for any purpose (the control group).

In the fourth group we have compared total enteric and control disease incidence during the two year effluent irrigation period with the enteric and control disease incidence during the non-irrigation period.

The results are analyzed according to the age distribution of the population, the irrigation season, the distance between irrigated tracts and residential areas and whether residential areas were predominantly upwind or downwind form the irrigated tracts, in addition to other factors.

Initial results indicate that for the total population no significant differences were found between effluent vs. non-effluent utilizing kibbutzim. However, a somewhat higher enteric disease rate may be present among the 0-4 year age group in kibbutzim practicing wastewater utilization.

Israel suffers from a severe shortage of water resources with over 95% of the total available water supplies being utilized currently. This situation has led to the extensive use of wastewater in Israel for irrigation of various crops and for enrichment of fishponds. Irrigation methods used in Israel today are mainly sprinkler and drip. About 90% of the total irrigation is done by sprinkling. The amount of sewage potentially available for exploitation in Israel is about 200 million cubic meters per year. About 20% was recycled in 1979, with the area of land irrigated with wastewater reaching 8,000 hectares.

The extensive use of effluent in agriculture creates a potential health hazard for field workers and the population adjacent to the fields. Humans possibly may be infected by aerosols containing pathogenic bacteria or viruses by inhalation of particles of 0.5-5.0 μm which may penetrate the respiratory system. Other modes of transport of pathogens from areas irrigated with wastewater are by direct contact with irrigation workers, contaminated crops, and fomites.

Hickey and Reist (1975) reported that pathogenic bacteria may be transported in aerosols from sewage to humans. In a later survey Clark et al (1976) emphasized the potential risk to sewage workers of enteric diseases from occupational exposure. Adams and Spendlove (1970) isolated microorganisms from aerosols at a distance of 1200m downwind from a sewage treatment plant. However, there is little proof of the degree of health risk from aerosolized sewage.

The Environmental Health Laboratory of the Hebrew University-Hadassah Medical School has succeeded in isolating coliforms at a distance of 350m and salmonella at a distance of 60m downwind from the wastewater irrigated fields (Katzenelson and Teltsch, 1976). Also, they succeeded in isolating enteroviruses at a distance of up to 100m downwind from effluent irrigation (Teltsch and Katzenelson, 1978).

One of the first retrospective epidemiological studies on the possible health risks associated with sprinkler irrigation with wastewater was carried out by our group in Israel (Katzenelson et al, 1976). The findings from this study indicated that in 77 kibbutzim (agricultural cooperative settlements) practicing mainly sprinkler irrigation with undisinfected oxidation pond effluent high in coliform count, the incidence of typhoid fever, shigellosis, infectious hepatitis and salmenolossis was 2-4 times higher than in 130 control kibbutzim not practicing any form of effluent irrigation.

This preliminary survey had serious methodological constraints in that the investigators based their findings solely on official communicable disease reports submitted to the Ministry of Health as required by law. These requirements do not include listing of all cases of relevant diseases and there are known serious defects in such reporting methods. Therefore, no conclusive findings may be based on this report. Consequently, a new research scheme, based on primary medical and environmental data collected at each kibbutz was needed to retest the hypothesis set forth in the initial retrospective study of wastewater irrigation in kibbutzim.

In this historical prospective study it is proposed to study in two stages the differences in enteric disease incidence in 81 kibbutzim spread throughout the country, by comparison of the disease rate in accordance with the following categories:

Stage I - Switch category: 13 switch-category kibbutzim defined as kibbutzim practicing two consecutive years of irrigation during which effluent sources were used for irrigation for at least part of the year and another two consecutive years in which only non-effluent water sources were used for irrigation, or vice-versa. Population: 4,860.

Stage II - Effluent vs. non-effluent utilization: 68 kibbutzim categorized as follows:

a. Thirty effluent-irrigating kibbutzim defined as kibbutzim which utilize both local effluent and/or that of nearby urban settlements for spray irrigation as welll as non-effluent sources and do not have fish ponds. Population: 13,531.

b. Twenty-eight non-effluent irrigating kibbutzim defined as kibbutzim which do not practice effluent irrigation and are not exposed to aerosolized sewage from effluent irrigated neighboring fields at a distance of 5 kms. from the residential area of the kibbutz, or to direct contact with sewage, and do not have fish ponds. Population: 11,096.

c. Eight kibbutzim utilizing effluent for fish pond enrichment in part of their fish ponds. The non-effluent fish ponds are enriched with chicken droppings and cow dung. Non-effluent sources only are used for irrigation. Population: 3,894.

d. Two kibbutzim defined as in (c) except that when both types of fishponds are drained twice yearly, the water is used for spray irrigation. Population: 1,111.

Using the above categories, we will examine the following working hypothesis:

- The wastewater borne (enteric) disease incidence rate - wastewater borne diseases are essentially enteric and will be referred to henceforth as enteric diseases (see Table 1 for list of diseases) in the irrigation season in effluent-irrigating kibbutzim is significantly higher than in non-effluent irrigating kibbutzim. On the other hand, we do not anticipate a higher enteric disease incidence rate in the non-irrigation season.
- The control disease incidence rate will remain uniform regardless of type of effluent. Control diseases are defined as diseases not associated with wastewater borne diseases.
- We anticipate that in the irrigation season the wastewater borne (enteric) disease rate will be higher for effluent-irrigating kibbutzim which use sprinkler irrigation and are exposed to aerosols than for effluent-irrigating kibbutzim not exposed to aerosols, i.e. kibbutzim utilizing effluent for fish pond enrichment.
- If the rates of enteric disease in kibbutzim utilizing effluent for fish pond enrichment are greater than the rates in those kibbutzim which are non-effluent irrigating, then it is possible to conclude that the increase is associated with the effluent-enriched fish ponds and that the pathogens are spread by direct contact with fish pond workers and/or by contact with or consumption of polluted fish.

Table 1.
Selected Diseases from Kibbutz Clinic Patient Records

Wastewater borne disease (Enteric diseases)	Non wastewater (Control Diseases)	Others
Typhoid fever	Measles	Influenza
Paratyphoid	Chickenpox	Food poisoning (bacterial)
Other Salmonella Infections	Accidents	
Bacillary dysentery	Rubella	
Gastroenteritis	Mumps	
Aseptic meningitis	Pneumonia	
Infectious hepatitis	Bronchitis	
Pathogenic coli; urinary tract infections	Upper respiratory infections	
Fever of unknown origin	Streptococcal throat infections	
Coxsackievirus infection		
Parasitic infections		
Amebiasis		
Leptospirosis		

The kibbutz in Israel is a good model which allows us to thoroughly investigate the above hypothesis. There are many kibbutzim which use sprinkler irrigation and in which the main sewage treatment is by oxidation ponds with no disinfection of effluent. In contrast, there are other kibbutzim which neither use effluent nor use effluent for fish ponds.

Each kibbutz is equipped with a clinic staffed by at least one nurse and serviced by a physician. In addition, there are regional laboratories which perform routine communicable disease examinations for the kibbutzim. Another advantage is the high level of consciousness among the kibbutzim. in health matters as well as their capability to record the environmental and medical data and their willingness to cooperate with a research effort of this kind.

Our study is based on data collected between 1974-77 from several sources:

1. Personal medical records of kibbutz populations. Data on 24 diseases was culled from these records for the period January 1, 1974 to December 31, 1977. Of the 24 diseases chosen 13 are considered wastewater borne (enteric diseases) while nine control diseases and the remaining two - influenza and food poisoning have been analyzed separately.

From the table it can be seen that in the total population no significant differences in enteric disease rates were found during the irrigation season between effluent irrigating and/or effluent enriching (Categories 1, 3 & 4) vs. non-effluent utilizing kibbutzim (Category 2 - control group). In addition, no differences were found in any enteric disease in this comparison. This holds true for all of the disease groups as well during the irrigation season. The same trend appeared during the non-irrigation season.

Tables 4 & 5 shows the distribution of the population by age group and sex and by type of population according to status, i.e. temporary resident, etc. No significant differences were found in the groups classified by age and sex, nor in those categorized by type of population.

TABLE 4. Age and Sex Distribution in (%) by Category

Age (*years*)	Category (1) Effluent Irr.	Category (2) Non-Effluent Irr.	Category (3) Effluent-Enriched Fish Ponds	Category (4) Effluent-Enriched Fish Ponds + Effluent Irr.
0-0.9	1.4 (0.7)*	1.9 (1.1)	1.7 (0.9)	1.6 (0.6)
1-4	7.7 (4.2)	9.0 (4.8)	9.5 (5.0)	10.1 (5.9)
5-9	10.1 (5.5)	9.8 (5.1)	10.4 (5.5)	9.9 (5.4)
10-18	15.2 (8.2)	13.8 (7.3)	14.5 (8.1)	13.8 (7.1)
19+	65.6 (32.8)	65.5 (33.2)	63.9 (32.2)	64.6 (31.5)
Total %	100.0 (51.4)	100.0 (51.5)	100.0 (51.7)	100.0 (50.5)

* The number in parenthesis indicate the % of males

TABLE 5. Population Distribution According to Type of Population by Category (%)

Type of Population	Category (1) Effluent Irr.	Category (2) Non-Effluent Irr.	Category (3) Effluent-Enriched Fish Ponds	Category (4) Effluent-Enriched Fish Ponds + Effluent Irr.
Members & Candidates	59.9	54.7	56.2	57.1
Children	31.9	32.5	36.1	32.6
Temporary Residents, Ulpan, Nahal	4.3	7.7	3.6	5.9
Others	2.9	3.9	2.8	3.3
Unknown	1.0	1.2	1.3	1.1
Total %	100.0	100.0	100.0	100.0

Table 6. presents a comparison of total incidence of enteric and control diseases according to distance of effluent irrigation fields from communal dining halls, while Table 7. compares total incidence of enteric and control diseases according to size of effluent-irrigated tract. No significant differences in enteric disease morbidity were found in the kibbutz groups distinguished by their distance from the communal dining room or by the size of the effluent irrigated tract.

TABLE 6. Comparison of Incidence of Enteric and Control Diseases According to Distance of Effluent Irrigation Fields from Communal Dining Halls. (Rate 1000/Month in Category 1 Kibbutzim)

Season		Up to 300 m.		301-600 m.		601-900 m.		901+ m.	
		Enteric Diseases	Control Diseases	Enteric Diseases	Control Diseases	Enteric Diseases	Control Diseases	Enteric Diseases	Control Diseases
Irrigation Season	Rate	33.3	82.0	57.2	169.0	62.2	160.0	47.2	149.2
	Ratio (enteric/ control)	0.41		0.34		0.39		0.32	
Non-Irrigation Season	Rate	19.1	108.6	29.9	157.3	28.7	125.5	31.0	157.0
	Ratio (enteric/ control)	0.18		0.19		0.23		0.20	
Population		2,823		6,792		1,198		2,429	
No. of Kibbutzim		7		14		5		4	

TABLE 7. Comparison of Total Incidence of Enteric and Control Diseases According to Size of Effluent-Irrigated Tract (Size of Tract - dunam) (Rate /1000/Month in Category 1 Kibbutzim

Season		Up to 50 dunam		51-400 dunam		401-750 dunam		751+ dunam	
		Enteric Diseases	Control Diseases	Enteric Diseases	Control Diseases	Enteric Diseases	Control Diseases	Enteric Diseases	Control Diseases
Irrigation Season	Rate	31.1	98.3	63.9	175.1	48.8	156.3	52.0	135.0
	Ratio (enteric/ control)	0.32		0.36		0.31		0.39	
Non Irrigation Season	Rate	11.8	63.7	35.4	181.9	27.8	143.5	31.1	148.7
	Ratio (enteric/ control)	0.19		0.19		0.19		0.21	
Population		2,250		4,835		2,410		3,747	
No. of Kibbutzim		6		10		5		9	

DISCUSSION

The results in this paper represent the first stage of data processing for this study. However, at this early stage it is clear that no significant differences have been found between effluent utilizing vs. non-effluent utilizing kibbutzim. There is some indication that differences will arise among the lower age groups, i.e. 0-4 years old, but further data computation must be performed before this can be proven conclusively.

This study suffers from several methodological problems associated with the limitations of an historical prospective study of this type. Medical data posed a problem in that servicibility of recording, defined as comprehensiveness, consistency and continuity of medical data recording was not uniform in all the kibbutzim. This includes the records of nurses, who play a decisive role in the clinic system in the kibbutz and those of physicians who usually visit the kibbutz twice weekly.

Furthermore, it was discovered that morbidity data for volunteers and temporary residents is not routinely recorded by the clinics. This group is especially critical to this type of study due to its possible high susceptibility to diseases related to effluent exposure. It is our feeling that such groups of previously unexposed people are the most vulnerable to change in health status and could provide critical information as to the degree of risk associated with wastewater irrigation during short periods of exposure.

Another difficulty which this study could not overcome was the problem of outbreaks. During the course of the study it became clear that vital information on outbreaks were not routinely recorded since the clinic staff on many kibbutzim was unable to deal both with the increased workload resulting from the outbreak and the recording of morbidity data during these periods.

We hope that many of these methodological problems can be overcome in our ongoing prospective study. In this study, supported by the U.S. EPA, special attention is being given to the servicibility of medical recording, care of volunteers, notifications and investigations of outbreaks as well as the obtaining of additional information on the health status of the kibbutz population from sources other than the clinics, i.e. checking work absenteeism for health causes and/or implementing a health diary. These checks on the system can aid in revealing the possible association between enteric diseases and effluent utilization.

ACKNOWLEDGEMENT

This study was financed by the U.S. EPA, Grant Number R-805174.* Special thanks to Professor Michael Davies for his guidance on questions of epidemiological methodology. In addition thanks must also be expressed to J. Applebaum and H. Gazit for supervision of field work and to F. Borensztajn, M. Barris & E. Sharon and Dr. Y. Wax for the statistical analysis. Without the willing and enthusiastic cooperation of several hundred kibbutz members, this field study could not have been carried out. Their active participation is appreciated.

*Mr. H. Pahren served as project officer (HERL EPA).

REFERENCES

Hickey, J.L.S., and P.C. Reis, 1975. Health significance of airborne microorganisms from wastewater treatment processes. Part II: Health significance and alternatives for action. Jour. Water Poll. Control Fed., 47:2741-2773.

Clark, C.S., A.B. Bjornson, G.M. Schiff, J.P. Phair, G.L. Van Meer, and P.S. Gartside, 1977. Sewage workers' syndrome, Lancet, 1:1009.

Adams, A.P., and J.C. Spendlove, 1970. Coliform aerosols emitted by sewage treatment plants, Science, 169:1218-1220.

Katzenelson, E., and B. Teltsch, 1976. Dispersion of enteric bacteria by spray irrigation. Jour. Water Poll. Control Fed., 48:710-716.

Teltsch, B., and E. Katzenelson. 1978. Airborne enteric bacteria and viruses from spray irrigation with wastewater, Appl. Environ. Microbiol., 35:290-296.

Katzenelson, E., I. Baium, and H.I. Shuval, 1976. Risk of communicable disease infection associated with wastewater irrigation in agricultural settlements, Science, 195:944-946.

Shuval, H.I. & Fattal, B. Epidemiological study of kibbutzim in Israel, Wastewater, Aerosols and Disease, Proceedings of a Symposium, September 19-21, 1979, Edited by H. Pahren and W. Jakubowski. U.S. Environmental Protection Agency, HERL EPA-600/9-80-028.

Developments in Arid Zone Ecology and Environmental Quality
Edited by H. Shuval

50

OCCUPATIONAL EXPOSURE TO DUST CONTAINING FREE-SILICA IN A CERAMICS FACTORY[a]

Rivka Kolton*, Elihu D. Richter** and Rafael Israeli***

*Hebrew University (Faculty of Applied Science and Technology - Environmental Sciences Program),**Hebrew University School of Public Health - Department of Medical Ecology, *** Ben-Gurion University for Health Sciences - Kupat Cholim Department of Occupational Medicine

ABSTRACT

In a ceramics plant (225 workers), analysis of respirable dust collected by personal samplers worn by exposed workers (analytic method =x-ray diffraction) documented exposures to free silica exceeding 0.050 mgm/m^3 - the Permissible Exposure Level (PEL) or Threshold Limit Value-Time Weighted Average (TLV-TWA) proposed by US National Institute of Occupational Safety and Health -- in 15 of 25 determinations. Excessive exposures were produced by dust with quartz content no greater than 8% by weight.

Such exposures to free-silica containing dust, if present in the past, could have accounted for x-ray findings compatible with silicosis described in a group of workers employed starting 5-23 years prior to this study. The exposures to free silica appeared to be a consequence of the failure to incorporate established principles for prevention of dust exposure when the plant was designed. Routine determination of free silica content, even at trace and low levels, is necessary to assess the potential hazard for silicosis in dusty workplaces.

[a] Full paper, in press, Israel Journal of Medical Sciences, 1981, based on M.Sc. thesis of R.K., 1976-1977.

Developments in Arid Zone Ecology and Environmental Quality
Edited by H. Shuval

51

THE RELATION BETWEEN WATER QUALITY OF SWIMMING POOLS TO THE INCIDENCE OF EAR INFECTION IN KINDERGARTEN CHILDREN IN ELEVEN KIBBUTZIM

Dr. D. Franklin+ and E. Simchen*

+Ministry of Health and *Hebrew University-Hadassah Medical School

ABSTRACT

Bacterial counts (total, coliform, faecal coli) and chemical examinations (chlorine or bromine) were done in the swimming pools of eleven kibbutzim in the upper Galillee region of Israel during the period May-September 1980. The pools were graded according to water quality.

The kibbutz nurses noted the incidence of ear infection in the above homogeneous population of four, five and six year old children during the same period. There was good correlation between the different bacterial indicators of poolwater quality but poorer correlation between them and chlorine/bromine residuals.

Infection rates were highest in the youngest age group, and also higher in divers than in non-divers. Crude infection rates showed no differences between swimming in pools of high and low bacterial quality. The differences became significant (on x^2 testing) when only non-divers were compared, and the relative risk further increased when the age adjusted rates were calculated.

Therefore we think that poor poolwater quality is an important risk factor in the etiology of ear infection in the swimmers of this population group in the above conditions.

Furthermore, we think that chlorine/bromine levels may not be a good indicator of bacterial quality of swimming pools. More intensive studies are indicated as follow-up, in respect of testing of other bacterial indicators, virological and fungal testing, incidence of skin, eye and intestinal infection, and infection rates in other age groups.

Developments in Arid Zone Ecology and Environmental Quality
Edited by H. Shuval

52

AIR QUALITY IN ISRAEL IN THE YEARS 1979 AND 1980

Michael Graber

The Environmental Protection Service, Ministry of the Interior, Jerusalem, Israel.

ABSTRACT

The Israeli National Air Quality Monitoring Network, which began to operate in December 1978, consists of seven stations placed in or near the main population and industrial centers of the country.

Each monitoring station was designed with its own data processor (a mini-computer type Computer Automation ALPHA-LSI), the purpose of which is to calibrate the air pollution and meteorology sensors, perform data quality control, average the measurements over times as required by Israeli standards and prepare half-hour and daily status reports of the air quality at each station. In addition, 5-minute averages of all air pollution and meteorology parameters monitored by the stations are recorded by the data processor on magnetic tapes.

Summaries of the results obtained by the network for the years 1979 and 1980 and a comparison of results with Israeli and other air quality standards are presented.

Although all annual averages of the pollutants monitored by the network were well below accepted air quality standards, violations of short-term (1/2-hour and 24-hour) air quality standards were observed during 1979 and 1980.

DESCRIPTION OF THE ISRAEL AIR QUALITY MONITORING NETWORK

The Israeli National Air Quality Monitoring Network, designed in 1974 (Jones et al.,1976) was put into full operation in December 1978, as the stations located in Ashdod, Beersheva, Greater Tel-Aviv and Jerusalem joined the two Haifa stations, which have been operating since September 1976.

The air quality monitoring network was established as part of the Israeli Air Quality Management Assessment Program, to support short and long-term needs of municipal and national authorities dealing with air pollution control and physical planning and to contribute to the development of programs to abate air pollution(Turpin and Hochstein,1979). In particular, the network was planned to:

1. Provide the data required to assess the air quality of the urban centers of Israel.
2. Provide information, in real time, regarding pollutant concentrations in the various regions of Israel, to enable alert and response actions in the case of an air pollution episode.
3. Provide the data base for long-range trend analysis of the air quality in Israel.
4. Generate the necessary data for calibrating existing air pollution dispersion models and those to be developed. These models are intended as tools for the air pollution control and planning authorities to achieve rational management decisions and policies.
5. Provide the general public with up-to-date information on air quality at the various parts of the country.

The national air quality monitoring network consists of seven continuous fully automated stations of which the six stationary stations are operative, while the mobile station is in the run-in period (It will be operated in the Greater Tel-Aviv area). The center of the network is located at the Israel Environmental Protection Service (IEPS), the Ministry of the Interior, in Jerusalem (Fig.1)

The stations which comprise the air quality network are individually operated and maintained by the Environmental Protection Units of the municipalities responsible for the regions in which the stations are located. This work is performed either by their own technical staff or through subcontractors.

A local firm, which is the representative in Israel of the American manufacturer of the monitoring stations, provides spare parts and services in the case of equipment failures or breakdown.

Central coordination amongst the monitoring stations and their operators, professional advice as well as the required financing were provided by or through the staff of the IEPS.

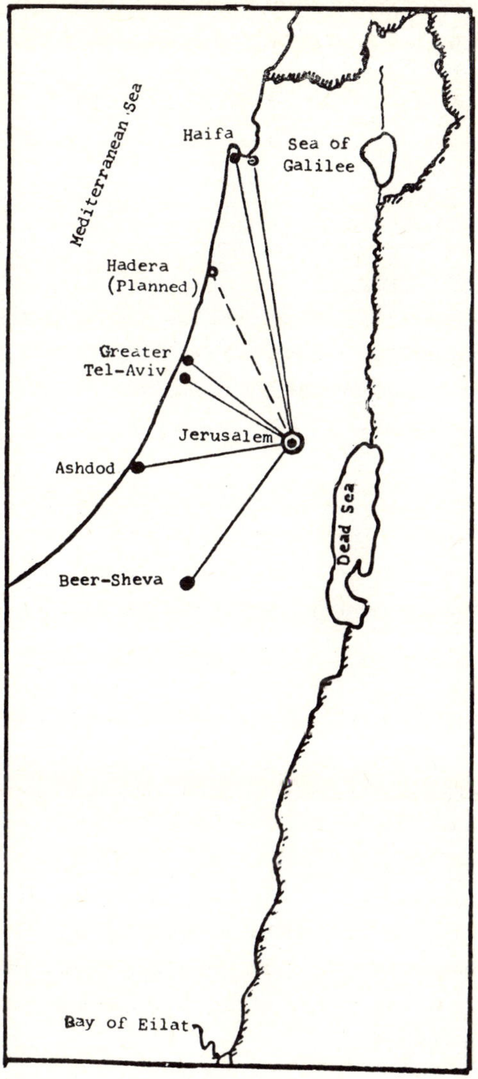

Fig.1. Israel National Air Quality Monitoring Network

A list of the individual pollutants and meteorological parameters which the network stations monitor is listed, together with the respective Israeli Ambient Air Quality Standards, in Table 1.

A detailed description of the Israel Air Quality Monitoring Network, including the design of its stations, the parameters monitored by them, type of equipment and special features and the criteria according to which the network stations were sited, was presented by Graber (1980).

SUMMARY FOR 1979 AND 1980 OF DATA FROM THE ISRAEL NATIONAL AIR QUALITY MONITORING NETWORK

A summary of the results obtained from the Israel National Air Quality Monitoring Network for the period 1st January - 31st December 1979 is presented in Table 2, and for the period 1st Jan. - 31st Dec. 1980, in Table 3. These summaries were condensed from the air quality status reports prepared and published each month by the Municipal Environmental Units of Ashdod, Beer-Sheva, Greater Tel-Aviv, Haifa and Jerusalem, which operate the individual monitoring stations of the network.

The monthly air quality status reports include the following information:

1. A verbal description of the air quality during the previous month, highlighting the extreme values measured; noting violations of the air quality standards if any; and commenting on the existence of, if any, weather conditions conducive to the build-up of pollution concentration. Table 1 is also included in each monthly report.
2. A table of daily averages of all parameters monitored at the station for each day of the month, underlining maximal values.
3. If any, a list of pollutants, with concentrations, date and time of occurance when levels exceeded the Israeli Standards for 99% and 100% of the time. *

* See footnote at next page.

TABLE 1: ISRAELI AMBIENT AIR QUALITY STANDARDS (1) AND PARAMETERS

Measured by the Israel National Air Quality Monitoring Network

No	Pollutant/Parameter	Symbol	Units (3)	1/2 – hour average 99%	1/2 – hour average 100%	2 – hour average 99%	2 – hour average 100%	8 – hour average 99%	8 – hour average 100%	24 – hour average 99%	24 – hour average 100%	Annual Average
1	Sulfur dioxide	SO_2	ppm (2)	0.3	0.6	-	-	-	-	0.1	0.15	-
2	Nitrogen oxides	NOx	ppm	0.5	1.0	-	-	-	-	0.3	0.45	-
3	Nitric oxide	NO	ppm	-	-	-	-	-	-	-	-	-
4	Nitrogen dioxide	NO_2	ppm	-	-	-	-	-	-	-	-	-
5	Oxidants	Ox	ppm	0.2	0.4	-	-	0.1	0.175	-	-	-
6	Ozone (4)	O_3	ppm	0.1	0.2	-	-	0.05	0.08	-	-	-
7	Total Hydrocarbons (4)	THC	ppm	7.5	15.0	-	-	-	-	3.0	4.5	-
8	Carbon monoxide	CO	ppm	30	60	-	-	10	17.5	-	-	-
9	Total suspended particulates (5)	TSP	ug/m^3 (2)	-	-	-	-	-	-	200	300	75
10	Total sulfur compounds	TS	ppm	-	-	-	-	-	-	-	-	-
11	Inhalable suspended particulates (5)	RSP	ug/m^3	-	-	-	-	-	-	-	-	-
12	Soiling Index (6)	COH	COH (2)	-	-	2.0	4.0	-	-	1.0	1.5	-
13	Temperature	TEM	°C	-	-	-	-	-	-	-	-	-
14	Wind Speed	WDS	m/sec (2)	-	-	-	-	-	-	-	-	-
15	Wind direction	WDD	degree	-	-	-	-	-	-	-	-	-

(1) Abatement of Nuisances Law, 1961: Abatement of Nuisance Regulations (Air Quality), 1971.

(2) ppm– parts per million; ug/m^3 – microgram per cubic meter; COH – Coefficient of Haze per 1000 linear foot; m/sec – meter per second.

(3) Israeli ambient air quality standards for all pollutants except for particulates and soiling index are stated in ppm. Tables of conversion to units currently in use are available

(4) Israeli tentative standards. (5) Not measured at the Haifa stations. (6) Measured only at the Haifa stations

Table 2: Summary of 1979 data from the Israel National Air Quality Monitoring Network

Pollutant	Averaging Time	Units	Ashdod (AD)	Beer-Sheva (BS)	Greater Tel-Aviv (TA)	Haifa Shuk Talpiot (ST)(4)	Kiryat Bialik (KB)(4)	Jerusalem (JM)
Sulfur dioxide (SO_2)	Annual	ppm	0.0035	0.0062	0.0110	0.0079	0.0043	0.0075
	24-hr max	ppm	0.029	0.019	0.048	0.064	0.034	0.023
	1/2 hr. max	ppm	0.351	0.127	0.203	0.610	0.173	0.080
Nitrogen oxides (NO_x)	Annual	ppm	0.021	0.037	0.052	0.034	0.027	0.016
	24-hr. max.	ppm	0.047	0.101	0.230	0.085	0.068	0.045
	1/2 hr.max.	ppm	0.313	0.626	0.660	0.415	0.297	0.193
Ozone (O_3)	8 hr. max.	ppm	0.078	0.070	0.072	0.081	0.050	0.057
	1/2 hr. max.	ppm	0.111	0.111	0.144	0.130	0.091	0.088
Total Hydrocarbons (THC)	24 hr. max.	ppm	3.7	3.3	4.5	-	-	1.4
	1/2 hr. max.	ppm	7.0	7.0	14.3	-	-	3.6
Carbon monoxide (CO)	8 hr. max.	ppm	8.6	7.1	11.8	-	-	2.4
	1/2 hr. max.	ppm	20.9	13.9	27.0	-	-	5.9
Total Suspended Particulates (TSP)	Annual (Geometric Mean)	ug/m^3	47	79	76	-	-	48
	24 hr. max.	ug/m^3	377	464	2604	-	-	290
	No. of 24 hr. measd. above $75ug/m^3$	-	13(1)	124	142	-	-	38(2)
Annual Uptime of Station (3)		%	73.3	81.6	76.2	66.9	29.9	59.5

(1) Because of equipment failures, only 125 24-hr. measurements were performed in the Ashdod station in 1979

(2) Because of equipment failures, only 218 24-hr. measurements were performed in the Jerusalem station in 1979

(3) This uptime expresses mainly failures of central components of the station such as the computer, the pump or power failures. The uptime of individual sensors is generally lower.

(4) THC, CO and TSP are not monitored by the Haifa and Kiryat Bialik stations.

Table 3: Summary of 1980 data from the Israel National Air Quality Monitoring Network

Pollutant	Averaging Time	Units	Ashdod (AD)	Beer-Sheva (BS)(2)	Greater Tel-Aviv (TA)(2)	Haifa Shuk Talpiot (ST)(1)	Kiryat Bialik (KB)(1)	Jerusalem (JM)
Sulfur	Annual	ppm	0.007	0.012	0.011	0.016	0.007	0.008
dioxide	24 hr. max.	ppm	0.106	0.026	0.053	0.065	0.070	0.022
(SO_2)	1/2 hr. max.	ppm	0.366	0.223	0.211	0.546	0.264	0.055
Nitrogen	Annual	ppm	0.020	0.031	0.049	0.033	0.028	0.022
oxides	24hr.max.	ppm	0.061	0.105	0.163	0.094	0.078	0.072
(NO_x)	1/2 hr. max.	ppm	0.202	0.633	0.524	0.427	0.338	0.285
Ozone	8 hr. max.	ppm	0.073	0.074	0.053	0.135	0.063	0.075
(O_3)	1/2 hr. max.	ppm	0.079	0.101	0.172	0.182	0.136	0.103
Total Hydrocarbons	24-hr. max.	ppm	3.0	2.5	2.8	-	-	3.8
(THC)	1/2 hr. max.	ppm	4.1	8.1	10.4	-	-	5.0
Carbon monoxide	24-hr. max.	ppm	6.2	1.9	10.0	-	-	9.8
(CO)	1/2 hr. max.	ppm	24.8	31.9	36.7	-	-	20.5
Total Suspended Particulate	Annual (Geometric Mean)	Mg/m^3	46	71	42	-	-	42
	24 hr max	Mg/m^3	925	974	222	-	-	948
(TSP)	No. of 24 hr. measd. above $75Mg/m^3$	-	25	100	16	-	-	44
Annual Uptime of Station		%	62.5	85.9	55.4	92.3	86.3	66.0

(1) THC, CO and TSP are not monitored at the Haifa and Kiryat Bialik stations

(2) Results based on data of January - November 1980

4. Maximal 1/2-hour averages of SO_2, NO_x, O_3, CO, and THC for each day of the month, including actual concentration and time of occurance, underlining maximal values.
5. If any, a list of all 1/2-hour values which exceeded the Israeli standards for 99% and 100% of the time.*
6. Maximal 8-hour averages of O_3 and CO for each day of the month, including maximal values.
7. If any, a list of all 8-hour values which exceeded the Israeli Standards for 99% and 100% of the time.

DISCUSSION OF RESULTS

A detailed statistical analysis of the measurements is required to state with confidence the conclusions derived from the results and to suggest ways (if necessary) of handling air pollution problems. This analysis should include computation of correlations of wind directions and other meteorological parameters with the measured levels of the pollutants, correlations of one type of pollutant with another, frequency analysis of results, comparison of recent results with previous or historic data, and, in the future, trend analysis.

At this stage, after two years of continuous monitoring but without detailed statistical analysis, the following facts and observations should be noted:

---

* The Israili Air Quality Standard determines both the permissible level of the pollutant in the air and the time that this level may be measured. For each pollutant, two threshold levels have been specified: The lower level (referred to as the standard for 99% of the time) may be exceeded for 1% of the time whereas the higher level (referred to as the standard for 100% of the time) may never be exceeded. Thus, the lower 24-hour standard may be exceeded 4 times per year, the 1/2-hour standard 176 times per year etc. (see Table 1).

During 1979 and 1980, annual averages of sulfur dioxide and nitrogen oxides for all the stations of the network were low compared to the U.S. standards (no Israeli annual standards exist for these pollutants).

During 1979, violations of the 1/2-hour, 8-hour and 24-hour standards have been observed for SO_2 in Haifa and Ashdod, for O_3 in Ashdod, Beer-Sheva and Tel-Aviv and for THC in Tel-Aviv.

During 1980, violations of the 8-hour tentative standard for O_3 have been observed in Ashdod, Beer-Sheva, Haifa and Jerusalem.

From the viewpoint of Israeli Standards, the most severe air quality problem is the TSP problem. During 1979, the Israeli annual TSP standard was clearly violated at 2 out of 4 stations measuring TSP. At the other 2 stations the annual TSP standard may also have been violated, but due to equipment failures, not enough data is available to obtain annual averages. The 24-hour TSP standard has been violated at all 4 stations measuring TSP.

During 1980, the Israeli annual TSP standard was not violated but the 24-hour TSP standard was violated at 3 out of the 4 stations measuring TSP.

It should be pointed out that because of the dry climate of Israel,and its closeness to dusty desert areas, these results of high dust concentrations are not surprising. However, the result showing an annual average during 1979 of TSP only slightly less for Greater Tel-Aviv than for Beer-Sheva could indicate that the contribution to the measured TSP might be from man-made sources as well as from natural ones. It should also be emphasized that currently the main concern with dust is the inhalable portion of it(i.e. the smaller particles which are fine enough to penetrate down into the lungs and cause damage to them). A special air quality standard for the inhalable dust is now being prepared in the U.S.(Miller,1979). The Israeli National Air Quality Monitoring Network is perhaps unique among monitoring networks in that it monitors both the total suspended particulates (TSP) as well as the inhalable portion of it. However, these measurements need further study and analysis before firm conclusions can be drawn from them.

5. The results from the Ashdod and Haifa stations raise the question regarding how well their location was chosen to represent the air quality in their vicinity. In Haifa, the air quality model developed for the Environmental Protection Unit in Haifa (MOH,1978) by the Israeli Meteorological Service has shown that it is unlikely that the two monitoring stations at their present locations will detect the extreme pollutant levels expected in that area. In Ashdod, in addition to the monitoring station of the Municipality of Ashdod, the Israel Electric Corporation is monitoring the air quality at several sites around the Eshkol power plant in Ashdod. The results from their Nir Galim monitoring station, as reported by the Municipality of Ashdod (MOH,1980) and the Israel Electric Corporation(Kovac et al., 1980), which is located about 2km from the Ashdod station, show that for SO_2, the Israeli 1/2-hour standard for 100% time has been violated during the period 7/78 -6/79 no less than 20 times. This is in contrast to findings of the Ashdod station where no violation of that standard was found. It should be mentioned here that the Eshkol power plant north of Ashdod lies near Ashdod's industrial area, which was sited there because it was considered the optimum location from air quality considerations as far as Ashdod was concerned. Results from Ashdod and Nir Galim stations prove that these considerations were correct to a certain extent. However, since the air quality standards in the Ashdod area were nevertheless violated, it is obvious that siting cannot provide the sole solution to Ashdod's air pollution problems.

ACKNOWLEDGEMENT

Establishing and operating the Israeli National Air Quality Network has been a cooperative effort in which many people have taken part. The Israel Environmental Protection Service wishes to express its gratitude to all the people of the Environmental Protection Units in the participating municipalities who operate the stations to produce the data, and to all personnel of the Environmental Protection Service who have taken part in this project. Special thanks are due to Dr.K.H. Jones (at that time at the USEPA), whose efforts in 1975, while serving as an UNDP/WHO air quality consultant to the Israeli Government, enabled the establishment of the Israel National Air Quality Network.

REFERENCES

Graber M. (1980) Data processing and analysis of the Israeli National Air Quality Monitoring Network,In: Selected Papers on the Environment in Israel, No.8, Publ. No.80-10, Environmental Protection Service, The Ministry of the Interior, Jerusalem.

Jones K.H., Zalman D., Graber M., Berger Holtz T. and El-Roy M. (1975) Air Quality in Israel - Selected Topics 1974-1975,Publ. 76-06, Israel Environmental Protection Service, The Ministry of the Interior, Jerusalem.

Kovac S.,Lahav D. and Balmor Y. (1980) Air Quality Monitoring Network for the Ashdod Area - A Review, ECD-251, Israel Electric Corporation Ltd., Haifa(in Hebrew).

Miller F.J. et al (1979) Size Considerations for Establishing a (USEPA) Standard for Inhalable Particles, J. Air Poll. Contr. Assoc., Vol 29(6):610-615.

MoH(1978) Annual Report 1977-1978, Municipality of Haifa, the Environmental Protection Unit, Haifa (in Hebrew).

MoH(1980) Annual Report - 1979, Municipality of Ashdod, The Environmental Protection Unit, Ashdod (in Hebrew).

Turpin L. and Hochstein S.(1979) The Environment in Israel, Publ.No. 79-01, Israel Environmental Protection Service, The Ministry of the Interior, Jerusalem.

Developments in Arid Zone Ecology and Environmental Quality
Edited by H. Shuval

53

STUDIES OF ACID PRECIPITATION AT THE "PENN STATE" SITE

R. G. de Pena, J.A. Pena, V.C. Bowersox* and J.F. Takacs

The Pennsylvania State University, University Park, PA 16802 and *Illinois Water Survey, Urbana, Illinois 61820

ABSTRACT

As part of the MAP3S program precipitation has been collected and analyzed for pH, conductivity, sulfate, nitrate, chloride, ammonium and other ions.

The collection was on an event basis. When trajectories were calculated sequential samples were collected in addition to the event ones.

Three years of data have been analyzed in detail and the results show that: There is a good stoichiometric balance of the four major ions: sulfate, nitrate, hydronium and ammonium. The principal contributor to the acidity in rain is sulfuric acid, while most of the acidity in snow is due to nitric acid. Hydronium and sulfate ions show a pronounced seasonal variation with maxima concentrations in summer and mimina values in the winter. The seasonal variation of sulfate is inverse to that of the gaseous sulfur dioxide in air. The SIV concentrations are in agreement with that of the sulfur dioxide, suggesting that equilibrium has been reached between the raindrops and the gas phase. The concentration of $SO_4^=$ in winter events is not influenced by the type of precipitation nor the trajectories. The concentrations of H_3O^+, NO_3^- and Cl^- show a continuous decrease with time except when there is an influx of fresh air from the N-NW.

Developments in Arid Zone Ecology and Environmental Quality
Edited by H. Shuval

54

TRANSPORT AND DUST DEPOSITION OF SAHARAN DUST IN THE EAST-MEDITERRANEAN

Yaacov Mamane* and Eli Ganor+

*Environmental Engineering, Technion, Haifa, Israel
+Ministry of Health, Tel-Aviv, Israel.

ABSTRACT

The production of natural dust from the Saharan desert was estimated to be between 200 and 260 million ton per year. These figures were based solely on the westward transport of Saharan dust. In this study estimates on the dust transport and deposition over the eastern Mediterranean are provided. The data were obtained from several studies concerned with dust storms in Israel including: ground and airborne particle mass concentration, vertical profiles, dust storm trajectories, particle deposition and the climatology of dust storms in the eastern Mediterranean.

INTRODUCTION

The transport of Saharan dust across the west coast of Africa, especially with the northeast trade winds, and estimates on dust production have been documented extensively (e.g. Carlson and Prospero, 1972; Prospero and Carlson , 1972; Bertrand et al. 1974; Ernst, 1974; Fraser, 1976; Jaenicke and Schutz, 1978; and Morales, 1979).

However information on the transport of Saharan dust to other sectors, such as Europe (Stevenson, 1969), the eastern Mediterranean (Ganor, 1975; Yaalon and Ganor, 1979), or various parts of Africa (Molski, 1966; Kalu,1979) is still lacking. In this study the production of Saharan dust transported to the eastern Mediterranean, the deposition of this dust and its contribution to urban particles are estimated.

THE NATURE OF DUST STORMS

Dust storms, moving from the Sahara and the surrounding deserts to the eastern Mediterranean, occur any time between October and May, but mostly from December to April (Lunson, 1950; Katsnelson, 1970; Ganor, 1975). They are associated with the passage of the warm sector of either a "Cyprus" cold low or a warm low ("Sharav") pressure systems. The latter affects the eastern Mediterranean during the spring season. The Cyprus low affects the region during the winter. As the fronts which define the borders of the warm sector move across, the strong westerly and south-westerly winds produce dust storms, often over a belt of about 1000 km wide and 1500 to 5000 m high (Yaalon and Ganor, 1979). Since the fronts may occur anytime, the dust storms too could be expected both at night or day. During the day however, they are helped by the vigorous convection over the desert. Some of these storms are very intense and could last for periods of three days or more. Visibility is reduced to few hundred meters or less and improves after the cold front has passed. In those days particle concentrations reach high values as shown in Table I. Table I summarizes the daily total suspended particle (TSP) concentration on Mt. Carmel, Israel, during several dust storms occurred in 1976-1977. These data which were obtained with a high volume sampler represents an average over a period of 24 hours (Shmueli 1979).

Table 1 - TSP concentrations ($\mu g\ m^{-3}$) during dust storms on Mt. Carmel.

Date of storm	Daily TSP Concentration ($\mu g\ m^{-3}$)					
	March 12	March 30	April 9	April 15	May 17	Nov. 23
Pre storm	59			92		
	51	57	43	100	32	41
	106	70	90	-	-	59
Storm day	329	208	214	1021	236	733
	70	192	161	-	193	166
	46	132	174	60	87	24
Post storm		49	92		40	

ESTIMATE OF THE SOURCE STRENGTH

Observed particle concentrations in the vertical and horizontal dimensions (Ganor, 1975) suggested that dust concentrations follow an exponential decrease in both directions. The mass concentration, C, at any point (x, z) could be therefore expressed by the following equation :

(1) $$C(x.z) = C_0 \exp(-hx)\exp(-kz)$$

where x is along the wind direction, C_0 is the mass concentration at the source, h and k are constants at any given distance. The above equation may also be a solution to a steady state transport model, in which diffusion and dry deposition are the dominant mechanisms, such as:

(2) $$\bar{u}\frac{\partial C}{\partial z} = -K_z \frac{\partial^2 C}{\partial z^2} - v_s \frac{\partial C}{\partial z}$$

where K_z is turbulent diffusivity, v_s the settling velocity of the particles, and, $\bar{u}$ is the average wind speed.
Substituting Eq. 1 in Eq. 2 yields an expression for h:

(3) $$h = \frac{k^2 K_z - k v_s}{\bar{u}}$$

In our case $h \sim k^2 . K_z . \bar{u}^{-1}$ since K_z is likely larger than 10^5 $cm^2 s^{-1}$ and v_s for micrometer-size particles is a fraction of 1 $cm.s^{-1}$.

The amount of dust transported through a vertical cross section, F_x, is given by integrating Eq. 1:

(4) $$F_x = \int_0^H \int_{-W/2}^{+W/2} C \bar{u}_z \, dy \, dz$$

where H is the thickness of the Saharan layer, W is the width of the dust plume and $\bar{u}_z$ is the average wind speed in the layer.

In order to estimate the amount of Saharan dust originated at the source the following assumptions and calculations have been done:

(a) k 0.77 km^{-1}. This was derived from the exponential decrease of mass concentration with height during the dust storm of April 1972: from ground levels of 1000 $\mu g\ m^{-3}$ daily average to 100 $\mu g\ m^{-3}$ at the top of the dust layer at 3000 m (Ganor, 1975).

(b) C_0 may be obtained from Eq. 1 assuming that K_z is about 10^5 cm^2 s^{-1}, $\bar{u}$ near the ground was measured during the same storm to be around 10 ms^{-1}, and particle concentration at a distance of 2000 km was found to reach 1000 $\mu g\ m^{-3}$, C_0 is then 3200 $\mu g\ m^{-3}$. This rather simplified calculations suggest that two thirds of the Saharan dust settle down during the first 2000 km, in agreement with Jaenicke and Schutz (1978) model calculations.

(c) For a typical dust storm of 1000 km wide, 3 km high and moving with a height averaged wind ($\bar{u}_z$) of 16 ms^{-1}, the amount of dust emitted per storm is approximately 70 tons s^{-1} as obtained from Eq. 4 and the above values.

(d) For ten dust storms per year (Katsnelson, 1970; Ganor, 1975) lasting each a period of around 30 hr., the amount of dust transported annually from the Saharan desert is about 70 million ton. About one third of this amount, i.e. 20×10^6 ton yr^{-1} reaches the east Mediterranean.

The uncertainty of the eastward transport of the Saharan dust is not better than a factor of 2. Our estimates were based on variable input parameters such as H, W, $\bar{u}_z$, C, the duration of the dust storm and number of storms per year. They may vary within the same storm, from storm to storm and from year to year. If each of the parameters do vary within 20 to 50 percent, than the estimated flux of material from the Sahara is correct within a factor of 2 to a factor of 3. Other source of error lies in the assumption that the Saharan region is a single point source, although other arid zones along the dust storm trajectories may also emit dust particles.

DUST DEPOSITION

If all the material which reaches the east Mediterranean is deposited over an area of 1000 x 1000 km^2 or even 500 x 1000 km^2, it would result with a deposition rate of 20 to 40 ton km^{-2} yr^{-1}.

The choice of an area of 500 or 1000 over 1000 km is based on Saharan plume dimensions as observed by satellites (Ganor, 1975). The above deposition rates are backed by independent measurements of settleable particles which were taken in Tel-Aviv during several years (Mamane et al., 1981). Figure 1 describes the monthly amount of the total dust settling in the Tel-Aviv urban area. It points clearly at two areas of relatively low deposition rates: the northwest and the south central regions - which are mostly residential quarters. In these locations, where the influence of industrial and commercial activities is limited, it was shown that

(a) One third of the total deposits is inorganic material insoluble in water. This corresponds to 50 ton km^{-2} yr^{-1} approximately.

(b) Most of this material, as analyzed by x-ray diffraction, show a mineral composition similar to that of the desert dust reaching Israel.

These findings are consistent therefore with our estimates of dust deposition in the range of 20 to 40 ton km^{-2} yr^{-1}.

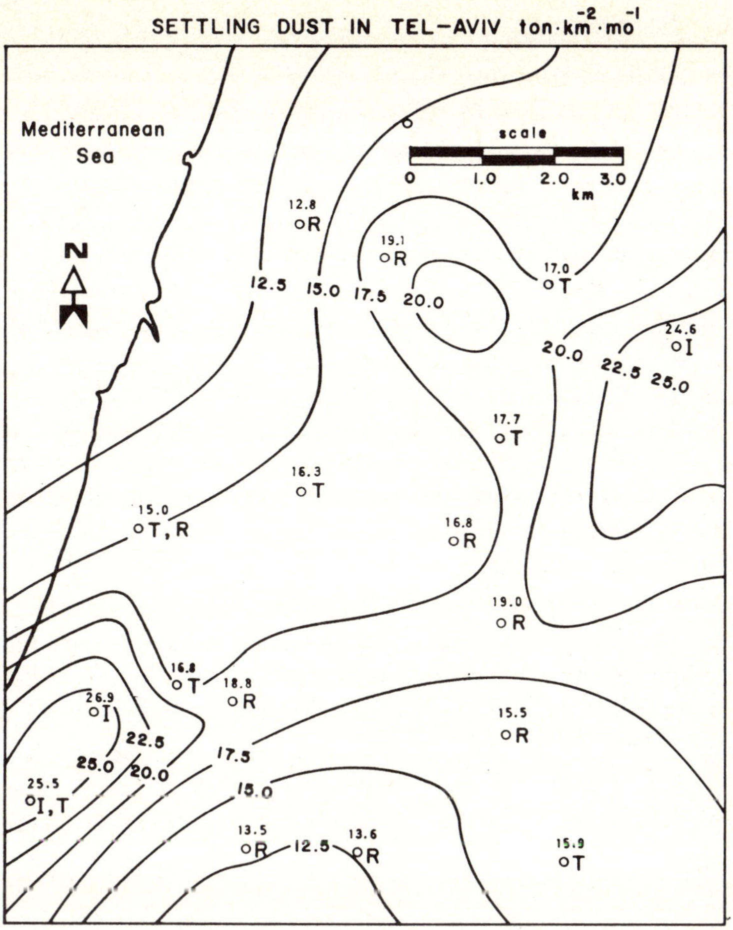

Figure 1 - Mapping of the total dust deposition in the Tel-Aviv area. The data are based on seven years of measurements, 1971-1977, at 16 sites. Dust was collected monthly with stainless steel cylinders (ASTM standards D 1739-70, 1973). Dustfall material was routinely analyzed for organic, inorganic water soluble and inorganic water insoluble content.

DISCUSSION AND CONCLUSION

Rough estimates show that the Saharan area may contribute annually 60 to 200 million tons of dust to the troposphere (Morales, 1979), an amount which may affect the regional and the global climate. Values of 260 million tons per year were reported by Jaenicke and Schutz (1978). Prospero and Carlson (1972) reported that 37 x 10^6 tons of Saharan dust are transported annually through the longitude of Barbados while 30 x 10^6 tons are deposited in the Atlantic Ocean. If one uses the model calculations of Jaenicke and Schutz (1978) in which 70% of dust particles settle down during the first 2000 km, than the source strength according to estimates of Prospero and Carlson (1972) would be about 220 x 10^6 tons yr^{-1} (the distance from the Sahara to the west Africa coast is about 2000 km).

All last three estimates were based on studies of the westward transport of the Saharan dust, and they do not consider the amount of dust carried annually to other sectors such as the East Mediterranean basin. These amounts are to be added to the "westward" estimates in order to obtain a more representative global figure. Therefore the total amount of desert particles generated by the Sahara is about 270 million ton yr^{-1}, based on Morales' summary (1979), 330 - based on Jaenicke and Schutz (1978) calculations, and 290 million ton yr^{-1} - based on Prospero and Carlson (1972) estimates. Not only the eastward transport but also the southern and northern transport are to be included in estimates of Saharan dust production.

REFERENCES

Bertrand, J., J. Baudet and A. Drochon (1974), "Importance of natural Aerosols in West Africa", J. Rech. Atmospheriques, 8 (3-4) 845-860.

Carlson, T.N. and J.M. Prospero (1972), "The Large-Scale Movement of Saharan Air Outbreaks Over the Northern Equatorial Atlantic", J. Appl. Meteorol. 11, 283-297.

Ernst, J.A. (1974), "African Dust Layer Sweeps into the Southwest North Atlantic Area", Bull. Amer. Meteorol. Soc., 55, 1352-1353.

Fraser, R.S. (1976), "Satellite Measurement of Mass of Sahara Dust in the Atmosphere", Appl. Optics, 15, 2471-2479.

Ganor, E. (1975), "Atmospheric Dust in Israel - Sedimentological and Meteorological Analysis of Dust Deposition", Ph.D. Thesis, 224 pp., Hebrew University, Jerusalem.

Jaenicke, R., and L. Schutz (1978), "Comprehensive Study of Physical and Chemical Properties of the Surface Aerosols in the Cape Verde Island Regions", J. Geophys. Res., 83. 3585-3599.

Kalu, A.E. (1979), "The African Dust Plume: Its Characteristics and Propagation Across West Africa in Winter", pp. 95-118 in Saharan Dust", Ed. C. Morales, John Wiley and Sons, New York.

Katsnelson, J. (1970), "Frequency of Duststorms at Beer Sheva" Israel Journal of Earth Sciences, 19, pp. 69-76.

Lunson, E.A. (1950), "Sandstorms on the Northern Coasts of Lybia and Egypt", Meteorol. Office, Prof. Notes No. 102, M.O. 5246, pp. 1-12.

Mamane, Y., E. Ganor, and A.E. Donagi, 1981, "Aerosol Composition of Urban and Desert Origin II - Deposition of Large Particles" - in preparation

Molski, B.A. (1966), "Tree Belts as a Protection Against the Harmattan in Nigeria", Paper III/6 presented at the International Clean Air Congress, pp. 54-57, London 4-7 Oct., 1966.

Morales, C., Editor (1979), "Saharan Dust", Scope 14, John Wiley and Sons, New York.

Prospero, J.M. and T.N. Carlson (1972), "Vertical and Areal Distribution of Saharan Dist over Western Equatorial North Atlantic Ocean", J. Geophys. Res., 77, 5255-5265.

Shmueli, S. (1979), "Daily and Weekly Cycles and Seasonal March of Suspended Particulate Matter Levels in Air at Technion City, Haifa", M.Sc. Thesis, 87 pp., Technion, Haifa, Israel.

Stevenson, C.M. (1969), "The Dust Fall and Severe Storms of 1 July 1968", Weather (London), 24, 126-132.

Yaalon, D.H. and E. Ganor (1979), "East Mediterranean Trajectories of Dust-Carrying Storms from the Sahara and Sinai", Chapter 9, from "Saharan Dust", pp. 187-193, Ed. C. Morales, John Wiley and Sons, New York.

Developments in Arid Zone Ecology and Environmental Quality
Edited by H. Shuval

55
ELEMENT CONTENT AND SIZE DISTRIBUTION OF AEROSOLS IN BEER-SHEVA, ISRAEL.

G. Shani, A. Kushelevski and A. Haccoun

Department of Nuclear Engineering, Ben-Gurion University of the Negev, P.O.B. 653, Beer-Sheva, Israel.

ABSTRACT

The aerosol quantity (TSP), elemental content and size distribution were measured in the city of Beer-Sheva. The measurements were performed for a period of one year under all kinds of weather conditions. Air was sampled on a Whatman 41 paper filter which was later activated by neutrons in a nuclear reactor and the gamma activity of the elements was measured for identification and quantity determination. A part of each filter was investigated with X-ray flourescence for elements undetectable by neutron activation analysis. Thirty elements were found. The size distribution was obtained by using a 6-stage cascade impactor.

The TSP level was classified according to meteorological conditions into three groups with the following average levels: High TSP level (890 ± 250 $\mu g/m^3$), usual TSP level (140 ± 30 $\mu g/m^3$) and low TSP level (59 ± 28 $\mu g/m^3$). The analysis of the elemental content of the aerosols was made according to cross-correlation of the abundance of the elements, meteorological conditions and size distribution. The various elements were divided into those naturally occurring - a part of the dust, or those man-made - emitted by industry and urban transportation. The identification of the sources of pollution was quite clear and there was a good agreement of the source identification by the various means - correlation and size distribution. There is a typical log-normal size distribution of the dust and an exponential fall of the urban pollution.

INTRODUCTION

Beer-Sheva is a relatively small city located on the border of the Negev desert. The industrial zone of the city is located on the east which is the most common wind direction. Because of the proximity of the desert, dust storms and Sharav (hot, dry, dusty weather) are relatively frequent in the city. The average annual rainfall is small - 200 mm, so that the air is not cleaned very often as it is north of Beer-Sheva.

The dust is the major air pollutant in Beer-Sheva, but the industrial zone, although not very large, is close to the city and it is mainly chemical industry producing bromine, chlorine and pesticides. The odors emitted from this industry are carried by the wind to the city and are a nuisance to the population. The combination of the industrial pollution and the dust complicates the situation further, since the industrial pollution is adsorbed by the dust and is carried over to the city.

In the present work, we have measured the quantities of aerosols under various weather conditions. We identified the elements in the aerosols and their abundance using neutron activation analysis and X-ray flourescence techniques. We also measured the size distribution of the aerosols and we were able to identify the origin of the elements - natural (dust) or industrial by size distribution.

EXPERIMENTAL TECHNIQUE

Total suspended particle levels were measured by passing a known volume of air through a filter. The difference in the weight of the filter before and after filtration divided by the volume of air gives the total suspended particle level.

In this work, air was sampled with a Staplex Air Sampler (TFIA-2) using 11 cm diameter Whatman 41 filters. Whatman 41 paper filters were chosen for their strength to withstand relatively high air flowrates, for their capacity to trap small particles with high efficiency and for their elemental purity (Dams, et al., 1972). Each sampling lasted for twenty-four hours with the sampler alternatively on for half an hour and off for half an hour. The airflow initially was 233 ± 6 liter/min. However, as the filter collected more and more particulate matter, the airflow rate dropped. Consequently, the airflow rate was measured at regular intervals during the sampling and the total volume of air flowing through the sampler was calculated accordingly.

The weights of the filters were obtained using a semi-micro balance after equilibrating the filter for twenty-four hours in a controlled humidity cell to obtain constant moisture content in the filter.

The filters were divided into two main parts. One was sent to the Soreq Nuclear Center for neutron activation analysis and the other one was ground and pelletted for XRF analysis using a Phillip X-ray Fluorescence Spectrometer Model PW-1410. The pellets were thick to simulate an infinite emitting medium. Numerical values were obtained in all cases with reference to standard solutions containing the elements to be analyzed. These standards were applied to clean filters and were sent simultaneously for activation and XRF analysis. In this way, errors due to changes in flux and spectrum were accounted for.

The well-known principles of instrumental activation analysis using a high resolution Ge(Li) detector and a multichannel analyzer were followed.

The spectra were analyzed using the computer program Jane to smoothen the spectra and to integrate the peaks (Schubeger, et al., 1975). To obtain the elemental concentrations, the following elements were studied: Br, Na, K, Zn, As, La, Sm, Eu, Sb, W, Au, Cr, Co, Fe, Se, Ce, Hg, Th, Hf, Ti, Rb, Mo, Yb, Ba, Cd, Lu, Ir, Cs and Ag.

Due to the distances between the Soreq nuclear reactor and the University labs (80 km), where samples were counted, short-lived radioisotopes were not analyzed. This prevented the measurement of some elements which produce only short-lived radioisotopes upon reactor neutron activation. This was remedied, to a small extent, by the use of XRF analysis. The following elements were studied by this method: Pb, Ni, Ca, Si, S, Ti, Fe, V, and Mn. Fe was measured by both methods and the agreement was within ± 10%. The lower detection limit was of the order of 0.01% mg/m^3.

The size distribution was obtained using a six-stage cascade impactor, APS-230 series. The 50% cut-off diameters of the impactor stages were 10 μm, 4.2 μm, 2.5 μm, 1.2 μm and 0.6 μm. A back-up filter was used to collect aerosols of diameter 0.4 μm and below. The filter paper with the dust distributed by the impactor was treated as described above, i.e. neutron activation analysis and X-ray flourescence.

RESULTS

Many measurements were made during the years 1977-1978. The results shown here are based on the analysis of 41 measurements made between June 1977 and May 1978. The analysis includes total suspended particle (TSP) level, elemental quantity in each sample and size distribution of aerosols containing each element.

The TSP levels were classified into three categories according to weather conditions on the day of measurement. High TSP levels were found on days with dust storms or Sharav. The average in this group is 890 ± 250 μg/m^3. Normal TSP levels were found on days which were not included in the first group and were not "clean" days due to rain. The average TSP in this group was 140 ± 30 μg/m^3. The third group was the low TSP level found on rainy days or immediately after rainy days. The average TSP level in this group was 59 ± 28 μg/m^3.

The average elemental content in the three groups is shown below in ng/m^3 and in percent (only the main elements are shown).

During normal meteorological conditions the following average elemental concentrations were found in ng/m^3:

Ca 17000 (12.2%), Si 16000 (11.5%), S 4300 (3.1%), Ti 300 (0.21%), Cl 3500 (2.5%), Pb 120 (0.08%), Fe 2700 (1.9%), Na 970 (0.7%), Br 275 (0.2%), K 1300 (0.93%).

During or following rainfall (low TSP) the following average concentrations in ng/m^3 were found:

Ca 7300 (12.6%), Si 8500 (14.6%), S 2600 (4.5%), Cl 2100 (3.6%), Pb 90 (0.16%), Fe 1400 (2.4$), Na 350 (0.6%), Br 180 (0.3%).

During Sharav conditions (high TSP) the following average concentrations were found:

Ca 150000 (16.8%), Si 160000 (18%), S 55000 (0.56%), Ti 2000 (0.22%), Cl 4600 (0.52%), Fe 2100 (2.4%), Na 3400 (0.38%), Br 240 (0.27%), K 10000 (1.1%).

The accuracy of the results is of the order of ±10%.

CROSS-CORRELATION BETWEEN THE ELEMENTS

The correlation between the occurrence of the various elements is shown in Table 1. In analyzing these correlations the following assumptions were made:

1) The major part of the TSP in the air is from natural sources. There is a certain contribution from industry and motor vehicles but as the population is relatively small and the industries still not very large, their contribution to the TSP is not large.
2) A high correlation between any elements and TSP is a sign that the element is part of the natural aerosol and dust concentrations.
3) If there is a very correlation between the TSP and the concentration of the element, i.e. when the TSP levels are very high or low, the concentrations of the element do not follow suit, then the source of this element is man-made or from minor natural sources not related to dust carried by the winds.
4) High cross-correlation between elements can either indicate that they are part of a chemical compound or that they both come from the same source.

The elements can be divided into three groups: One group with high correlation with TSP and high cross-correlation between its members. This group consists of Fe, Co, Se, Lu, Sm, Mn, Ti, V, W, K, Ca, Si and Na. A second group with about 50% correlation between the members of the group and TSP and about the same level of correlation between its members. This group includes S, Pb and Cl. The third group includes those elements with very low or negative correlations with TSP and other elements of the table. This

TABLE 1. Correlation Factor between the Elements and TSP (x 100)

	Co	Sc	Hg	La	Sb	Sm	K	Br	Na	Fe	Mn	Ni	V	Pb	Cl	Ti	S	Si	Ca
TSP	98	99	-30	95	-24	99	99	4.3	90	99	99	91	99	62	56	99	58	99	99
Ca	96	96	-37	92	-26	99	99	4.8	91	98	98	89	98	58	54	98	55	99	
Si	99	98	-38	96	-26	99	99	4.9	89	99	99	92	99	60	55	99	56		
S	72	18	-16	78	22	69	68	22	61	57	58	51	60	58	46	59			
Ti	99	95	-42	97	-	-	99	.8	88	99	99	92	99	61	55				
Cl	63	11	-29	61	43	84	63	-10	75	56	58	53	55	29					
Pb	85	-5	-28	86	28	31	74	21	55	60	68	70	62						
V	99	93	-47	96	-	99	99	3.2	88	99	99	92							
Ni	99	48	-	96	-	93	93	-1.1	77	93	34								
Mn	99	94	-	98	-	99	99	0	89	99									
Fe	99	98	-32	96	-24	99	99	3.4	89										
Na	94	74	-21	83	4.3	98	97	-1.7											
Br	4.5	-14	-44	32	2.9	68	17												
K	99	-	-	96	-	99													
Sm	-	-	-	-	-														
Sb	-	-	-	-															
La	99	-	-																
Hg	-27	-																	
Sc	97																		

group includes Sb, Br and Hg.

Regarding the first group, the very high cross-correlation within the group probably indicates that they are primarily crustal in origin, hence all correlating excellently with TSP and consequently with each other.

The elements of the second group probably owe their origin to industrial and crustal sources. This explains the partial correlations with TSP and with other crustal elements such as Si and Ti of the first group. The fact that S, Cl and Pb belong to this group is not surprising since S and Cl are used extensively by the local chemical industry as well as being crustal and Pb undoubtedly comes at least partly from motor vehicle exhausts. Regarding Cl, it is noteworthy that the correlation with Na is only 15%, indicating that slat (NaCl) is not the only source of Cl in the air.

The third group includes Hg, Sb and Br. Here the correlation is nearly zero or even negative with TSP and other major crustal elements in this table. The fact that when TSP rises there is no corresponding rise in these elements and in the case of Hg and Sb it tends to decrease, shows that the source is probably not crustal, but from local industry emissions of these elements which are random. A negative correlation can be explained by assuming a local source which is diluted by the weather conditions conducive to high TSP.

The correlation fits in very well with the fact that very large quantities of mercury are used in the electrolytic processes of the production of Cl_2 by the Makhteshim Company.

The evidence that bromine in the TSP is industrial in source also comes from comparisons of the Br level in Beer-Sheva and Shivta, which is 30 km from Beer-Sheva to the southwest. Levels two and three times higher are found in Beer-Sheva than in Shivta. In all cases where the predominant wind was from the direction of the bromine plant, the levels rose dramatically.

AEROSOL SIZE DISTRIBUTION

It was found that the size distribution of the aerosols can be divided into two kinds. One distribution is the log-normal distribution where the shape of the size distribution is more or less Gaussian on a log scale.

The size distribution is given by

$$f(d) = \frac{1}{\sqrt{2\pi}\,\log\sigma}\exp - \left[\frac{(\log d - \log \bar{d})^2}{2\log^2\sigma}\right]$$

where d is the aerosol diameter, $\bar{d}$ is the average aerosol diameter and σ is the standard deviation of the distribution.

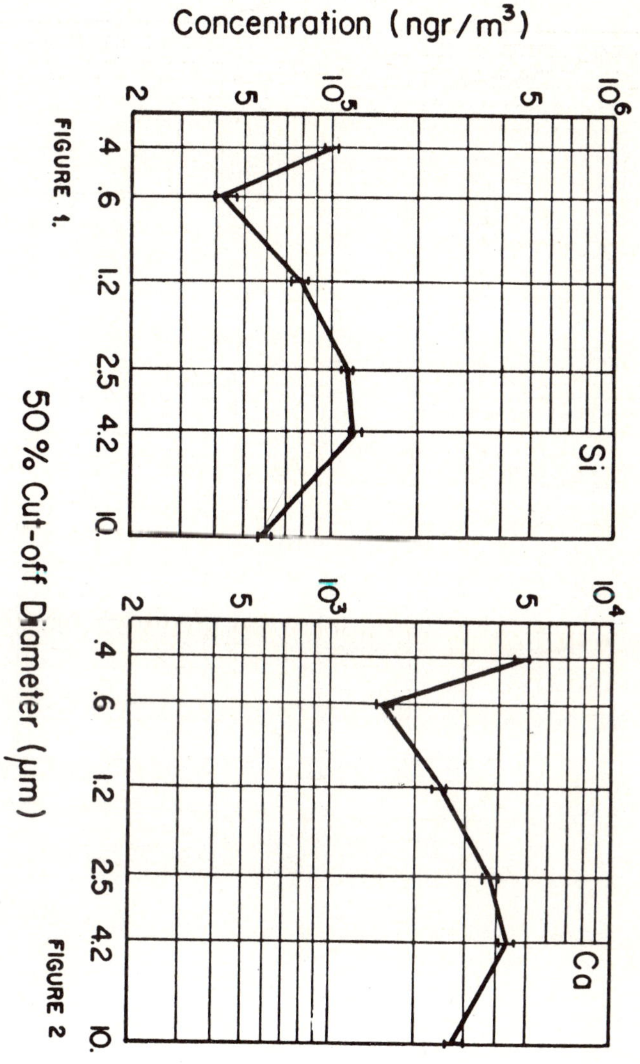

FIGURE 1.

FIGURE 2

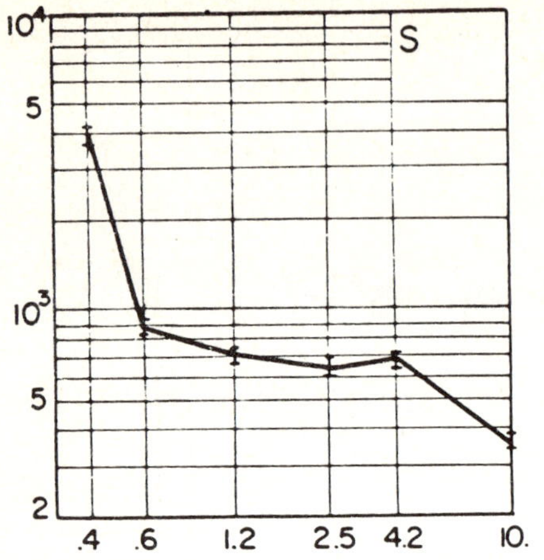

FIGURE 3

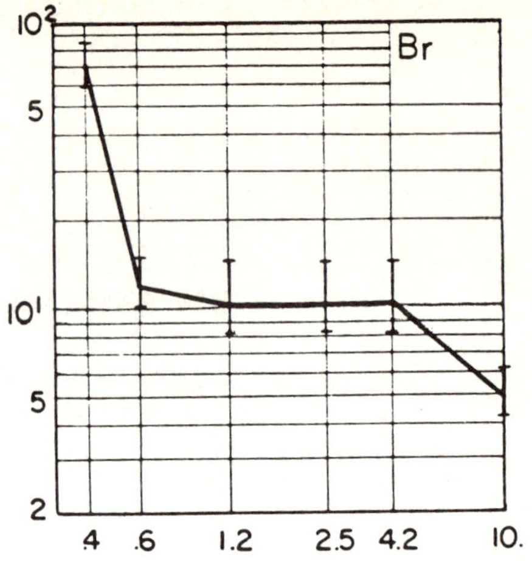

FIGURE 4

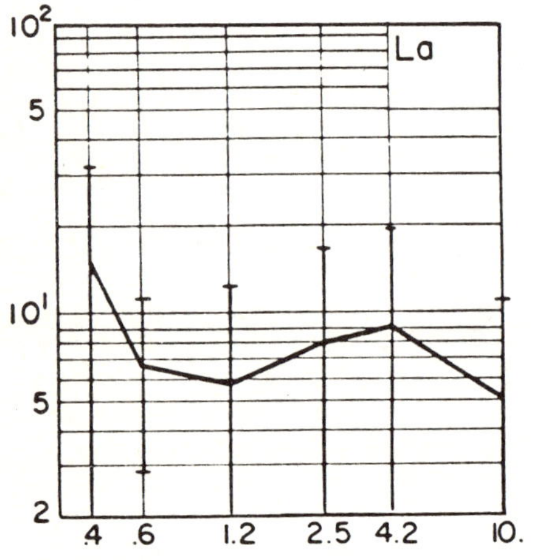

FIGURE 5

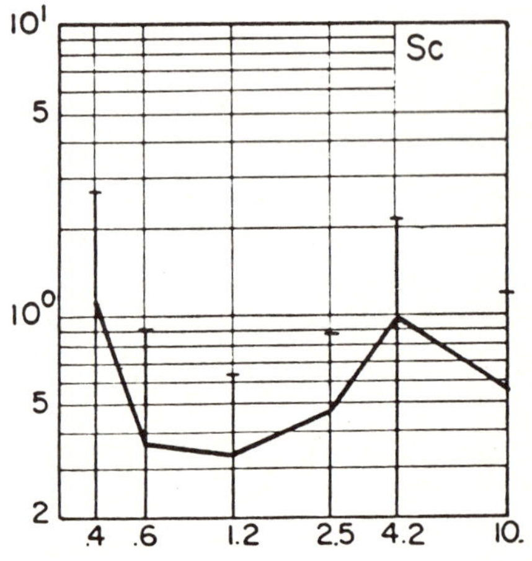

FIGURE 6

Concentration (ngr/m^3)

50% Cut-off Diameter (µm)

The second distribution obeys the rule $n(r) = cr^{-\beta}$ where n is the number of aerosols per unit volume, r is the aerosol radius, c is a constant and β is a number between 3 and 4. The distribution in this group is a rapidly falling curve as the radius increases.

It was found that dust from the desert collected on a windy day or during a dust storm has a log-normal distribution around 4.2 μm while the typical urban air pollution caused by industry and transportation has the exponential fall distribution. Elements such as Ca and Si which were identified as parts of the dust coming from the Negev desert as calcite ($CaCO_3$) and quartz (SiO_2) showed the typical size distribution.

Elements such as sulphur emitted from car exhausts and elements emitted from industry such as Br and Cl have the rapid fall distribution. This is a typical urban pollution distribution. The size distribution can be used for identification of the pollution source.

Some elements might come from both sources and the distribution is the combination of the two, i.e. La and Sc. Examples are shown in Figs. 1 - 6.

REFERENCES

Dams R., Rahn K.A. & Winchester J.W. (1972) Evaluation of filter materials and impactation surfaces for non-destructive neutron activation analysis of aerosols. *Env. Sci. & Tech.* 6, No. 5, 411.

Schubeger P.S. *et al.* (1975) JANE - An easy to handle computer program for different levels of qualitative and quantitative γ-ray spectra analysis. *J. Radioanalytical Chem.* 25, 141.

Developments in Arid Zone Ecology and Environmental Quality
Edited by H. Shuval

56

CHARACTERIZATION OF SUSPENDED DUST PARTICLES IN THE JERUSALEM AIR

Yigal Hirsch, Mordechai Peleg and Menachem Luria

Environmental Sciences Division, School for Applied Science and Technology,
The Hebrew University of Jerusalem, Jerusalem, Israel.

ABSTRACT

Continuous measurements of particle mass concentration, particle size distribution and other environmental measurements were conducted at a monitoring station in Jerusalem, Israel, during the 1978/79 winter season in order to determine the processes which occur in suspended dust particles in the metropolitan area.

This investigation has revealed that the main source of man-made suspended dust particles in the city center is vehicular traffic. It was also found that, as expected, the total particles mass concentration and particles density were higher by day and lower by night as anticipated, although the diurnal variations in the mass of the respirable fraction were minor.

Most of the man-made aerosol mass are initially emitted to the air in the form of small particles (sub-microns), but as a result of atmospheric processes, larger particles are evolved. The rate constant which characterizes this coagulation process (2.8×10^{-8} cm^3/sec) was estimated from the field measurements. This rate constant indicated that urban particles coagulate on almost every collision, and that the speed of this process is the main mechanism responsible for the sharp decrease in particle number density shortly after midnight.

The sub-micron particles, number, surface area & volume distribution, as well as the diurnal variation of particle mean diameter were evaluated. The influence of climatic factors (dust storms and precipitation) on suspended dust particles in Jerusalem air are also discussed.

Developments in Arid Zone Ecology and Environmental Quality
Edited by H. Shuval

57

ENVIRONMENTAL IMPACTS AND CONTROL ASPECTS OF FLUIDIZED BED COMBUSTORS OF POWER PLANTS

Ekrem V. Kalmaz

The University of Tennessee, Department of Engineering Science and Mechanics, Environmental Impact Research
317 Perkins Hall, Knoxville, Tennessee 37916 U.S.A

Abstract

Combustion of coal in the fluidized bed combustor systems results in heat transfer and heat release rate coefficients that is one order of magnitude higher than in the conventional boilers. The fluidized bed combustion boiler is an enclosed vessel containing a distributer plate which supports a bed of solid sorbent material through which combustion air is passed at a rate approximately 0.60-4.50 m per second. The fluidized bed combustors are used to generate steam for production of electricity. Several environmentally significant pollutants are emitted from these combustors.

In this paper, the entry of these pollutants into the air, water and terrestrial environments is discussed in relation to their production, discharge and disposal. Chemical and physical properties and environmental significance of these pollutants are discribed.

Developments in Arid Zone Ecology and Environmental Quality
Edited by H. Shuval

58

ANALYSIS OF SYNOPTIC CONDITIONS IN THE EASTERN MEDITERRANEAN (March 1979 TO FEBRUARY 1980) THAT LED TO ELEVATED AIR POLLUTION CONCENTRATIONS IN ISRAEL

Uri Dayan* and Michael Graber**

**Department of Atmospheric Sciences, Hebrew University, Jerusalem*

*** Israel Environmental Protection Service, Ministry of the Interior, Jerusalem.*

ABSTRACT

The Israel National Air Quality Monitoring Network, which began its full operation in December 1978, supplies continuous air quality data from its stations located in Ashdod, Beer-Sheva, Jerusalem, Haifa and Tel-Aviv.

From the data provided by the network during the period beginning March 1979 and ending February 1980, air pollution spells have been identified. Air pollution spells have been defined as incidents in which elevated concentrations of two or more primary gaseous air pollutants were observed at two or more of the network stations, at the same time.

A detailed synoptic analysis of the weather situations prevailing during those spells , and their classification has been performed.

It is suggested that this classification can serve as an Air Pollution Potential (APP) for the Eastern Mediterranean Region. Also, this classification can serve as the basis for a reliable 24-36 hour forecast of the expected air quality.

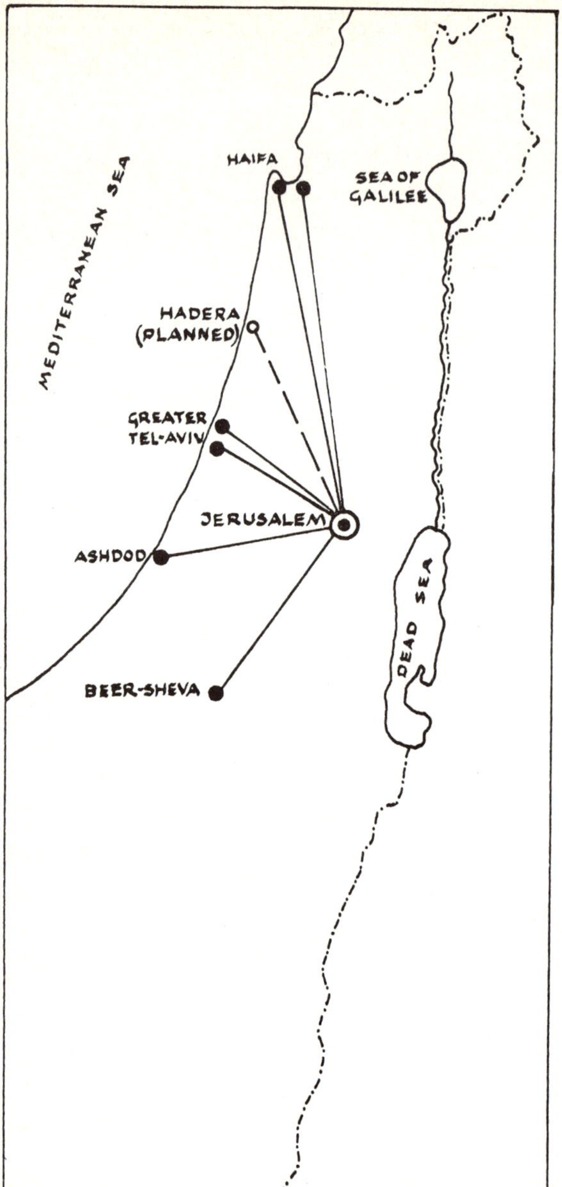

FIGURE 1: ISRAEL NATIONAL AIR QUALITY MONITORING NETWORK

INTRODUCTION

Air pollution (APP) is defined as an atmospheric condition where the efficient dispersion of pollutants in the air is limited and large quantities of pollutants emitted into it under these conditions would cause the build-up of elevated air pollution concentrations.

The similarities between the prevailing climate and the geographical configuration of Israel's Mediterranean coastal plain and those of Southern California have been noted by several meteorologists in connection with the well-known high APP (and its high concentrations of actual air pollutants) of the Los-Angeles basin (California, U.S.A.) (Jaffe, 1969; Halevy and Steinberger, 1974; Tadmor and Manes, 1973).

From the annual variation in Israel of the meteorological parameters which determine the APP (namely wind velocity and the depth of the mixing layer - see e.g. Holzworth, 1974) Rindsberger concluded (1974) that the occurrances of spells of high APP should be spread out more or less evenly over the whole year with a higher frequency of APP spells expected during the winter months (Rindsberger, 1976).

Neumann (1973) on the other hand, based on the persistence during summer months of the meteorological phenomena termed "subsidence temperature inversions", concluded that in Israel the high APP is expected from about May to October.

Until now no air quality data were available on a country-wide scale which would enable the validation of a proposed APP against actual air pollutants trends.

The Israel National Air Quality Monitoring Network, which began its full operation on December 1978, supplies continuous air quality data from its stations located in Ashdod, Beer-Sheva, Jerusalem, Haifa and Tel Aviv (see Graber, 1980).

At an early stage of the operation of the network, it became apparent that elevated concentrations of the primary gaseous air pollutants (i.e. SO_2,NO_x, THC and CO) were usually registered simultaneously at more than one monitoring station and often for more than one pollutant.* This finding

---

* An exception to this pattern is exhibited by the total suspended particulates (TSP). Elevated concentrations of TSP were observed during dust storms and hazy conditions, and were not correlated with the incidences of elevated concentrations of the gaseous pollutants.

indicated that regional factors (such as the APP), and not local factors (such as the strength of emission sources), are dominant when incidents of elevated air pollution concentrations occur.

In this study we have identified the incidents during the period beginning March 1979 and ending February 1980 in which elevated concentrations of two or more primary gaseous air pollutants with levels higher than 15% of the Israeli air quality standard (Graber 1980) were registered at several stations at the same time. We shall hereby define these incidents as air pollution spells. A detailed synoptic analysis of the weather situations prevailing during these spells and their classification has been performed. In our conclusions (section 3) it is suggested that this classification of the weather situations can serve as an APP for the Eastern Mediterranean region, including Israel. It can also serve as the basis for a reliable short-term forecast (24-36 hours) of the expected air quality.

SYNOPTIC ANALYSIS

A list of the dates when air pollution spells were observed in Israel (according to data from the Israel National Air Quality Monitoring Network) during the period beginning March 1979 and ending February 1980 is given in Table 1. The synoptic conditions prevailing during each of these spells have been analysed and classified. From Table 2 it can be seen that the synoptic patterns leading to elevated air pollution concentrations in Israel are limited to five types, occuring most frequently during spring, less frequently in autumn and winter and infrequently in the summer.

The following five types of synoptic patterns, defining five classes of air pollution spells, occured during the period stated:

AIR POLLUTION SPELLS CLASS 1 - WARM SECTOR OF A COLD DEPRESSION

Prior to the development of the synoptic configuration which led to Class 1 air pollution spells, our region was usually influenced by easterly winds. With the falling of atmospheric pressure in the Eastern Mediterranean basin due to the advance from the west of a cold cyclone, these easterly winds veer gradually to westerly winds. Class 1 spells reached their climax when our region was under a south-westerly flow and their end was prompted by the westerly shift of the winds, once the cool depression has moved eastward.

TABLE 1. SPELLS OF COUNTRY-WIDE ELEVATED AIR POLLUTION CONCENTRATION IN ISRAEL (MARCH 1979 TO FEBRUARY 1980) AS MONITORED BY THE ISRAEL NATIONAL AIR QUALITY NETWORK, AND THE CORRESPONDING SYNOPTIC PATTERNS.

Date of Spell			Monitoring Station Reporting Elevated Air Pollution Levels						Synopt Clas
Month/Yr.	Date	No. of days	AS	BS	HI	HII	JE	TA	
3/79	2	1.0		X	0	0	X	X	IV
3/79	5-6	2.0	X	X	X	0	X	X	IV
3/79	21-23	3.0	X	X	0	0	X	X	IV
3/79	28-31	3.5	X	X	0	0	X	X	I
4/79	1-3	2.5	X	X	X	0		X	I
4/79	10-13	3.5	0	0	X	0	X	X	II
4/79	25-30	5.5	X	X	X	0	X	0	IV
5/79	10-12	2.0	X	X	X			X	I
5/79	23-25	2.5						X	I
6/79	21-25	4.5	X		X	X		X	V
7/79	15-17	2.5				X			III
7/79	24-26	2.5	X	X	X			X	III
9/79	28	1.5	X	X	X	0			IV
10/79	22-23	1.5		X				X	I
10/79	29-30	1.5	X	X	X			X	V
11/79	5-6	1.5			X	0		X	I
11/79	25-26	1.0		X		0		X	II
12/79	2-5	3.0	X	X		0	X	X	II
12/79	19-20	1.5	X	X	X	0			I
1/80	1-4	3.0	X	X	X	0	X	X	I
1/80	6-10	4.0		X		0	X		I
1/80	22-30	8.0	X	X	X	0		X	I
2/80	2-5	3.0	X	X	X	X	X		IV
2/80	7-9	2.0			X	X		X	IV

The 500 millibar synoptic charts showed a pattern of transition from a synoptic configuration influenced by an anticyclone, or a weak zonal flow, to one dominated by a trough type flow, which caused a dynamic instability cooling of the atmospheric layers accompanied by strong south-westerly winds which brought the air pollution spells to an end.

AIR POLLUTION SPELLS CLASS II: WARM SECTOR OF A WARM AND SHALLOW DEPRESSION.

During air pollution spells of Class II, our region was within the warm sector of a Khamsinic depression (Joseph et al. 1973). Air pollution levels began to rise while our region was still within the high pressure system. The warm cyclone, formed at that time over the Gulf of Gabes, Tunis, moved within 24-36 hours eastward along the Mediterranean coast and reached the delta of the Nile in Egypt. The arrival of this warm depression in our region was accompanied by a shift to the south of the surface winds, by a steep increase in the temperature, a sharp decrease in the relative humidity and rising concentration of air pollution.

Class II spells ended with the movement of the Khamsinic depression along the Israeli coast to the North. This caused the winds to change to cool and moist westerlies arriving from the Mediterranean Sea.

The 500 millibar synoptic charts obtained for Class II were similar to those obtained for Class I spells. A gradual replacement of the upper-air ridge with a trough moving eastward caused a lowering of temperatures through the whole atmosphere and a build-up of an atmospheric instability which led to the dispersion of the elevated pollution concentrations.

AIR POLLUTION SPELLS CLASS III: SHALLOWING OF THE PERSIAN TROUGH, OR PRESENCE OF A WEAK PRESSURE GRADIENT

These synoptic patterns are typical of the summer season (see Table 2) when the extensive North-African upper-air subtropic anticyclone advances to our region and causes subsidence and stability of the atmospheric profile over our area. (Meteorological Office 1962). In the first case, the air pollution spells ended with the gradual deepening of the shallow trough. The decrease in air pollution concentrations was also influenced by the deepening of the marine turbulent boundary layer. In the second case, the synoptic pattern in time developed into a Cyprus low characterized by light variable winds, and generating a weak cold front. Class III spells ended after the weak cold front swept our area and the winds changed subsequently to cold and moist north-northwest flows.

AIR POLLUTION SPELLS CLASS IV - MIGRATION OF THE AXIS OF THE RED SEA TROUGH

Air pollution spells Class IV are characterised by a synoptic pattern where a ground-level pressure trough extends towards the north from the Red Sea low with the north-south axis of the trough lying over Israel. It was observed that these air pollution spells usually last for only short durations (i.e. 12 - 24 hours). Class IV synoptic pattern brings to our region regimes of light and variable winds, resulting in the increase of air pollution levels. Class IV spells usually end with the eastward movement of the axis of the trough and the penetration into our region of cool and moist marine air from the north western sector of this trough, accompanied by low-level clouds. See Fig. 8a and b for ground-level synoptic analysis of a typical Class IV spell. Inspection of the temperature and humidity profiles on thermodynamic diagrams (tephigram) revealed that subsequent to the marine air penetration into our region, the inversion base was lifted. This led to the reduction of air pollution concentrations, due to improved dispersion conditions. At the 500 millibar level, a replacement of the ridge-type flow by a trough type flow was observed towards the end of Class IV spells.

AIR POLLUTION SPELLS CLASS V - LOCAL SURFACE ANTICYCLONE ACCOMPANIED BY A WEAK UPPER AIR RIDGE

This situation occurred mainly in the beginning of the summer season and also, to a lesser extent, during the spring and autumn seasons. It developed usually as a result of shallowing of the climatological trough extending from the Persian Gulf, which led to a synoptic pattern lacking a significant pressure gradient in our area. Under these conditions a high pressure ridge, which extends from the Balkan peninsula to the delta of the Nile develops into a closed high pressure system and moves eastward to Israel. Such a ground-level anticyclone is usually accompanied by an upper-air ridge. Following the subsidence of the system, a warming and drying-up of the region is observed, accompanied by rising air pollution concentrations. Class V spells end either a) as a result of the incoming from the west of a Khamsinic depression which causes the ground level winds to veer, or, b) as a result of the deepening of the Persian Gulf trough. The development in either case causes the elimination of the local ground-level anti-cyclone and the subsequent termination of the Class V air pollution spell. See Fig. 10a, b, c, and d for synoptic analysis of a typical Class V pattern, ended by a penetration of a Khamsinic depression.

Key:
X - elevated air pollution concentration registered
0 - no data available for that period
AS - Ashdod; BS- Beer-Sheva; HI- Haifa Shuk Talpiot; HI - Haifa Kiryat Bialik;
JE - Jerusalem; TA - Tel Aviv.

TABLE 2: SUMMARY OF SYNOPTIC PATTERNS LEADING TO SPELLS OF ELEVATED AIR POLLUTION LEVELS IN ISRAEL (MARCH 1979 - MARCH 1980)

Season Synoptic Class	Spring	Summer	Autumn	Winter	Total	%
I	4	-	2	4	10	42
II	1	-	1	1	3	13
III	-	2	-	-	2	8
IV	4	-	1	2	7	29
V	-	1	1	-	2	8
No. Spells/Season	9	-	5	7	24	100
%	38	12	21	29	100	

DISCUSSION AND CONCLUSIONS

Rindsberger (1969) has proposed criteria for issuing meteorological warnings for air pollution episodes in the Greater Tel Aviv area, based on a modification of the Holzworth APP criteria (Holzworth, 1974). Based on these criteria, all air pollution episode days (28% of the time) were observed in 1976 and 169 days (46%) in 1977 (MOH, 1979). However, air pollution measurements for the same period reported in Tel-Aviv by the Ministry of Health (MOH), show that the actual air quality in Tel Aviv, although far from being good, is definitely not as bad as one would tend to believe from the above-mentioned high frequency of episode days.

For comparison, one should note the frequencies reported for the occurrances of APP in various countries (Holzworth, 1974): U.S. (West) - 13%, U.S. (East) - 5%, Canada - 4%, USSR(West) - 7-10%, Central Europe - 4%, Japan - 5-6%.

The Israel Air Quality Monitoring Network provides real-time information on the status of air quality in most of the densely populated and industrial areas in Israel. The air pollution monitored by this network in 1979 showed that the occurrances of elevated concentrations was not limited to one locality but in most cases was observed simultaneously over the whole country. Further analysis of these results showed that these air pollution spells were related to five well defined classes of synoptic weather patterns. We propose that this weather classification should serve as the basis for the Israeli APP rather than the modified Holzworth criteria.

Since the meteorological parameters leading to stable atmospheric conditions influences both formation of pollution spells and formation of fog (mainly light wind conditions) it is not surprising that our synoptic classification is supported by the findings of Levi (1967) regarding the synoptic conditions in the Eastern Mediterranean which are favorable for fog formation in Israel. It should be mentioned that the use of a synoptic weather classification to define an Israeli APP, as suggested here, led us to a relatively large percentage of time with high APP (about 18% of the time in 1979). However, accumulation of more air quality data in the coming years will provide the data necessary to define more accurately those weather situations in which high air pollution concentrations tend to occur as well as to refine and validate the classification proposed by us.

Our main concern in this work is to develop a system which will provide a reliable forecast of the air quality in Israel. It is a relatively easy task to predict from synoptic data the persistence, or change in the future (24-36 hrs.) or a given weather situation in our region. Thus, the real-time air quality data, provided by the Israel National Air Quality Network, in conjunction with reliable forecasts of the synoptic situation, will enable the presentation to the public of a forecast of the air quality expected 24-36 hours in advance.

ACKNOWLEDGEMENTS

The operation of the Israel National Air Quality Monitoring Network, which is providing the air quality data necessary for this work, is a cooperative effort in which many persons from several institutions are taking part. The authors wish to express their gratitude to all those persons, and, in particular, to the staff of the environmental protection units in the participating municipalities.

The authors also wish to thank the Israel Meteorological Service for providing the synoptic data, particularly Mr. Salomon of the archives, and Mr. Z. Kaplan, Division of Planning, Ministry of the Interior, for drawing the synoptic charts presented here.

REFERENCES

Graber, M. (1980) Data Processing and Analysis of the Israel National Air Quality Monitoring Network, presented at CODATA - the Israel-France Interdisciplinary Symposium, 5-6 March 1980)publ. 80.03, IEPS, Ministry of the Interior, Jerusalem).

Halevy C. and E.H. Steinberger, (1974) Inland penetration of the Summer Inversion from the Mediterranean Coast in Israel. Israel J. Earth-Sci. 23:47-54.

Holzworth G.C. (1974) Climatological Aspects of the Composition and Pollution of the Atmosphere World Meteorological Organisation Technical Note No. 139 WMO No. 598 Geneva, Switzerland.

Jaffe Sh. (1969) Synoptic Conditions Leading to Air Pollution Potential in the Greater Tel-Aviv Area. Meteorological Letters No. 17, Series E, The Israel Met. Service, Bet-Dagan (in Hebrew).

Joseph J.M. and A. Manes and D. Ashbel (1973) Desert Aerosols Transported by the Khamsinic Depressions, and their Climatic Effects. J. Appl. Met. 12 (5):792-97.

Levi M. (1967) Fog in Israel Israel J. Earth Sci. 16:7-21, Met. Office 1962. Weather in the Mediterranean 2nd edition Vol. I M.O. 391, Air Ministry, Met. Office, London.

MOH (1979) Summary of Air Pollutants in Israel in 1976 and 1977 publ. No. APR-50 the Research Inst. for Env. Health Nuisances, Min. of Health, Tel-Aviv (Hebrew).

Neumann J.(1973) Some Meteorological Aspects of Air Pollution in Israel, presented at the 3rd world conference of Engineers and Architects, Tel-Aviv, 17-24 Dec. 1973 (Contribution No. 73-22, Dept. Atmospheric Sciences, Hebrew University, Jerusalem).

Rindsberger M. (1974) Analysis of mixing depth over Tel-Aviv Israel, J. Earth Sci. 23:13-17.

Rindsberger M. (1976) Air pollution potential in Greater Tel-Aviv Area Israel J. Earth Sci. 25:127-132.

Tadmor J. and A. Manes (1973) Photochemical Smog Potentiality of Urban Areas in Israel. Israel J. Earth. Sci. 22: 93-105.

Figs. 2: Surface Synoptic Chart

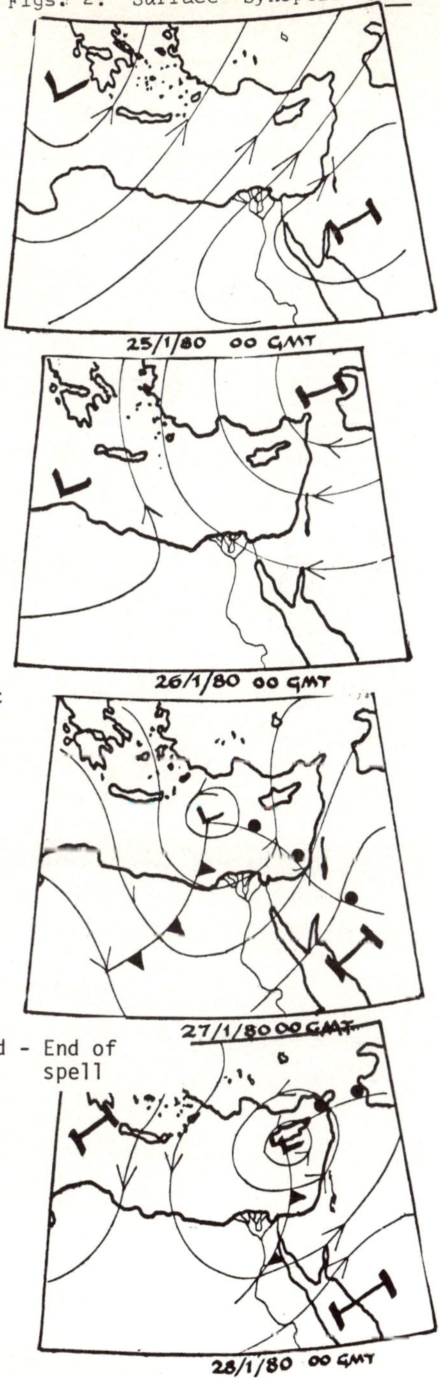

Figs. 3: 500 mb-Synoptic Chart

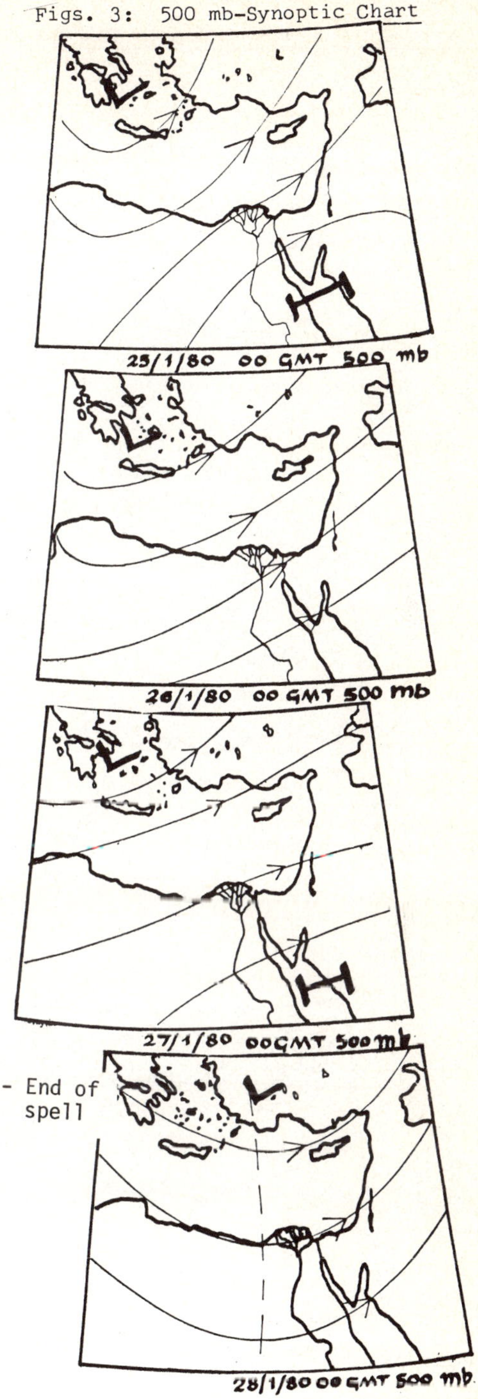

Figs. 4: Surface Synoptic Chart

a

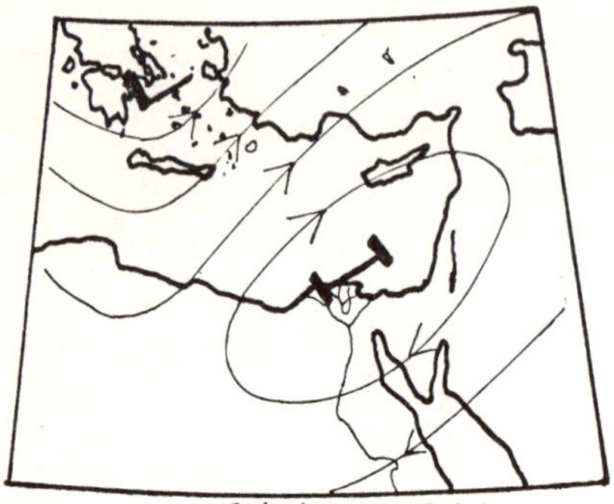

21/3/79 00GMT

b

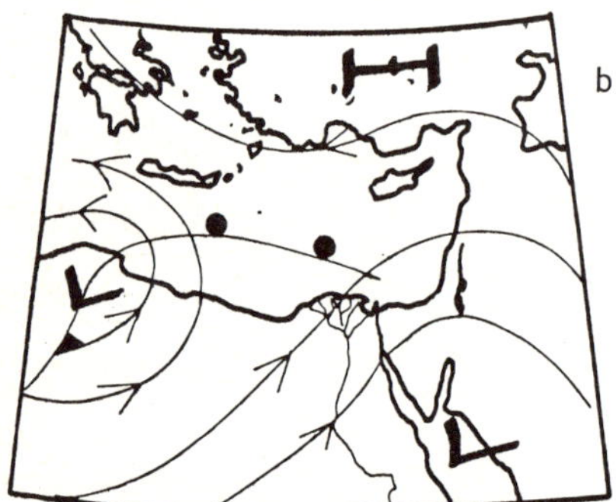

22/3/79 00GMT

c - End of spell

23/3/79 00GMT

Figs. 5: 500 mb Synoptic Chart

a

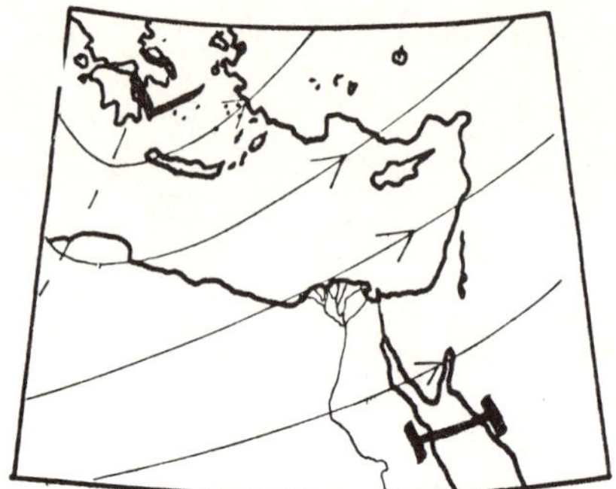

21/3/79 00 GMT 500 mb

b

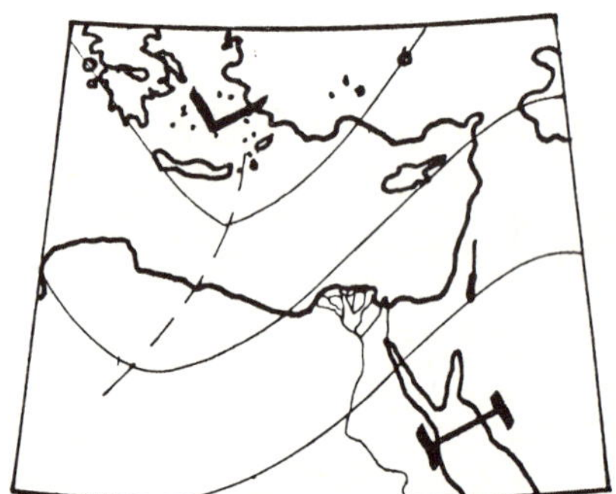

22/3/79 GMT 500mb

c - End of spell

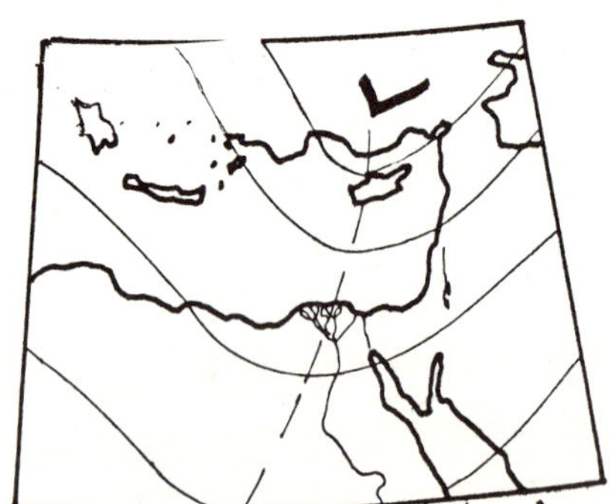

23/3/79 00 GMT 500mb

Figs. 6

a - Date of maximum concentration

b - End of spell

Figs. 7

a - Date of maximum concentration

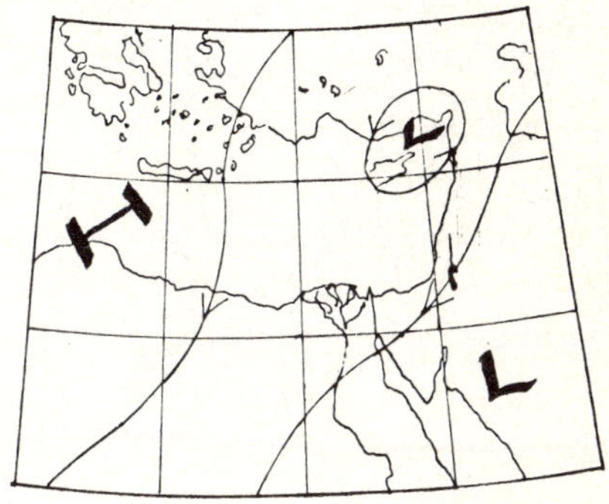

b - End of spell

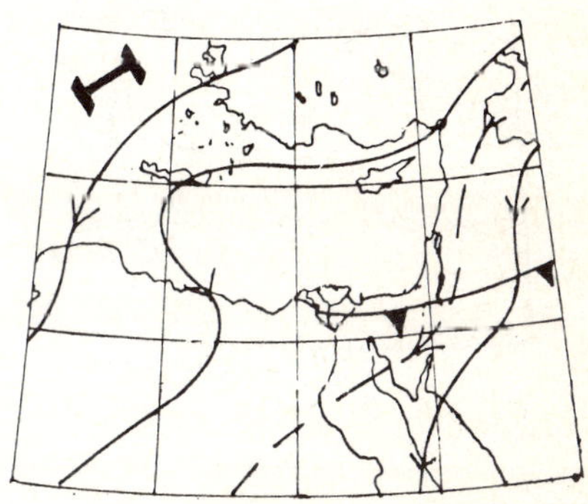

Figs. 8: Surface Synoptic Chart

Figs. 9: 500mb Synoptic Chart

a

2/3/79 00 GMT

a

2/3/79 00 GMT 500 mb

b - End of spell

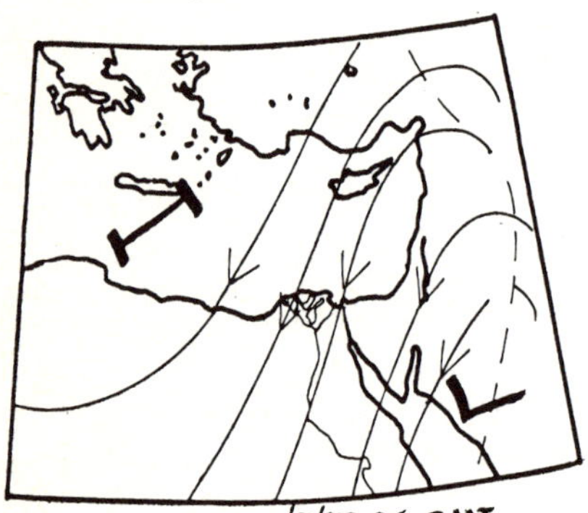

2/3/79 06 GMT

b - End of spell

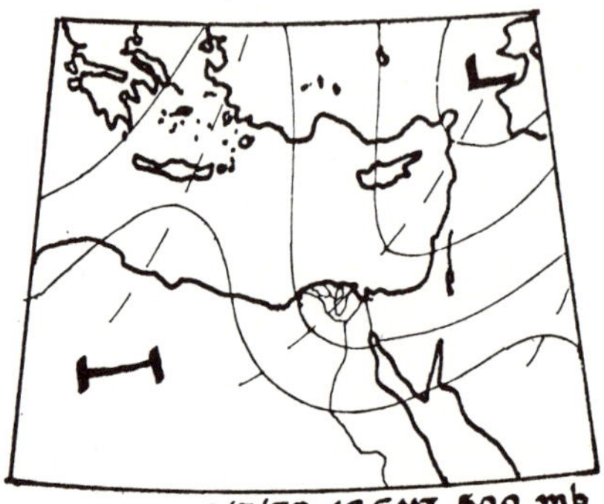

2/3/79 12 GMT 500 mb

Figs. 10 Surface Synoptic Chart

a

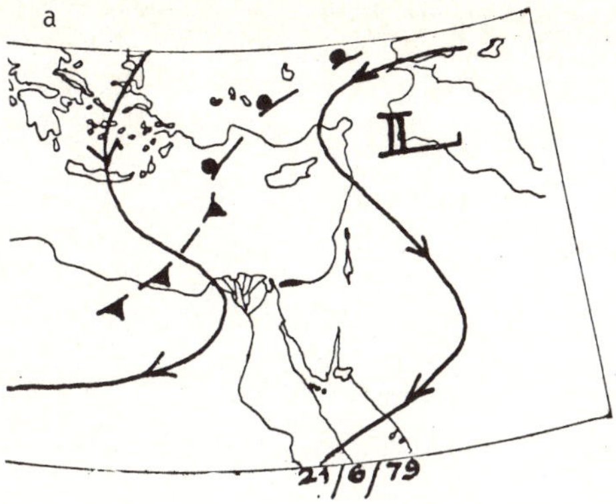

b

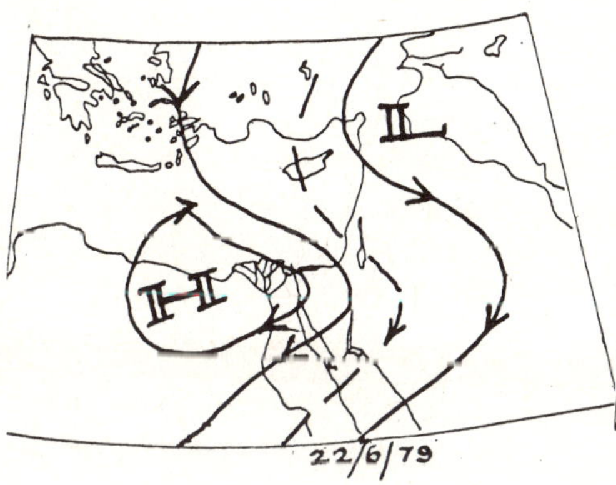

c

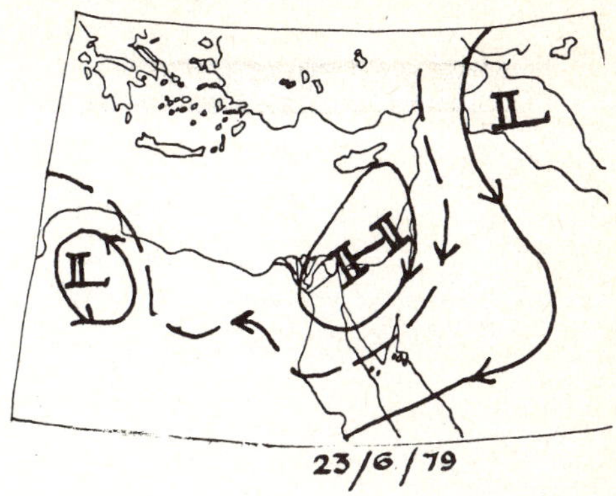

d - End of spell

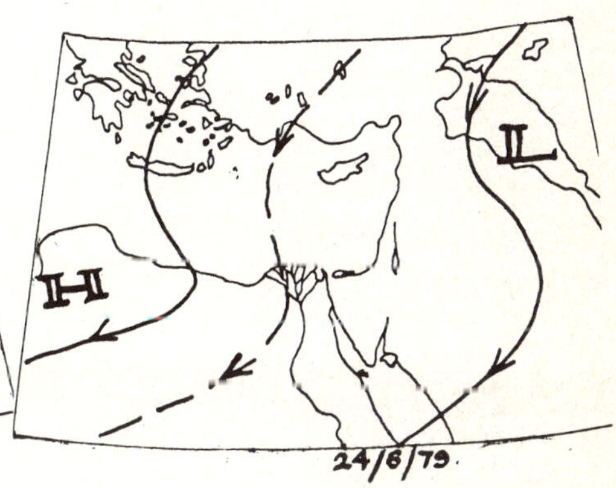

Developments in Arid Zone Ecology and Environmental Quality
Edited by H. Shuval

59
SOME ASPECTS OF VEHICLE POLLUTION ESTIMATION IN ISRAEL*

M. Gutman**   I. Hocherman**   A. Stotter**

***Transportation Research Institute, Technion, Israel Institute of Technology*

ABSTRACT

The present paper is resulting from a study on pollutant emissions from motor vehicles, and they are a function of driving patterns and vehicle characteristics. In the first phase, the driving characteristics in Israeli cities were investigated and compared with known characteristics of driving cycles in the world. For this purpose, characteristic routes were chosen in Tel-Aviv, Jerusalem and Haifa. Speed and distances were continuously measured with a fifth whell instrument. It was found by comparison that the chosen driving cycles in Israeli cities were very similar to the European cycles. From each of the cycles, average parameters were calculated and may be used for calculation of the pollution inventory from vehicles in Israel.

INTRODUCTION

The combination of rapid population growth, extensive urbanization, industrialization and rapid motorization has, in the last two decades, created an increasing problem of air pollution in Israel.

Motor vehicles are the main source of pollution in urban areas everywhere. In Israel the number of motor-vehicles increased from 70,000 in 1960 to 260,000 in 1970 and to about 500,000 in 1980, a seven-fold increase in 20 years, while the development of the road infra-structure continually fell behind. The result is congestion in city centers and on main urban arteries, which further contributes to the increase in air pollution.

The major pollutants emitted from motor vehicles are carbon monoxide (CO), Nitrogen oxides (NOx), hydrocarbon (HC), lead and smoke. The exhaust products are a result of combustion conditions in the engine under high temperature and pressure at an air-fuel ratio that is often less than stoichiometric. The amount of pollutatns emitted from a vehicle is a function of the type and quality of fuel, engine size, age of vehicle and level of maintenance, e.g., (Stern,1977, Kent et al.,1978). HC emissions can in-

---

* The study was sponsored by the Environmental Protection Service of the Ministry of the Interior, Israel.

crease by a factor of 10 as a result of a simple malfunction in the ignition system. CO emissions are mainly a factor of fuel-air ratio. Malfunctions in the curburation or fuel injection systems may induce a two fold increase of CO emissions (Wiener & Grogger,1979). For the same vehicle, the amount of pollutants varies with road and weather conditions and with driving patterns. Parameters such as speed, acceleration and deceleration and proportion of standstill time all affect the emission rate. Figures 1 and 2 (Kent,1979) and Table 1 (Bond et al.,1974) illustrate the effect of average speed on levels of CO and HC emissions.

TABLE 1. Emission factor for automobile exhaust for various routes

Route type	Average route speed mph	Pound per Vehicle Mile HC	CO
Business	10	0.023	0.35
Residential	18	0.015	0.21
Arterial	24	0.013	0.17
Rapid Transit	45	0.0085	0.10

DRIVING CYCLES IN VARIOUS COUNTRIES

In order to set standards for vehicle pollutant emissions under realistic driving patterns, the notion of "driving cycle" was developed. A driving cycle is a graph of speed vs. time, used to simulate real operating characteristics on a dynamometer. Several countries have developed driving cycles which are used mainly for emission tests for new vehicles (Simanaitis,1977). The first driving cycle was developed in California during the fifties. This cycle had evolved into the currently used UDDS75 which is based on a 12 mile route in downtown Los Angeles (Klingenberg et al.,1977).

The European drivind cycle, ECE-15, is based on a study which was carried out in several major cities in Europe in 1969. Table 2 presents some characteristic parameters of the American, European and Japanese driving cycles (Klingenberg et al.,1977). The parameters differ among cycles as they reflect the different urban driving patterns in the various countries.

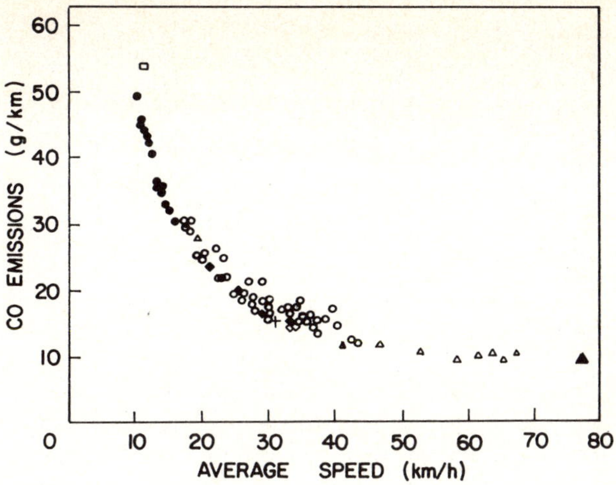

FIG. 1. CO emissions vs. average speed.

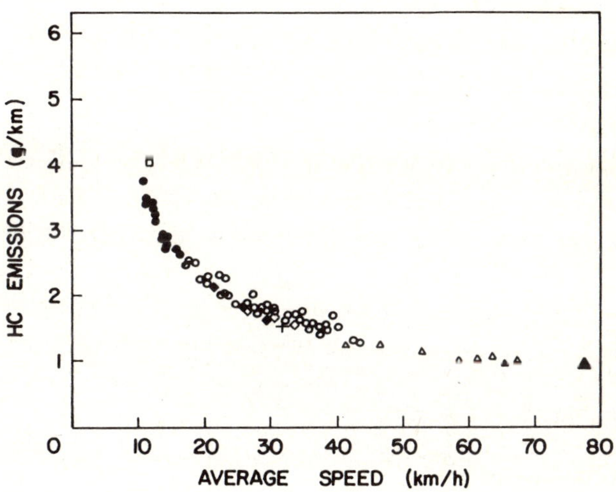

FIG. 2. HC emissions vs. average speed

TABLE 2. Assessment Criteria Figures of some well-known Driving Cycles and our results for Israel.

No.	Criterion	Abbreviation	Unit	Driving Cycle						Israel's Mean driving pattern
				Europe	Calif.	US 72	US 75	Japan 10 Mode	Japan 11 Mode	
1	Average speed of the entire driving cycle	$\bar{V}_1$	km/h	18.8	35.5	31.5	34.1	17.7	28.9	21.8
2	Average speed during actual driving (Standstills excluded)	$\bar{V}_2$	km/h	26.4	41.7	38.1	41.2	23.7	38.4	29.9
3	Average acceleration of all acceleration phases ($\dot{V} > 0.1\ m/s^2$)	$\bar{\dot{V}}_+$	m/s^2	0.64	0.64	0.58	0.59	0.63	0.52	0.54
4	Average deceleration of all deceleration phases ($\dot{V} < -0.1\ m/s^2$)	$\bar{\dot{V}}_-$	m/s^2	0.71	0.65	0.69	0.70	0.62	0.58	0.56
5	Proportion of standstill time ($V \leq 3$ km/h), $\lvert\dot{V}\rvert < 0.1\ m/s^2$)	s	%	28.7	14.6	17.3	17.4	25.4	24.8	27.40
6	Time proportion of constant speed ($\lvert\dot{V}\rvert < 0.1\ m/s^2$)	K	%	30.3	21.2	20.0	20.5	22.2	11.9	15.50

RESEARCH PURPOSE

The Transportation Research Institute of the Technion, with fundings from the Environmental Protection Service is conducting a study on pollutant emissions from motor vehicles. The major aim of this study is to provide means for measuring a pollution inventory. This is needed to obtain estimates of air pollution levels in specific sites, as well as for estimating total amount of pollution in a region.

As was mentioned before, pollutant emissions from vehicles are a function of both driving patterns and vehicle characteristics. Accordingly, the study consists of two stages. In the first phase, a representative Israeli urban driving pattern was established. In the second phase those findings will be used for getting type, engine-size and age specific coefficients which will enable the calculation of the amounts of pollutants emitted from a vehicle.

METHOD

In order to define the Israeli urban driving pattern, representative routes were chosen in the three major cities: Tel-Aviv, Jerusalem and Haifa. The routes were constructed as to include a path through the city center as well as major arterial roads leading to residential areas. It is known that about 50% of urban traffic is concentrated on 2% of the streets. The routes were chosen to represent these main streets. The length of the routes in each city is 10-12 km.

Driving pattern measurements along these routes were carried out using a single instrumented vehicle which followed other vehicles driving along the route ans simulated their behavior. The data of 45 test drives were made, all between 6:30A.M. and 7:00 P.M. From these recordings parameters that characterize the driving pattern were calculated; namely, average speed, average net speed, mean acceleration, mean deceleration, percent standstill time and time proportion at constant speed.

These parameters were defined as the weighted average of the parameters calculated for the three cities. The weights were 2 for Tel-Aviv and 1 for Jerusalem and Haifa. The reasoning behind the weighting is that about 50% of the vehicles in Israel are registered in the Tel-Aviv area, while Jerusalem and Haifa were taken to represent all other cities.

TABLE 3. Test results of urban driving pattern in Israel.

Abbreviation	Tel-Aviv Route 1 $\bar{X}_1$	σ_1	Haifa Route 2 $\bar{X}_2$	σ_2	Route 3 $\bar{X}_3$	σ_3	Mean	Jerusalem Route 4 $\bar{X}_4$	σ_4	Route 5 $\bar{X}_5$	σ_5	Mean	Mean* $\bar{X}$	Standard Errors** $\hat{\sigma}_X$
$\bar{V}_1$	22.3	2.72	18.43	1.92	23.33	2.45	20.88	23.76	2.62	19.68	2.68	21.72	21.80	0.491
$\bar{V}_2$	30.76	3.00	24.11	2.00	31.32	2.06	27.72	32.90	3.31	27.52	2.24	30.21	29.90	0.524
$\bar{V}_+$	0.548	0.054	0.51	0.073	0.54	0.03	0.522	0.54	0.033	0.57	0.04	0.555	0.54	0.009
$\bar{V}_-$	0.582	0.053	0.54	0.022	0.55	0.03	0.545	0.55	0.041	0.55	0.035	0.550	0.56	0.008
s	28.53	4.49	22.53	5.53	26.12	5.94	24.335	27.71	3.30	28.63	4.42	28.17	27.40	1.492
K	12.86	3.98	19.26	3.88	14.72	3.78	16.99	18.63	2.56	20.27	2.09	19.45	15.50	1.289

* $\bar{X} = 0.5\,\bar{X}_1 + 0.125\,(\bar{X}_2 + \bar{X}_3 + \bar{X}_4 + \bar{X}_5)$

** $\hat{\sigma}_* = (0.5^2\,\frac{\sigma_1^2}{\eta_2} + (0.125)^2\,[\frac{\sigma_2^2}{\eta_2} + \frac{\sigma_3^2}{\eta_3} + \frac{\sigma_4^2}{\eta_4} + \frac{\sigma_5^2}{\eta_4}])^{1/2}$

$\bar{X}_i$ - mean results; σ_i - calculated standard error

i - route number; η - tests number

RESULTS AND DISCUSSION

Table 3 summarizes the results of phase 1 of our study.

Table 2 presents a comparison of the parameters of some well known driving cycles and the Israeli driving pattern. The Israeli parameters are similar to those of the European driving cycle, the ECE-15, especially in the most important parameters, namely average speed and percent standstill time.

This result will enable planners in Israel to use European results of emission levels for new vehicles. For used vehicles, adoption of European results is not necessarily valid because of different maintenance standards as well as different weather and road conditions.

Stage 2 of the study which is now ongoing will provide data of emissions from used vehicles in Israel.

REFERENCES

Bond R.G. et al. (1974) Handbook of Environmental Control Vol. 1: Air Pollution. CRC Press, U.S.A.

Kent J.H. et al. (1978) A Driving Cycle for Sydney. Transpn. Res. Pergamon Press, London, 12, 147-152.

Kent J.H. et al. (1979) Motor Vehicle Emissions and Fuel Consumption Modelling. Transpn. Res. Pergamon Press, London, 13, 395-406.

Klingenberg H. et al. (1977) Analysis of the Measuring and Testing Procedures for Motor Vehicle Exhaust Emissions - Regulated in Europe and in the U.S.A. Volkswagenwerk AG, Forschung und Entwicklung, Wolfsburg.

Simanaitis D.J. (1977) Emissions Test Cycles Around the World. Automotive Engineering, 8, 34-43.

Stern A.C. (1977) Air Pollution Vol V. Academic Press, New York.

Wiener R. & Grogger P. (1979) Car Emissions as Function of Vehicle Characteristic. Environment, 7, 39-41.

Developments in Arid Zone Ecology and Environmental Quality
Edited by H. Shuval

60

OZONE DESTRUCTION PROCESSES AT NIGHT AND EARLY MORNING.

E.H.Steinberger*, E. Ganor+ & A. Donagi+

Department of Atmospheric Sciences, The Hebrew University, Jerusalem, Israel
+ Research Institute for Environmental Health, Sackler School of Medicine, Tel-Aviv University, Ramat-Aviv, Israel

ABSTRACT

Continuous ozone measurements in Tel-Aviv and Jerusalem during the last 5 years have shown high concentrations of ozone during the night. This high nocturnal ozone decays rapidly during the early morning hours.

The processes which can destroy ozone are as follows:

(1) $NO + O_3 \longrightarrow NO_2 + O_2$

(2) $O_3 + h\nu \longrightarrow O_2 + O$ $\quad 290 \leq \lambda \leq 780$ mμ

(3) $NO_2 + O_3 \longrightarrow NO_3 + O_2$

(4) $O_3 + HC \longrightarrow$ products

The rate constants for reactions (3) and (4) are at least two orders of magnitude smaller than that of reaction (1), while reaction (2) is active only during sunlight. Therefore only reaction (1) has to be considered.

Ozone decay rate can therefore be written as follows:

(5) $$-\frac{dO_3}{dt} = kO_3 \cdot NO.$$

Where k is the rate constant, O_3 and NO the ozone and nitric oxide concentrations.

Ozone decay rates during the night and early morning hours were analyzed for 1979 in Tel-Aviv, and compared with the NO concentrations.

It was found that there is a difference between ozone decay on Saturdays and ozone decay on weekdays.

On Saturdays the high levels of the night remained constant and no decay was observed in most cases. At the same time NO levels also stayed constant and low. This can be explained by the low NO emission (since the morning rush-hour traffic is absent).

On weekdays ozone concentrations started to decrease from 04:30 L.S.T., reaching minimum values at about 07:30 L.S.T..At the same time interval NO concentrations increased, reaching maximum values, at about 08:00 L.S.T..

Developments in Arid Zone Ecology and Environmental Quality
Edited by H. Shuval

61
TRAFFIC NOISE IN AN URBAN AREA

R. Levi+, O. Arnon* and N. Moses*

+Division of Human Environmental Sciences, Hebrew University and *The Environmental Protection Service, Ministry of Interior

ABSTRACT

The Geometrical parameters of traffic noise in an urban area were investigated in and around buildings. The vertical and horizontal distances from the street as well as other geometrical parameters were measured. The dependence of noise levels on the geometrical parameters was analyzed using statistical methods. The dependence was examined both for the internal and external noise levels, and also for the differences between them. This paper describes a prediction model for internal noise levels in buildings, based on their geometrical parameters. A similar prediction model is described for the differences between external and internal noise levels.

Developments in Arid Zone Ecology and Environmental Quality
Edited by H. Shuval

62

STREAMS IN THE DESERT -- WHAT DO THEY HAVE IN COMMON?

F. D. Por

Department of Zoology, Hebrew University of Jerusalem

ABSTRACT

Stream ecology has been widely studied in temperate and sometimes in wet tropical areas. The arid environment streams have largely been neglected.

Severely fluctuating hydrographic conditions, such as water volume and current velocity, temperature and dissolved oxygen content are the dominant limiting factors. Fluctuation is both seasonal and secular. Increased salinities may appear during the summer crisis. Waters are usually hard and alkaline.

As a rule there is a reduced tree canopy and therefore insolation is high. The importance of plant detritus in the streams is relatively reduced and detritus is mainly of herbaceous origins. Turbulence is usually great.

A marked reduction in the biotic diversity results. The biota of these streams are eurythermic and euryhaline organisms, resistant to low oxygen concentrations. The importance of the hyporheic and hygropetric refugia for the reinstatement of the fauna after the drying out or severe shrinking in the summer, is considerable. The relative role of the insecta and of other amphibious organisms increases. Likewise, the importance of the passively dispersed organisms is considerable.

Peculiar colonization events, "stepping stone" dispersal as well as water-divide shifting are very common in the arid climate streams. Relicts are relatively frequent in isolated or endorheic basins. The annual and secular fluctuations of the environment are expressed in uphill shortening and consequent downhill extension of the rivers. In many cases the rivers break up during the summer into a series of semi-stagnant pools.

Due to high temperatures and insolation, the rate of biological purification is rapid. However, human interference with the velocity, volume and even seasonality of these streams, the so-called "hydraulic pollution", is very considerable. Also, being frequently slow-flowing streams, with small altitude differences, mechanical oxygenation is reduced. High summer temperatures aggravate the oxygen problem. Sewage and agricultural runoff may turn intermittent streams into permanent flowing rivers.

Because of the scarceness of the streaming waters in the arid and semiarid regions, the problems of management are further aggravated. The number of conflicting interests related to every stream is very high. The same river may be used, and at the same time, as an open sewer, as a collector of agricultural runoff, as a recreational facility and be a last survival area for an exceedingly important faunal element.

Developments in Arid Zone Ecology and Environmental Quality
Edited by H. Shuval

63

DECOMPOSITION OF ALLOCHTHONOUS PLANT LITTER IN ISRAELI COASTAL STREAMS.

Seth R. Reice and Gershon N. Herbst

Department of Zoology, University of North Carolina and IES Laboratory, Nature Reserves Authority and Department of Zoology, The Hebrew University of Jersualem.

ABSTRACT

Leaf litter decomposition was studied in three Israeli streams which flow into the Mediterranean Sea. Decomposition has never previously been studied in Israeli running waters. The streams differ in level of pollution and salinity.

Nahal Daliya (Tout) is a clean freshwater stream. It is a winter stream, which is typically dry from June until October. The mean discharge during the months of this study (January-March) is 568 ± 227 l/sec. ($\bar{x} \pm \sigma$). The winter chlorinity is low ($\bar{x}$ = 59 mg/l) and dissolved O_2 overages 143% of saturation. Nahal Tanninim is a clean, slightly saline stream. It is permanent with peak flow in the winter months. The mean discharge for the study period is 2498 ± 180 l/sec. Dissolved O_2 averages 107% of saturation. The chlorinity averages 911 mg/l. Upstream N. Tanninim is fresh (chlorinity 24-126 mg/l) and its discharge is much lower (winter $\bar{x} \pm \sigma$ = 764.5 ± 235 l/sec.). Its flow is quite comparable to the Hadera and the Daliya. It goes dry from late August until early October. N. Hadera is a polluted stream. It has no dissolved oxygen, presumably due to high BOD. It is essentially dry during the summer (May-October) when it consists mainly of sewage and drainage water. During the study period the mean discharge is 569 ± 206 l/sec. The chlorinity averages 248 mg/l placing it intermediate to the Daliya and the Tanninim.

Decomposition of leaves of two common riverine plants was studied. One is an introduced tree (*Eucalyptus rostrata*) and the other is a reed (*Phragmites communis*). The *Eucalyptus* sheds its leaves and many fall into the streams. The *Phragmites* often falls in as a complete plant. Both species are present in the watersheds of all three streams.

Replicate 5 g. leaf packs of each species were introduced in January 1981 and sampled biweekly until March 1981. All leaf packs were placed on sandy substrata. Dry weight and ash free dry weight were determined for each sample date. Benthic macroinvertebrates associated with the leaf packs were identified to species whenever possible and biomass was determined. The relationships between animal biomass and leaf decomposition rates were also investigated. One major comparison was made between the Hadera and the other streams (the Daliya and the upstream site of the Tanninim). This contrast permits the understanding of the influence of extreme pollution on decomposition in Israeli coastal streams. The comparison between the upstream and downstream stations of the Tanninim allows examination of the effects of discharge and salinity on decomposition. The effects of the salinity gradient from Daliya to Hadera to Tanninim were also analyzed.

Developments in Arid Zone Ecology and Environmental Quality
Edited by H. Shuval

64
THE AQUATIC INVERTEBRATE FAUNA OF THE UPPER JORDAN RIVER AND FACTORS EFFECTING THEIR DISTRIBUTION[1]

R. Ortal, G. Herbst and H. Schnur

IES Laboratory. Nature Reserves Authority and Department of Zoology, The Hebrew University of Jerusalem

ABSTRACT

Biological and physical data were collected at approximately monthly intervals from 7 stations along the Upper Jordan River, from the Benot Yacov Bridge to Lake Kinneret, between May 1979 - April 1980. No significant differences were observed in temperature and dissolved oxygen content among stations, however, differences in current velocity, depth and sediment type were observed. The summer of 1979 was the driest in more than 50 years.

More than 70 invertebrate taxa have been identified. The majority of taxa were typical lotic forms, however, Cladocerans represented a large faunal component (12 taxa). Three taxa, *Leydigia acanthocercoides*, *Ilyocryptus sordidus* and *Macrothrix laticornis* appear to maintain resident populations within the river. The remainder of Cladoceran taxa enter the river as (catastrophic) drift.

Changes in community structure were observed along the river and three different regions are recognized: an upstream area, a middle transitory region and a downstream delta area. Station 7, characterized by silty sediments and reduced current velocity had the lowest species richness and lacked typically lotic taxa such as *Hydropsyche* (Trichoptera) and *Prosopistoma* sp. (Ephemeroptera).

Classification of samples by current velocity indicated that Simulid larvae, *Hydropsyche*, *Prosopistoma* sp. and *Dugesia biblica* (Turbellaria) were found significantly less often than by chance at current velocities of less than 20 cm/sec. Leeches (Hirudinea) were found more often than expected by chance at these low current velocities.

Differences in species richness among stations with similar current velocities, water temperatures and dissolved oxygen content may be explained by differences in turbulence or in sediment structure. The potential significance of depth, current velocity, sedimentation and turbulence in making management decisions on minimum flow requirements for the Upper Jordan River is discussed.

[1]Financial support was provided by the National Water Planning Company.

Developments in Arid Zone Ecology and Environmental Quality
Edited by H. Shuval

65

ECOLOGY OF DESERT RAINPOOLS

Ch. Dimentman

Department of Zoology, Hebrew University of Jerusalem

ABSTRACT

Temporary waters of various climatic zones and especially of arid regions are characterized by the instability of the abiotic conditions prevailing in them. Consequently, a fast biological succession develops.

The speed of maturation and the duration of life cycle of temporary water inhabitants are influenced by temperature. The occupants of the Sudan rainpools, which fill during the hot summer months, have a faster development and shorter life span than related species in rainpools of the Central Negev which fill with cold water during the winter.

Energy input in rainpools of arid and semiarid regions seems to be mainly allochthonous. This energy transfer may be carried out by several agents, such as runoff, wind and animals. In many of these pools, the secondary production exceeds the primary production.

The number of species, species diversity and complexity of food webs in arid region rainpools is low compared with those of semi-arid regions. Similarly, the above parameters are low in temporary pools from the mountains, in rock pools and short-lived pools by comparison with pools from the coastal plain, flooded basins and longer lived pools.

Survival of dry periods is carried out either by the production of various resting forms or by emigration to "refugia." The former phenomenon is common amongst protozoans, rotifers and lower crustaceans. It is also known in sponges, coelenterates, flat and round worms, annelids, bryozoans and molluscs, as well as some species of Dipteran larvae and even fish. However, only a few representatives from these groups are able to withstand the extreme dessication of desert pools.

In the desert rainpools of Israel and Sinai, a resting stage is utilized by various crustaceans: Anostraca, Cyclopoida, Cladocera and Ostracoda, and, in a few seasonal lakes, also by Notostraca and Conchostraca. Emigration to refugia in these deserts is common amongst various groups of insects such as Odonata, Hemiptera, Coleoptera and Diptera, as well as the amphibian Bufo viridis.

Developments in Arid Zone Ecology and Environmental Quality
Edited by H. Shuval

66
ECOLOGY OF THE SOLAR LAKE AND ITS HYDROPHILID PREDATORS

Jonathan Spira

Oceanography Program, The Hebrew University of Jerusalem; Israel

ABSTRACT

The ecology of the hypersaline "Solar Lake" adjacent to the Gulf of Eilat and situated 18 km. south of Eilat, is investigated. The lake is characterized by extreme fluctuations of temperature and salinity which have an important influence on community dynamics.

Species diversity is reduced and the food-web is shortened and simplified. Predators are limited to Hydrophilide (Coleoptera) larva.

Throughout the winter the fauna is limited to a shallow epilimnion (0.5 - 1m) overlaying a flat algal mat. A single predator, larval *Berosus* sp., actively preys on the planktonic brine shrimp *Artemia salina* (L.) and on two species of benthic Harpacticoid Copepods.

During the summer (May - Sept.), a new biotope develops as the water level recedes and the flat algal mat is exposed to desiccation. A limited fauna of Ephydridae larva and terrestrial Arthropods inhabit this supralittoral belt. They are predominantly preyed upon by larval *Anacaena* sp. (Hydrophilidae) which also utilizes *Artemia* cysts as a food source.

Algal distribution is determined solely by physico-chemical factors while invertebrates are limited by a combination of both physico-chemical factors and the distribution of algae.

The ecology of the "Solar Lake" is discussed with emphasis on the role of the two Hydrophilide predators.

Developments in Arid Zone Ecology and Environmental Quality
Edited by H. Shuval

67

SEASONAL SUCCESSION OF ZOOPLANKTON IN A SEWAGE RESERVOIR

Heather J. Bromley

IES Laboratory. Nature Reserves Authority and Department of Zoology, The Hebrew University of Jerusalem

ABSTRACT

The zooplankton fauna of a sewage reservoir at Na'an, Central Israel was studied for 5 months.

Zooplankton was first evident in 1980 only in March, and was represented by a very small number of Epiphanes senta (Rotifera). By April, the rotifer fauna was well developed, with about 4000 specimens/litre, and consisted of Brachionus angularis (dominant), Asplanchna cf. amphora, Brachionus calyciflorus and Filinia longiseta. The only crustacean fauna consisted of a small number of copepodites.

In May, there were about 300 rotifers/litre, dominated by B. calyciflorus. Numbers of B. Angularis were reduced and Asplanchna had disappeared. The Microcyclops minutus (Copepoda) population had increased (21/litre), and there was a sudden appearance and peak in numbers (230/litre) of the cladoceran Moina micrura, including males and ephippial females.

The rotifer numbers had declined to 1550/litre in June and Brachionus budapestensis and B. angularis were dominant. Asplanchna had reappeared and numbers of B. calyciflorus were reduced. The crustaceans were dominated by Microcyclops (155/litre) while numbers of Moina had fallen (15/litre).

In July, the zooplankton was dominated by the crustaceans Microcyclops (38/litre) and Daphnia magna (25/litre) with the virtural disappearance of Moina and all the previously recorded rotifers. The rotifer Hexarthra sp. was the only species present in significant numbers (16/litre).

Some possible explanations of the seasonal succession as observed for the different zooplankton groups and of the species within each group are put forward, and some observations on the predator-prey relationships within the rotifer population are made.

Developments in Arid Zone Ecology and Environmental Quality
Edited by H. Shuval

68

AQUATIC ENTOMOFAUNA OF A SEMI-TROPICAL DESERT OASIS - EN GEDI (DEAD SEA AREA)

David G. Furth

Inland Water Ecological Service, Nature Reserves Authority

ABSTRACT

Collections and surveys of the aquatic insects of the Dead Sea oasis En Gedi were made irregularly between 1968/1969 and 1981. Most samples were taken with an aquatic insect net (475 microns) during 1980-81. Information from the literature, unpublished data from various specialists, as well as data from several local collections has been added to the author's collections to form a comprehensive faunal list (Table) of the aquatic insects known from En Gedi. The known distribution in Israel for the En Gedi aquatic insects is also provided in the Table as well as their zoogeographical affinity based on general distribution. There is a distinct predominance of tropical African (Ethiopian) elements in the aquatic entomofauna of En Gedi which generally correlates with studies of other groups of the fauna and flora of this oasis.

Appropriate general information about each aquatic insect family found in En Gedi is given as well as more specific details of the species existing there. An entomological community analysis is then provided for the seven recognized biotopes in this isolated desert oasis, also represented in most springs in the arid zones of the Middle East. A second Table (community analysis), based on the author's field observations, lists the presence or absence of the most typical members of each of the seven biotopes. From this community analysis certain species have been assumed to be broadly or narrowly (eurytopic or stenotopic) tolerant of factors in their respective biotopes and may be used as indicators of the water quality in such environments.

Kibbutz En Gedi was founded in 1953 and since that time the modern human influence on the natural aquatic system has increased greatly due to growing usage of water for agriculture, resident population, and tourism. Pesticide pollution has also played a role in the history of this fauna. The potential limiting factors of the aquatic entomofauna are reviewed and related to a possible improved management of the En Gedi Nature Reserve towards protection of the entomofauna representatives of the various aquatic environments.

Developments in Arid Zone Ecology and Environmental Quality
Edited by H. Shuval

69

ZOOPLANKTON IN THE FOOD WEB OF LAKE KINNERET

Dr. Moshe Gophen

Kinneret Limnological Laboratory
P.O.B. 345, Tiberias, Israel (14102)

ABSTRACT

The study of the zooplankton as a part of the food web in Lake Kinneret has been carried out by the author since 1968.

The biomass of zooplankton is composed of 58% Cladocera, 35% cyclopoid copepods and 7% Rotifera. Species composition, based on biomass in percent of Cladocera and Copepoda, is as follows: Cladocera: Ceriodaphnia reticulata 38%, Diaphanosoma brachyurum 30%, Bosmina longirostris 21%, Ceriodaphnia rigaudii 6%; Copepoda: Mesocyclops leuckarti 90%, Thermocyclops dybowski 10%. The number of Rotifera species is 19.

The annual cycle of biomass variations of Cladocera shows a peak during March-May, due mostly to Ceriodaphnia reticulata, while during July-August cladocerans are intensively preyed upon by zooplanktivorous fish, mainly Mirogrex, and their biomass rapidly decreases. The annual pattern of variations of Copepoda biomass is smaller, compared to Cladocera. The copepod population is more stable than the cladoceran community because of two reasons: 1) The escapability of copepods is higher than that of cladocerans, consequently copepods are preyed upon by fish to a lesser extent. 2) The range of food sources of copepods is wider than the range of cladoceran food sources.

Filter feeder zooplankters feed mainly on small nannoplanktonic algae. These herbivorous organisms are preferred by zooplankton predators. Analyses of interactions between food web components indicate a summer paradox in the Kinneret ecosystem: food requirements of zooplanktivorous fish and herbivorous zooplankters are high because of the high water temperatures but food sources for both components are limited. Moreover, the feeding activity of filter feeder zooplankton can help in the prevention of summer blooms of nannoplanktonic algae caused by abnormal nutrient conditions in the lake and may influence water quality.

AUTHOR INDEX